THE CHALLENGE
OF SCIENCE EDUCATION

The CHALLENGE OF SCIENCE EDUCATION

Edited by

JOSEPH S. ROUCEK

Essay Index Reprint Series

 BOOKS FOR LIBRARIES PRESS
FREEPORT, NEW YORK

507.1
R752c

INTERNATIONAL STANDARD BOOK NUMBER:
0-8369-2070-8

LIBRARY OF CONGRESS CATALOG CARD NUMBER:
77-128305

PRINTED IN THE UNITED STATES OF AMERICA

Dedicated to

DR. DAGOBERT D. RUNES

for his contributions to all aspects of education

PREFACE

When Sputnik was announced to a startled world in October of 1957, it appeared almost unbelievable that the Soviet scientists could have gone ahead of the United States in the scientific race and produced this space satellite—an accomplishment which was only in the planning stage in the United States at that time. The American reactions, as well as of all other nations, were immediate, ranging all the way from the denunciation of the USSR to radical proposals for revolutionizing the American educational system by copying that of the USSR. Ever since, the development in the satellite programs of the United States and Soviet Russia has been rapid.* More recently, the Soviet Union's determination to win world leadership in science and technology has been evidenced again, this time in the atomic power field. Moscow's disclosure that it had built and was operating the largest atomic power plant in the world was timed to coincide with the Geneva atoms-for-peace meeting in 1958. (But some secrecy was maintained even then about the plant, especially where it was located. And the Soviet press failed to report adequately the earlier British and United States feats in this same area). In October, 1958, three Soviet scientists won the 1958 Nobel Prize in physics for building a better atomic "mousetrap." One of their traps—a cosmic-ray counter—was circling the globe in Sputnik III; others were in nuclear laboratories around the world. (The winners were P. A. Chervenkov, I. M. Frank, and Igor E. Tamm, all Moscow professors; they were cited jointly by the Royal Swedish Academy "for the discovery and interpretation of the Cherenkov effect," the basis of a novel cosmic-ray counter). They were the first Soviet physicists ever to receive a Nobel prize, but, far from being unrepresentative, they were only typical members of a large group of remarkably talented and trained workers in their country. The post-war work of Soviet physicists has contributed substantially to such major achievements as the development of Soviet nuclear weapons, the intercontinental ballistic missile and the successful earth satellites. Among Soviet physicists

* F. J. Krieger, *Behind the Sputniks. A Survey of Soviet Space Science* (Washington, D. C.: Public Affairs Press, 1958) ; Commdr. David C. Holmes, U.S.N., *What's Going on in Space* (New York: Funk & Wagnalls, 1958) ; M. Vasiliev, translated by Mervyn Savill from the Italian, *Su Sputnik Nel Cosmo,* introduction & notes by Wilham Beller (New York: The Dial Press, 1958) .

themselves, however, the greatest honors have been given to those who have made fundamental contributions to basic problems of modern theoretical physics; and these contributions, to the average observer, sometimes have little or no immediate practical application. It was significant in this respect that the Soviet physicist most honored in 1958 was Academician N. N. Bogolyubov, a Ukrainian, generally considered as one of the world's outstanding theoreticians in mathematical physics; and the most publicized Soviet physicist had been Academician Igor V. Kurchatov, generally believed to have been the scientific director of the Soviet nuclear weapons program.

Then, on January 3, 1959, the Moscow radio carried this announcement: "On January 2, 1959, a cosmic rocket was launched toward the moon in the U.S.S.R. . . . The launching . . . again demonstrates to the world the outstanding achievements of leading Soviet science and technology."

The rocket—"Lunik," the Russians are calling it—was unquestionably the greatest achievement of the Space Age, and its psychological impact upon the world was profound (it also emphasized the steady widening of the horizons of the cold war). Lunik made plain too that the Soviets still hold the lead over the United States in the space competition.

At any rate, these world-shaking scientific advancements produced endless discussions and proposals, whose main theme has been the sudden realization of the importance of scientific education to American life and safety. The once-derided egghead was suddenly called upon to come out of his ivory tower and furnish the brains to restore American primacy in scientific development.

The present volume is a systematic effort to synthesize the cross-currents of thinking and the evaluation of educational practices in the field of science by the distinguished specialists in this field. A definite effort has been made to cover the tested past experiences and to indicate what are the possibilities, and even the necessary needs, to strengthen the role of science education in order to discharge the challenge presented to it by the threat of world crisis and international tensions, in which science plays such a dominant role.

It must be stressed that the present volume does not (and actually cannot) cover all aspects of the history of science teaching, nor of all its ramifications in modern society. Strictly speaking, there is no aspect of education which is not influenced by or uses scientific principles. For instance, modern educational buildings are being built "scientifically" and their modernization contributes to the improvement of science thinking and practices. Yet, this field cannot be covered in this volume. The same applies to several areas of subjects, such as astronomy, archeology, agriculture, meteorology, etc.

Most of our contemporary problems in all aspects of life will be determined increasingly by developments in science and technology, and the graduates from all educational institutions, if they are to continue to play a full part in the determination of our course of history, must be informed about the methods, scope and limitations of the sciences; and this can be done only by a well-functioning educational mechanism, on all levels and in all its ramifications.

Today, science (and technology) is everybody's concern, not only Americans but all human beings in all corners of the globe. To Americans, this concern implies that a greater pool of scientifically-minded citizens and of scientists is needed to enable the free world to regain and then maintain a position of scientific superiority in the cold war era. To this end, this book has been written, on the assumption that a survey of what has been done and what can be done is one of the most reliable "scientific" approaches.

JOSEPH S. ROUCEK

University of Bridgeport (Conn.)

TABLE OF CONTENTS

SELECTED AREAS

COLLEGE AND UNIVERSITY EDUCATION

AUXILIARY ASPECTS

Framework

THE CASE FOR AND AGAINST SCIENCE AND SCIENTISM

JOSEPH S. ROUCEK
University of Bridgeport

The miracles accomplished by science and technology have produced a widespread belief in America—to which many academicians subscribe— that the scientific method can solve all problems. In fact, most Americans think that we are all "better off because of science." But there are also those who have certain dark suspicions. Most of them want to know more about science, according to recent surveys.[1]

Specifically, some of the answers to the Science Writers—N.Y.U. survey produced these results: 83% of the 1,919 people questioned said "the world is better off because of science," citing increased standards of living, medical progress and so on; but half, when asked, reflected fears of atomic power and its implications. For the ill effects of science, 12% blamed scientists, 12% blamed politicians, 12% blamed "evil persons," 8% blamed foreign powers, 3% blamed the military. To the question whether "it is all right for scientists to work on anything they want to, even if their discoveries conflict with religious teachings," 34% said no; 48% said yes, 9% had no opinion, 4% had varied answers. (Scientists' chief characteristics were rated as brilliance and studiousness, oddity, lack of religious convictions; some mentioned good balance and dedication to work and to humanity; but only 6% thought scientists were characterized by curiosity.[2]

What Is Science?

"What is Science"? In 1947, Yale asked that question by inviting Dr. James Bryant Conant (then President of Harvard University) to deliver the annual Terry Lectures at New Haven. Dr. Conant agreed to deliver the lecture, but not on that subject, because he considered it impossible to define science successfully or adequately. (The subject he finally agreed to speak on was "On Understanding Science").[3]

In his lectures,[4] Dr. Conant tried to allow science define itself by describing the story of what scientists had done in a few basic lines of research when science was passing into its modern phase from one to three centuries ago. The examples were focused mainly with the attempts of the scientists to discover the nature of fire and chemical activity, the nature of the atmosphere and the properties of air and the nature and properties of electricity. The scientists interested in those fields are often criticized today in the light of present-day knowledge, especially since "they were very stupid when compared with the bright boys of today."[5]

Broadly speaking, science can answer questions about phenomena which can be investigated. (But not all phenomena can be investigated, such as heaven or hell). But not all questions can be also answered, for many of the data dealing with all phenomena are non-objective, non-quantitative, and non-measurable, especially those dealing with the problems which involve basic questions of right and wrong. Is democracy superior to communism? Does might make right? Is Buddhism preferable to Mohammedanism as religion?

In 1951, Dr. Conant was courageous enough to try a definition of science: "Science is an interconnected series of concepts and conceptual schemes that have developed as a result of experimentation and observation and are fruitful of further experimentation and observations."[6] This is far from the dogmatic certainty that is expected of science. Dr. Conant admits that it seems "to equate scientific activity with a form of madness." Yet science is not a quest for certainty. It is a speculative enterprise, successful only to the degree that it is continuous. And the important word in this definition is "fruitful."

This conception of science makes it dynamic. Science is an activity. It avoids philosophic dogma as to "reality." It permits contradictory theories to persist side by side, as in the concepts of light as wave motion and as particles, for instance. It distinguishes science from "accumulative knowledge." And, above all, it separates scientific inquiry from progress in the practical arts.

The conceptual scheme which is both the origin and the result of scientific activity commonly develops from broad speculative ideas. And imaginative speculation depends on intuition and inspiration. It "rarely if ever" comes from an examination of the facts and the careful use of logic. (It is this point which has been overlooked by the exponents of the coldly calculating "scientific method.") But the conversion of a speculative idea into a fruitful conceptual scheme (which others would call a theory) depends wholly on experiment and observation. It is the experimental testing of ideas and deductions that is the major activity of science at work.

The relation of science to common sense is vital to the understanding of science. The practical experiments done each day by the artisan, cook, or the engineer, which at first sight seem akin to experiments in science, are not science, because they are empirical, "cut-and-try." They concern some special device or dish or process with no interest in generalization. The degree of empiricism is an index of the degree of science, in reverse. The chef at the Waldorf may experiment endlessly, but his tests are empirical and thinking is just common sense. On the other hand, the surveyor in the field is at the opposite extreme and depends wholly on the great concepts of mathematics.[7]

More broadly speaking, then, science is a set of mental habits, an organization of modifiable knowledge, which permits man transform his environment. It seeks to describe its subject matter and point out such uniformities of occurrence as are found to exist. Scientists are interested in the recurrent rather than the unique. They use the method of description to portray what they have observed. In this description, they use classification as a way of breaking down the complex structures and processes which they study, in order to organize them into concepts which can be meaningfully interrelated. They try to discover uniformities in the many complex phenomena which they study in order to reduce them to a smaller number of principles. When such principles have become precisely stated and sufficiently verified, they are considered scientific laws.

Sciences differ in the extent to which they have been able so far to place in completely quantitative form all the data studied and the relationships discovered. Astronomy and physics have gone far in this direction. Sociology trails far behind. However, the physical sciences developed gradually from purely descriptive, non-quantitative statements to a consistent system of concepts which could be defined precisely and whose interrelationships could be measured with great precision.

Science is often confused with the experimental method, a method under which some of the sciences have made remarkable advances. The experimental method is an extremely important aspect of such sciences as physics and chemistry, but is of relatively little usefulness in sciences like astronomy and paleontology. In sociology, it has had limited, but increasing usefulness.

The conception of science as a method of gaining knowledge has become increasingly widespread in recent decades. According to this conception, science consists not only of a body of verifiable facts and their demonstrated interrelationships, but also of the method by which these facts are gained and verified. The key point of scientific method is objectivity, the use of methods which as far as possible eliminate the personal bias of the observer and are available to others who would check his findings. In all the sciences, there is always a healthy skepticism and a tradition of criticism of methods and findings which do much to eliminate the possibility of acceptance of statements which have not been verified through rigorous methods of investigation and checking.

To summarize (and popularize) briefly the outstanding characteristics of sciences, we can say that they are insisting on: careful stating of assumptions; careful definition of terms; employment of objective methods of observation; classification of phenomena observed; suggesting and testing new hypotheses; expression of uniformities in quantitative terms wherever possible; development of a body of theory which relates different aspects of the findings in meaningful ways but which is subject to modification in the light of newer findings; and continuous criticism and re-examination of generalizations which, however well established they seem to be, are constantly open to critical re-examination in the light of new findings. Notice also that, when dealing with any scientific problem, such a study must use both the scientific and philosophical method. It is true that science is truly international. Physical laws are the same in Japan as in America, astronomy is identical in Soviet Russia as in England. The natural laws are simply in themselves, and men discover them. Yet no science can be discovered unless the cultural base is in agreement with scientific aims.[8] No science can survive without being supported by cultural base. Equally true is the fact that cultural biases stamp the very theory of scientific method; even the language is involved, and the Indo-European language mentality "has determined the course of science, and when other cultures wish to incorporate science they borrow from this great bank. They coin terms for the concepts invented in this type of language and bend their grammar and logic to conform to the pattern already set. Thus, although science is perhaps permanently fixed within the mold of Western culture, other cultures can take it over."[9]

One of the basic tenets, tacitly assumed by all scientists, has been that man can control his future in his interests. When disturbing consequences crop up following scientific discoveries, the assumption is made that the remedy for the consequences of science is more science. But this is only faith, "and its inseparable commandment that all value-judgments except the supreme scientific value-judgment must be renounced in the scientific role is one commandment. Science of any and every sort is in this sense one science."[10]

Hence, science is not just a dead accumulation of collected facts but an approach to unsolved problems. Historically, we differ from our forefathers and their predecessors by having science which has been changing our dependence on nature to one of mastery, the replacement of guesswork and "common sense" by authentic facts, and by building up civilization according to plans rather than haphazardly.

Science, when truly appreciated and applied, can really do wonders.[11]

How many people realize, for instance, that our knowledge of the external world through pictures (motion pictures, historical records, newspapers and periodicals, TV), depends on mirrors, lenses, microscopes and telescopes? "The limitations of the eye are obvious. It would be difficult to evaluate the advancement of human knowledge since the invention of optical instruments. This only emphasizes that a lens is more than just glass."[12] Science, furthermore, can produce material abundance and learn how to produce it. It can do so by applying a genuinely scientific discipline to the study of all possible causes—psychological, social, economic, biological—and by elaborating the probable wide variety of means which are necessary to solve them. Just as mathematics is an intellectual tool for accomplishing, when rightly used, certain particular operations that would be impossible without it, so is the whole body of science the tool now available to society for producing, if it is rightly used, the free and desirable life which the "age of plenty" has made theoretically possible. Yet, science cannot solve *all* the problems, since it is necessary for the scientist to deliberately limit his field of inquiry to those that can be cultivated by their own tools and methods; he must be able to observe and experiment. (Wrote William Harvey in 1628: "It were disgraceful, therefore, did we take the reports of others upon trust, and go on coining crude problems but of these, and on them hanging knotty and captious and petty disputations. Nature herself is to be addressed; the paths she shows us are to be boldly trodden; for thus, and whilst we consult our proper senses . . . shall we penetrate at length into the heart of her mystery.")

Scientific Facts and Ethics

The impact of scientific discovery on Western culture has been evident since the Renaissance. There is nothing new about the use of scientific findings, irrespective of their origins, in military warfare and industry. The scientific knowledge and skills of Archimedes were pressed into service by Hiero in the defense of Syracuse. Galileo's work on "violent motions" and projectiles were undoubtedly put to use by the Venetian armorers. But from the Renaissance period, science has been playing an ever increasingly important part in human warfare and welfare. In more recent decades, especially with the discovery of the atomic energy, scientists have acquired the knowledge, and therewith the power, to destroy or preserve all life on earth—and possibly the very earth itself.

What is really new in the world is not so much the role of science as the role of the scientist. The traditional conception of the scientist interested only in the pursuit of truth and not worried about the uses to which that truth might be put, is rapidly becoming obsolete. More and more scientists have been concerned over issues which have become acute largely in consequence of their professional activity. Even before Hiroshima, but especially since, scientists in democratic countries have stepped out of their laboratories and organized committees and associations, trying to influence domestic legislation and foreign policy. In America, scientists have been advisers to the President, to Congress, to various departments of the government, and to the American representatives at the United Nations.

Thus the scientist is persistently reminded that he is human, that he also has a conscience, which only man, of all the animals, has invented. Hence, every scientist is at one time or another jolted out of his objectivity to reflect whether he is doing the right thing by society. And he is also persistently attacked by the prophets of doom who accuse him of helping to destroy religion, of inventing devilish ways of destroying man's desire for simple life, and of inventing machinery which has unlimited destructive capacities.

But is the burden of responsibility really that of the scientist? Leonardo da Vinci wrote in his secret notebook in stubby mirror script, "I do not describe my method of remaining under water for as long as I can remain without food . . . on account of the evil nature of men who would practice assassinations at the bottom of the seas . . ." Alfred Nobel, more trustful of his fellow men, presented them with half a dozen new high explosives, and then used a large portion of his profits to persuade them not to blow each other to pieces.

The distrust of the scientific attitude is especially striking in the field of social philosophy. Problems of the natural world may be solved by scientific method, but not problems of men. There is no scientific formula which can solve the ever-recurring series of human problems.

But this reasoning indicates some misunderstanding of scientific thinking and action. No formula is a solution to a problem. Solutions involve commitment to a course of action. Action transforms situations; it literally re-solves them by eliminating the specific difficulties that provoke inquiry. The resolved situation may of course become problematic in turn. But by what canons of logic or common sense does it follow that because one problem is followed by a second we cannot intelligibly speak of having solved the first? A physician does not promise to cure his patient once and for all time by a specific form of therapy. The problem of health must be solved every day anew. Cure-alls are for quacks, not for scientists. The scientist continually examines his own tradition for the purpose of action, for the purpose of deciding on tomorrow's experiment.

Basically, the scientist in his laboratory is not concerned with good and evil; as an individual he can do little to solve the social problems he creates, and only the application of the scientific method to society offers him a way out. Science is a technique, not a way of life. It is a self-contained, self-perpetuating discipline. Its procedures are designed to enlarge understanding and increase knowledge; they are not concerned with promoting the happiness of man. If the scientist abandons his criteria and turns to social, economic and political standards, he solves no problems. He has abandoned science. The scientist is thus absolved of formal responsibility for the effects of his work. But does this relieve the mind of the biologist who sees his typhus antitoxin turned into an instrument of bacteriological warfare? Or of the chemist whose vitamin researches blossom into a campaign of intimidation of the public for profit?

The proof of the pudding is that the assassinations that Leonardo feared are practiced now from the bottom of the seas to the roof of the skies. Self-imposed censorship is futile, for discovery is not a unique stroke of genius that flashes once and disappears. Again and again, there have been simultaneous discoveries in science. (The periodic arrangement of the elements was conceived by Mendeleev and Lothar Meyer, independently, within a few months of each other, while Darwin and Wallace arrived at the fundamental concepts of organic evolution at almost the same time) . It has been suggested also that the scientist be given the control he now lacks. Some, in fact, are willing to place the steering wheel entirely in his hands. Since planning is essential in the modern world, who in this vale of mechanical tears is better suited than he to make and carry out the plans? The technocracy of the early thirties was definite as well as crude in such proposals. And there are several well-meaning sociologists, intrigued by the precise measuring instruments and neat mathematics of science, who think aloud wistfully of applying the same formulas to society.[13] Unfortunately (or even, fortunately!), the scientist has little advantage, if any, over the layman in matters of politics and government. And the politician and the statesman is not enveloped in an aura of omniscience and infallibility. The statesman, if and when he acts "scientifically" and wisely, must think and operate literally in two worlds. He must know things as they are (and here the scientist can help him); he must, however, also know things as they ought to be (the field of ideologies). Scientifically grounded wise statesmanship consists in possessing scientifically verified factual theory concerning what is the case, and scientifically verified normative social theory of what ought to be the case, and then achieving as much of the ideal as possible changes in the factual will permit. That form of society is the good one which embodies in the emotions of men a sensitivity to nature in its aesthetic aspect and orders its education, its intellectual outlook, and its social institutions in the light of the latest verified scientific theories.[14]

Since the beginning of the 17th century, scientific discovery and invention has advanced at a continually increased rate. This fact has made the last three centuries basically different from all previous ages, creating a gulf dividing man from his past, generation from generation, and finally decade from decade. Science, which was only a rebellious force in the early 17th century, is now integrated with the life of every individual and community. We can expect the rate of scientific advances to increase.

The application of science has led to industrialism which has also provided a certain way of life and a certain outlook on the world. Even in America and Britain, the oldest industrial countries, these changes have provided certain breaches of continuity and social and psychological stresses. But the difficulties have especially arisen in regard to the "underdeveloped" countries, upon which industrialism and science has burst violently, as something foreign, and demanding imitation of enemies and leading to disruption of ancient national folkways, mores and ideologies. Although this shock has been endured, in varying degrees, by Germany, Russia, Japan and partly in India, elsewhere it has been causing upheavals. The very attractiveness of Marxism shows how an alien ideology could find followers in the countries whose very way of life is entirely alien to the conditions which had given rise to Marxism. Obviously, Marxism is especially suited to countries where industrialism is new. But so the other forms of contemporary madness—nazism, imperialism, and Stalinism—are the natural result of the impact of science "on nations with a strong pre-scientific culture." And the end of this ideological upheaval is not yet to be seen.[15]

How to handle these explosive forces that might bring on another World War III? But such a question takes us beyond the sphere of science into that of ethics. But the trouble is that a scientific outlook tends to make many parts of traditional moral codes appear superstitious and irrational; furthermore, science now often creates a new environment with duties which might have been already discarded. Science is also related to power, and especially the power of destruction. But science is not responsible for the responsibility of power holders.

The most influential sciences have been physics and chemistry; biology is beginning to rival them. From the standpoint of human welfare, the scientific aspects of psychology, anthropology, and social psychology, with sociology, are becoming rapidly social sciences from the standpoint of human welfare. Yet science, by itself, cannot supply us with ethics. It can show how to achieve a given end, and it may show us that some ends cannot be achieved. But among ends that can be achieved our choice must be decided by other than purely scientific considerations. Hence the survival of humanity does not depend upon science but rather upon what human beings desire and decide. When science can continue, and whether, while it continues, it can do more good than harm, depends upon the capacity of mankind to make wise decisions.

What is, then the scientist's influence in social and political affairs? At best, scientists have been most influential when recommending policies that depend most intimately upon the technical knowledge which they possess. But there have been also "babes-in-the-woods" scientists, who believe in political miracles—for example, that the USSR, in the short time they say we have left, will accept some form of world government or atomic control. The scientist is peculiarly conditioned by his training to be partisan of freedom. But this also leads to trouble when the most idealistic proponents become convinced that political idealism is more important than the passing aspects of power politics.[16]

These debates usually boil down to two irreconcilable points of view: (1) one extreme point of view relates to "what ought to be," and the other (2) clearly separates "what is" from what "ought to be."

The first point of view reasons that scientists are "responsible" for making scientific discoveries, and thus also for the uses made of them. This assumption also involves the idea that scientists are responsible for the uses made of scientific discoveries and that each scientist is morally responsible for seeing to it that these societal uses are moral. But this point of view singles out the scientist out of other human responsibilities. Is, for instance, the miner of ore responsible when the scrap iron is sold to Japan and then used against his country? "Such an extension of responsibility would be absurd because of the impracticality of it, and in particular would make impossible that specialization and division of labor which is one of the foundation stones of our modern civilization."[17] Furthermore, how can the scientist make his contribution without being free? And even, how can scientists foresee, direct and control the uses made of discoveries? Above all, such an approach assumes that society cannot deal with the issues raised by scientific discoveries except by forcing the scientist to do something uncongenial for which he might not be qualified anyhow. In fact, the applications made of scientific discoveries are seldom made by the scientists themselves—and this is usually done by politicians, statesmen and industrialists. In short, "the cry for responsibility is often no more than the cry of a lazy man to get someone else to do for him what he ought to do for himself," reports Bridgman.

The basic aims of pure science, unlike those of applied science, are neither fast-flowering nor pragmatic. The quick harvest of applied science is the usable process, the medicine, the machine. The shy fruit of pure science is "Understanding." "The grand aim of all science," Einstein once remarked, "is to cover the greatest number of empirical facts by logical deduction from the smallest number of hypotheses or axioms."

Objectivity Versus Subjectivity

Inseparable from the whole question of the relation of ethics to empiric research is the never-ending debate between the proponents of the ideology that all research in science must be "useful," and those who claim that "ought to be" must be eliminated from scientific research in order to accomplish the non-evaluating task.[18]

It was Karl Marx who undercut the entire concept of scientific objectivity, providing the theory that men's ideas and even their sciences are intimately dependent upon the matrix in which they develop. (Freud did the same in a somewhat different sense some years later). But more recently we have changed the terms for the ideological framework. We now use "cultural" rather the class matrices. Our way of looking at the world is determined by our culture as a whole, as well as by the one aspect of it, our language. We now use the term "definition of the situation," describing the process which people go through in perceiving, evaluating, and interpreting what goes on about them. Even men of science whose integrity cannot be questioned, frame their research problems and interpret their findings in accordance with these basic preconceptions. Furthermore, the perception of problems is even more culture-bound, both objectively and subjectively, than the perception of data. Science develops according to the questions asked by the people, and these questions depend in turn upon their culture. In short, scientific inquiry can never be independent of its cultural matrix, which calls attention to some kinds of problems and ignores others; it is also related to the scientist's own values, biases and the situation in general.[19]

How this came about can be seen from the following survey of the attitudes toward science.

The Rise of Science as an Institution

The backbone of science is an attitude, a scientific attitude which accepts authority not as the source of the truth but only as the bearer of knowledge which can be effectively applied to the solution of human problems. The foundation of such reasoning is offered by fresh conceptual experience, experimentation and by three levels of theory: (1) classification; (2) explanation; and (3) formulation of general laws. But, and this is important, these "laws," these theories, are never considered as definite or absolute (thus differing from the claims of ideologies), only hypothetical, which are used to create additional theoretical consequences which can be subjected to empirical tests.[20]

The difference between ideologies and science has also led to the eternal struggle between proponents and opponents of scientism, and has excited the concerns of the man of religion, of letters, the critics, the historians, and the philosophers, as well as the man of science.[21]

Fundamentally, man is interested in natural phenomena for two reasons: he has an interest in comprehending and explaining nature on the one hand, and an interest in controlling nature on the other; in other words, he has intellectual and practical interests. Yet, the most common definition of science as "a body of organized knowledge which has been obtained in a certain manner, and which implies a certain attitude toward this knowledge and its means of derivation,"[22] tends to regard science to be of recent origin; and very frequently science is confined to "Western" culture, and restricted to a few persons within this culture; and even more frequently, science is equated with contemporary laboratory practices.

Actually, contemporary science, as a body of knowledge, methods, and attitude, is rooted in the antiquity of the species.[23]

One of the most widely accepted notions of the origin of science is that of James G. Frazer: that religion developed out of magic, and science developed out of religion.[24] While Frazer did not state a date of the appearance of science, one is led to believe it had its tentative beginnings in the ancient civilizations of the Near East.[25]

Frazer's view of the evolution of science is superior to that of Auguste Comte's theory which is also widely held: that thought has passed through three successive stages: the religious, metaphysical, and the positive or scientific. According to Comte, the religious stage was displaced by the metaphysical. The latter reached its peak during the middle of the 17th century, and prepared the way for the scientific stage, which in Comte's day was superseding the metaphysical. Frazer's conception is more acceptable since it makes science indefinitely more ancient than does Comte's view and also makes scientific development a more tentative and crescive process.

If we define science simply as "being that system of social rules and behavior by means of which man achieves control over his environment,"[26] then science is neither recent, nor confined to the "West," nor practiced by a few. Some degree of science and scientific behavior can be found throughout all human history, and to some degree in every society, regardless of the technological level of its culture. (Turgot: ". . . the resources of nature and the prolific seeds of science are found wherever there are men") .[27]

Anthropology usually shows that man invented first tools (the ax, the plow, the wheel, etc.) , and then used them for building (as shown in the pyramids of Cheops). But important in the history of science is the attitude toward it. The Egyptians appear to have been interested in the control of natural phenomena and hardly in comprehending its intellectual aspects. The Greek culture, on the other hand, was mostly enticed with the search for explanation, seeking order and unity in nature.[28]

In fact, our mechanical and industrial revolution was made possible by the methods of thinking of the Greek thinkers, with whom science was above all contemplative; they formulated every major problem in science and every major mode of attack on it. The Romans, in turn, were absolutely uninterested in the search for explanation. These two different attitudes toward natural phenomena are still with us today. Man's search for explanation is known as science; man's efforts to control nature is known as technology, applied science (also: the science or study of the practical or industrial arts) .

The difference between science and technology reappeared during the Middle Ages, when technology sprung up. For instance, in the 9th century, the horse collar was invented; about the same time, the nailed horseshoe was used—both helping to increase the productivity of agriculture. Complex mining machinery, often using horses, increased the productivity of mines. Then came the great Gothic cathedrals, and movable type of printing.

In the 12th century, the West started to get acquainted with Greek manuscripts and thus with Greek science, with astounding results.[29] The schoolmen absorbed the ancient Greek sciences contained in the works of Aristotle, Ptolemy, and others (together with the additions and comments made by Arabic scholars) ; but they added little of their own. Natural science was considered, when it was considered at all, within a theological and metaphysical frame of reference. The habitual reverence for authoritative texts induced them to accept the errors as well as the useful knowledge contained in the works of the ancient scientists, and their training made them depend too exclusively on deductive logic (despite the experimental work of Roger Bacon and others at Oxford in the 13th century and some of the Paris Nominalists in the fourteenth century) . They were interested in abstracting the essence of things rather than in measuring physical properties, and this tendency was strengthened by the spell of Aristotelian physics (which explained the action of natural phenomena by the qualities or "dispositions" supposed to be inherent in them) .[30] Thus even medicine suffered from this approach, which was suited to theology and philosophy, but not sciences which have to deal with concrete physical things.

The Marriage of Science and Technology

Modern science developed, interestingly enough, independently of technology, and technology was quite unrelated to science at the beginnings.

Modern science can be traced to the opening of the modern era when the medieval thought began to be replaced by new philosophical crosscurrents which revolted against the dogmas, Aristotelian and Ptolemaic. This revolution of the 16th and 17th centuries did far more than transform man's concepts of the physical universe and remake the entire texture of human life. It changed the character of man's habitual mental operations. Not only the external culture pattern but the internal processes of thought were revolutionized. Here is the origin of both the modern world and of the modern mentality. Historically it outshines everything since the rise of Christianity and reduces the Renaissance and the Reformation to mere episodes within the system of medieval Christianity.

The greatest hurdle that the human mind ever had to overcome—and the first great step toward the modern age—was the overthrow of Aristotle's incorrect concept of motion and the substitution of the present concept of inertia. To make this change did not require better observation or more precise measurements. All the facts were at hand. But to understand the facts a different kind of thinking was needed, a transposition in the mind of the scientist himself.

The classic year of 1543 saw the appearance of the books of Copernicus and of Vesalius—the one a triumph of theory, the other a victory of facts.[31] The master thinkers of the 16th and 17th centuries created an instrument —Baconian experimentation and Cartesian rationalism—which gave rise to an astounding technology that has surpassed the instrument itself and has made nearly everything possible. Yet, up to about 1800, the search for control of natural phenomena (technology) and the search for explanation of nature (science) went on their separate ways.[32]

But at the turn of the last century, an important link started to develop between them; the contributions of the scientists began to be used by the engineers and technologists. For instance, the scientific contributions of Franklin, Coulomb, and others was promoted by the scientist Michael Faraday into a general theory of electro-magnetism, which, in the hands of such technologists and inventors as Siemens and Edison, has created our modern electric power industry. As scientists started to think of the relations of things in terms of mathematical order (numbers, ratios, and geometrical figures), their advances in the mathematical sciences led to the scientific revolution. The mathematical approach to science required accurate measurement to bring satisfactory results; this led to the improvements of the known instruments of measurement and to the invention of new ones.

Science still remains, today, an intellectual activity, a search for explanation; the results of this search are bodies of knowledge that the technologist uses in his search for control over nature. Consequently, today science and technology are inseparable, although they both remain diverse activities. Science is an intellectual activity, technology a practical one; science deals chiefly with ideas, technology chiefly with things.

Thus, because of these basic differences, different cultures react differently to science and to technology. To the Soviets, science, placed at the service of the proletariat, is believed to be a tool of nearly unlimited power. To the people of the United States, this is a dangerous conception, since, in the hands of aggressive Communism, it might be the most deadly weapon. At the same time, this concept is destroying the American outlook which is also fundamentally materialistic, but interested primarily in technology rather than in science (while the Soviets have been able to stress the inseparable aspects of technology and science).[33]

Apologists and Critics

The marriage of science with technology has had the most far-reaching implications for modern life. The achievements of science had attracted the interest of educated men even before the 16th century, and many people found them disturbing in their implications. The Copernican picture of the universe was a shock not only to men's religious beliefs but also to their ideas about the world in which they lived. Kepler and Descartes, both religious men, were convinced that in demonstrating mathematical order throughout the universe they were glorifying the supreme rationality of God. But some men began to feel that there was no need for Divine Providence in a mechanical universe, which operated like a gigantic clockwork. But the full impact of the new science and technology was felt only in the eighteenth century, when it produced another intellectual revolution, the Age of Reason.[34]

Machinery, the most characteristic product of this union of science and technology, forced men to modify their philosophical outlook. (Since reason had produced machinery, the philosophers were convinced that reason could solve every problem).

Machinery, an apparatus made out of iron, having no organic life, began to produce goods which formerly had been the work of the hands of living people guided by their minds. The human sources of energy (hands, muscles, plus a little of brain) started to be less and less important and more useless. While, in "good old times," most of the production was handled by specially trained master craftsmen, the utilization of machinery allowed the use of unskilled labor (at the beginning of the Industrial Revolution at least), and even of children and women. During the second half of the 18th century, James Watt, by perfecting the steam engine, made available an almost inexhaustible source of cheap power; the steam engine was soon adapted for travel by land and sea. Another series of inventions made possible the spinning and weaving of cloth by machinery, and Whitney's cotton gin reduced the cost of the chief raw material used by the textile trades. The use of coal in the smelting of iron led to the rise of processes by which iron and steel were made available at low cost and in almost unlimited quantities. The development of machine manufacturing swept away the hand industries.

The traditional institutions (the family, the associations of farm and shop, the parish church) were transformed by the Industrial Revolution. The training offered under the old system by the shop was completely disrupted; children working in factories were no longer apprentices but rather laborers; parents employed in shop or mine were separated from each other and their children much of the time; the parishes of the manufacturing centers had little resemblance to rural parishes.

At the beginning of the 19th century the condition of the laborers (and particularly of the laboring children of the poor in England) was miserable in the extreme. It was true that the quantity of goods was rapidly rising, but there was also displacement of labor and the increase in the quantity of poverty. Since these problems were most glaring in England at first, the English thinkers began to cogitate on these conditions of the laborers. "I have seen them; famine had thinned them to skeletons, and they were dumb with despair," proclaimed Lord Byron (in his only speech in the House of Lords) .[35]

Adam Smith, an Englishman, published in 1776 (the year of the American Declaration of Independence) *An Inquiry into the Causes of the Wealth of Nations.* He demanded freedom of competition, free and unrestricted trade within a country, and freedom of trade between nations. Let each country produce those goods which are best suited to its soil, climate and natural resources. Let things alone! Let the government keep hands off! Let freedom reign!

Smith preached his doctrine at a time when the Industrial Revolution was getting under way, and the new factory owners and capitalists welcomed the ideology which gave them a free hand and allowed them to make what they pleased and how they pleased and to employ their workmen—women and children as well as men—for long hours and low wages. His ideas carried the day.

Yet, the ideas of *laissez-faire* failed to solve the problem of the growing number of superfluous persons produced by the application of science to technology, whose child, technocracy tried to be the redemption doctrine by declaring that machinery will be able to fulfil the ancient wish-dream of a superfluity of goods when man will be able to escape the drudgery of existence by having most of his exhausting labor performed for him by the machinery. How deep was the growing resentment was shown by the movement of "General Ludd," a village idiot (Ned Ludd from Leicestershire) , which tried to handle the problem by destroying the spinning machinery by mass action.[36]

On the intellectual front, the efforts to handle the problem of machinery were not so much against science as against planning the applications of science, especially from the standpoint of religious and philosophical theories. While the *laissez-faire* theorists, following Adam Smith, proclaimed that machines were accomplishing the great harmony of society and producing more and more wealth for everybody ("the greatest happiness of the greatest number"), there arose a group of Utopian and Christian Socialists (Owen, Fourier, Saint-Simon) who were convinced that the troubles were caused by the mistakes in planning and in utilizing religious and humanitarian values in society, although, in the end, better days and better men were destined to appear.[37] Although these men were popular, their popularity was soon submerged by the attractiveness of the theories of Karl Marx, who propounded *scientific* socialism and gave a rousing welcome to the increasing misery produced by the machinery since it would, in the end, release the labor masses from misery by taking it over, conceived the capitalistic system as a systematized conspiracy, and, in contrast to the utopians, was convinced that the end of the rainbow could be found in the paradise of the machines for the labor class accepting his doctrine of dialectic materialism.[38]

Notice, however, that Marx glorified science, while condemning the results of the industrial applications of science.

"Marxism regards laws of science—whether they be laws of natural science or laws of political economy—as the reflection of objective processes which take place independently of the will of man. Man may discover these laws, get to know them, study them, reckon with them in his activities and utilize them in the interests of society, but he cannot change or abolish them. Still less can he form or create new laws of science."[39]

It is paradoxical fate of Marx's thought to have exerted an influence upon the modern scene comparable to that of a new faith. Paradoxical, because Marx started with a repudiation of the religious consciousness as "the sigh of the oppressed creature." Unlike such socialist predecessors as Saint-Simon, who coupled his doctrine with the call for a new Christianity, Marx proclaimed the criticism of religion as the first great step in a "merciless criticism of everything existing" which was supposed to disclose, indirectly, the principles of a new world. The task, as he came to see it, was to be the first to transform socialism "from utopia into science." Nevertheless, "history will remember him not as the author of a scientific theory but as the apostle of a secular creed."[40]

Marx's assumption was that his theories were "scientific." But the tragedy of Marxian thought has been, of course, that Marxism, just as most of the "scientific" theories designed for social action, is an ideology par excellence.[41]

At any rate, the eternal trouble between "what is" and "what ought to be," which Marx tried to solve, is still with us.

While the Soviet rulers have made a fetish of science, there have been many influential and vocal critics of science in the Western countries accusing it of being a sacred cow.

Existentialism and Science

One of the paradoxes of modern times is that with the increase of scientific knowledge there has developed a growing skepticism of science as a mode of knowledge. The more power scientific knowledge commands, the more profound in many quarters is the distrust of the possible uses of scientific method in social affairs. Evidence of this skepticism and hostility can be found in many contemporary philosophical, theological and literary movements, of which existentialism is the most agonized but not the most significant.

This attitude is not based upon the falsity of specific conclusions about human behavior advanced by scientists. Nor does it arise from the failure of concerted, systematic attempts by social scientists to grapple fruitfully with problems of men and society on a scale comparable to the activities of the natural scientists in their field. For few such efforts have been made. Rather does it flow from a philosophical premise which contests the relevance and validity of the scientific approach to man in his distinctively human capacity. It is alleged that judgments of value, which are central to the human enterprise, are forever beyond the competence of scientific inquiry.

Existentialism, one of the most popular ideologies of the day, is a protest against the submersion of the individual in a mass society. It expresses the realization that there are "subterranean forces of life" with which pure reason cannot deal and that reason itself has its roots in these irrational forces. It wants a new conception of human thinking, which recognizes that existence is primary and logic only secondary. It asks, in short, for an intellectual revolution, for a new set of standards by which the works of the mind can be judged, believing that the intellect in the modern world has become an inhuman gadget and that organized reason has given our civilization unprecedented powers, which are used without taste or moral insight. This field of thinking has been headed by Kierkegaard, Nietzsche, Heidegger and Sartre, and has influenced such varied products of modern culture as the painting of Cézanne and the cubists, the writing of men like Joyce and Faulkner, and the Principle of Uncertainty in modern physics. All of these developments reveal the growth of a new conception of human experience and a common conviction that the traditional categories and ideals of abstract reason are insufficient to place men in touch with reality.[42]

The favorable judgment of existence as good, the creative dynamism of Western culture and its gradual trend toward democracy all have their roots in the Judaeo-Christian tradition. According to the Biblical book of Genesis, all things are originally good. Life is therefore worth living, and man is encouraged to exercise his talents for the conquest of subhuman nature. God is the Father of all, and through the sacraments every man has access to His presence. This tradition was once favorable to the development of free personality. Now, however, its intellectual foundations have been undermined, its life solidified in rigid institutions and its actual influence on life and culture negligible. Waldo Frank attributes this significant development to the objective methods of science, which have shattered the ancient picture of an ordered cosmos, and to the advances of modern technology, which have rendered ancient social patterns obsolete.[43] These changes should be welcomed, for if they were rightly used, they might provide us with unparalleled opportunities for the free exercise of personal existence. But they have not been rightly used. Scientific thought has destroyed the old philosophies, but it has given us no superior global view to take their place. Like these philosophies, it has viewed man objectively as a thing, ignoring his free existence as it is lived from the inside. It has neglected the concrete world of lived experience, which goes with such existence, and has tried to squeeze this world into a universe of objective facts that is really abstracted from it.

In this abstract universe of science, no legitimate place can be found for personal life or for anything transcending man. Such diverse modern points of view as that of Hegel, Marx, Spengler and contemporary positivism all agree in subordinating the erratic, individual subject to stable activities, which Hegel called "objective spirit." This widespread anti-personalism is now illustrated in those familiar patterns of mass conformity and mechanical organization that suggest the gloomy predictions of Aldous Huxley's *Brave New World* and George Orwell's *1984*. As against these objectivist trends, Frank maintains that the most essential human values are to be found only in personal choice and fulfillment. A social organization must be judged not merely in terms of its functional efficiency but also in terms of the free existence it supports. Such existence, of course, cannot be produced and organized mechanically. The individual must win it for himself. But he may be aided by accurate descriptions of the concrete world in which he lives from day to day and its wider cosmic dimensions. In this way he may be led not only to a deeper understanding of himself but also to the direct and immediate awareness of a being which both transcends him and touches him very deeply.

This awareness of ultimate mystery can be developed without falling back on traditional religious forms and with full acceptance of the trusts of science. Each individual must work out his own mode of reflection, but here again he may be aided by the experiences of others. (This secular approach is near certain current forms of Protestant thought, and his account of experience on which it is based, is markedly similar to that of such living European thinkers as Karl Jaspers and Gabriel Marcel.[44]

The Fear of Scientism

By "scientism" is meant the transmutation of science into a modern form of logic. The practitioners of scientism are largely popularizers of science and advertising agencies; they have tended to develop the stereotype of the scientist as a person of distinguished appearance who is disembodied, absent-minded, brilliant, infallible, and slightly grey at the temples, and who comes from behind the King George beard he wears while touting "medical" products to appear clean-shaven in his role of salesman for other gadgets.

On the higher level, the faith in this new gadget, with its associated myths, values, and rituals as the scientific ideology has been attacked by numerous thinkers and philosophers (Brooks Adams, Sorokin, Spengler, Lamprecht, Paul Barth, Niebuhr, and others).

The years since 1900 have been marked by social restlessness and by the extraordinary advances made in science and technology. There is a connection between the restlessness and this scientific advance. Most of our economic and labor and ideological problems are the direct result of technologic innovation, and technologic innovation in turn springs from scientific research. Airplanes, automobiles, electron tubes, radio-telephonic communication and broadcasting, television, radar, synthetic rubber, the antibiotics, hormones, vitamins, the release of atomic energy—we saw these evolve before our very eyes.

Of special interest is electronics, a new branch of electrical engineering. It not only gave us a new means of mass communication in the form of broadcasting and television but, as the remarkable computers which have been introduced of recent years indicate, it promises to give us automatic machines which will do much routine office and factory work. The theoretical scientist, still regarded as an impractical dreamer, pointed the way that led to electronics. He tore the atom apart and showed that electronics enters into its structure. Nor should relativity be overlooked. For it told the physicist exactly how much energy can be released from an atom and therefore proved to be of practical value in developing the atomic bomb.

Is it an accident that science is as confused as the world at large today? The cocksureness of the Victorian scientists, to whom all nature was only a machine, has evaporated. So has the cocksureness of statesmen and economists and sociologists. There is the "principle of uncertainty" in atomic physics, which, though it has a special meaning, has a symbolic application in science as a whole and even in the affairs of the world. "The universe is no longer the comfortable infinite expanse that it was before Einstein made our heads swim with new ideas about closed space and about energy that turned into mass when velocities were high enough."[45]

Montaigne thought that a nation with uneventful annals was blessed. The late Dr. Alfred N. Whitehead rejoiced in restlessness and insecurity; for out of them came great scientists, great social reformers. Two terrible world wars have been fought in the years since 1900, yet these same years witnessed great social reform, new scientific benefits, disease conquered, less drudgery, more comfort. Indeed, science has come as much into its own that when today we speak of "progress" we usually refer to scientific and technological progress.

Yet, at the same time, this very course of development in our times has had a decidedly chastening effect upon the former ideas of "progress." The certainty of the late 1890's that science would fertilize a brave new world devoid of disease, poverty and brutality has been transformed into the dread that it might annihilate our old world.[46] White magic has turned to black and the vision of scientists bearing gifts to mankind has been replaced by the specter of Dr. Einstein speaking on equations and logarithms, his language being incomprehensible but terrifying. The age of electricity which we welcomed so rapturously is about to be replaced by the era of atomic energy which may turn our earth into a paradise or into a wasteland. Man's faith has become man's fear.[47]

Is Science a Sacred Cow?

Bishop Wilberforce, debating evolution with Thomas Huxley, used to lay them in the aisles by asking whether Huxley was descended from a monkey on his mother's or his father's side. The world of the 1870's roared. Mr. A. Standen, a chemist by profession and a medievalist at heart, continues the great Wilberforce tradition. As he has a nice wit, he gives the modern reader a laugh or two, if not earth-shaking roars.[48]

He goes at "Science," scientists, pseudo-scientists and the scientific method with dash and vigor, in the best Oxford debating club manner. John Dewey and Lancelot Hogben have their heads briskly knocked together, while relativity is not nearly what it is cracked up to be. He uses all the tricks in the debater's bag: logic chipping, reasoning by analogy, quoting out of context, two-valued judgments, spurious identification, wild extrapolation. For instance, his ultimate conclusion from the strict application of the scientific method, is: "We can say good-bye to freedom."

Standen complains that there is a modern priestcraft building up "Science" into a towering new cult, if not religion, and bids us beware. If there is such a danger, we certainly should beware, states Stuart Chase, but "the way to meet it, one suspects, is by more popular knowledge, not less."[49]

Standen insists that science is becoming an absolute rather than an honest problem-solving tool. It competes with other absolutes, which are immeasurably superior. The proper study of mankind is the pursuit of the True, the Beautiful, and the Good, not vulgar matters like survival. "The first purpose of Science is to learn about God"—a sentiment to which Bishop Wilberforce could take no exception. If one would try to summarize Standen, it appears that he is against free investigation, and advocates investigation limited by Authority, preferably pretty ancient authority when man had a loftier moral sense. He especially deplores the use of the scientific method to help solve some of our desperate political and social problems—even though many of them, such as the factory system and mechanized warfare, are the direct result of science.[50]

It is true that there is a need to evaluate the effects of science and search for ways to use it more intelligently. There is room for severe criticism of its misuses. Quacks and hucksters have sold us many a nostrum with a picture of a white-coated servant, squinting down a microscope and proclaiming that science approves the advertiser's product. The methods by which science has been taught can be also criticized, giving the student intellectual indigestion rather than enlightenment. We can also question the social scientists unable to work out techniques for their own problems, or inventing a fearful, technical patois to cover their delays in finding facts. But Standen would even undermine the citadel of science, since he is apparently against the scientific method for solving human problems. He sternly advocates the older methods of appeals to Authority, to the supernatural and to intuition. He would throw all social scientists bodily out of the window as mischievous meddlers, castigate the psychiatrists, and even call the biologists on the carpet. Only the physicists receive his somewhat frosty nod, provided they stick to matter and energy.[51]

Scientific Philosophy

As we have already noted, there are many interpretations of science. If astrology, magic and religion have played their part in the development of science, and if today it is realized that the fundamental concepts of physics are abstractions framed by the mind to bring order and simplicity into an apparent chaos of phenomena, there are reasons. Most historical relations of science with philosophy have to be constantly revised, and especially in recent decades when the revision again became necessary because first-class theoretical physicists like Einstein, Heisenberg, Schrödinger and de Broglie are not only scientists but also philosophers, because the epistemological writings of A. S. Eddington, Bertrand Russell and others have strongly influenced scientific thinking. The old distinction between matter and energy still holds good in engineering quarters and in everyday life, but it has disappeared in atomic physics. We have also the new era which has been ushered in by the splitting of the uranium atom and the release of enormous amounts of energy. These advances in physical science have their more pallid counterpart in those made by the biological sciences. The leaders of the semantic movement are far from backward in their proposals to clear up all the big philosophical issues about science and human needs by analyzing the language habits underlying our personal and social confusions. Is science merely a tool, or can it be a way of life? What is the moral and philosophical significance of relativity physics? Why do people disagree and how can they reach agreements on moral, political, and religious questions? Is scientific method neutral to ethical questions, or does it have a morality of its own?[52]

The present century has seen the rise of a notable movement in philosophy, which has set about to criticize the ideals of intellectual certainty and moral absolutism, and has undertaken, in a systematic way, to provide a new and more sober conception of the meaning of rationality. In America this movement has been represented by the pragmatists and instrumentalists, whose orientation has been primarily moral and social. In Europe, Bertrand Russell and Ludwig Wittgenstein at Cambridge, the "logical empiricists" of the Vienna Circle, and a school of "scientific empiricism" at Berlin, led by Professor Hans Reichenbach, have approached this task from the point of view of developments in the natural sciences and symbolic logic.[53]

The school believes, in general, that traditional philosophy, from Plato to Kant, raised important questions, even though it formulated them in confusing ways. But it raised these questions before it had the necessary materials for answering them. Their grandiose systems—though presented as products of reason—are little more than rationalizations of non-rational beliefs, of more interest to the psychologist than the philosopher. By "philosopher" (according to Reichenbach) is meant the modern empiricist who has dismissed metaphysics as meaningless, and recognizes as knowledge only the truths obtained by science. Thus Aristotle, Kant, Spinoza, Hegel and Marx had disguised emotional preferences. In the light of the new orientation of this philosophy, moral laws cannot be considered true or false since they are expressions of personal desire and directives for the behavior of others. "Science tells us what is, but not what should be." Most ethical debates are thus over means, rather than ends, and in this realm science can give us genuine answers.

Basically, science does not and cannot determine the judgment of values. Science may help mankind to get what it wants, and clarify the actual alternatives between which culture patterns may choose, show the "costs" of the choices and predict the consequences. But the decision cannot be scientific; hence no ethical and moral code can be provided by scientific method.

Yet, science helps to be build philosophic values, if the culture pattern in which the scientist operates, wants such help. An interesting proposal in this respect has been presented by F. S. C. Northrop, according to whom every culture in the past has had a philosophical basis (a theory of the nature of man and what is good for him).[54] Setting out to prove it in 435 pages of closely reasoned analyses of the histories of Mexico, the United States, Britain, Germany, Russia and the Orient, Northrop concentrates on science, religion and art, but ranges all over the cultural map. One of his basic theses is that the present forms of all Western cultures belong to the past because the assumptions behind them no longer square with scientific and hence philosophic truth. Indeed, they never did; the last system of thought that accounted coherently for all the facts of science known at the time was the Aristotelian system of St. Thomas Aquinas. Aristotle's scientific and self-consistent thought merged with Christian teaching in the 13th century, and provided "interconnections" between all departments of knowledge—something which the Humpty Dumpty modern world has desperately needed. (This is the present appeal of Tomism!). Plato and St. Augustine, for whom the world of sense was almost as unreal as it later was for John Locke, were superseded in St. Thomas' system by Aristotle, biologist and logician, who knew what was "real." But since the discoveries of Galileo and Newton, Western societies have been based upon a succession of scientific philosophies, each of which added its own mistakes in correcting its predecessors. Finally science itself took a turn, with the physics of Einstein, that knocked the props from under even the inadequate philosophies of Locke, Kant and Hegel. "The traditional modern world is as outmoded as the medieval world." In fact, it is worse than that. With no philosophical coherence at the top, faulty and contradictory Western ideologies have been at war, like the societies that cling to them. We need a worldwide philosophical formula that will synthesize the best of East and West.

The answer does not lie in the West going off the deep end into the mysticism of the East (as Aldous Huxley, Christopher Isherwood and other Anglo-American intellectuals seem to have done). Nor the East should drop its own culture for Westernism. "Nothing can do more harm to democracy than the thesis, so popular with many contemporary moral and religious leaders, that science is neutral, if not positively evil, with respect to human values . . ." Scientific method is satisfactory, the glory of the West; but the "modern" views of the world constructed on it were flawed by a basic error, found in Locke, whose 17th century philosophy contained the premises of Jeffersonian democracy, the Declaration of Independence, and the United States Constitution. The error consisted in the theory that "physical substances" (space, planets, flowers) are definable only in Newtonian terms (extension, mass, volume), and thus have no sensuous qualities (depth, heat, fragrance) but are supplied with them by the "mental substance" of the observer. With this "three-termed relation" as the villain of the piece, philosophy has been in trouble for nearly 300 years. As a consequence of it, Locke saw no purpose in government except the protection of private property; and in consequence of that, the United States government has failed to this day to understand the Mexican and other foreign governments. As another consequence, Protestant Christianity conceived the human soul differently from Catholic Christianity, in consequence of which Protestantism has shared the poverty and confusion of "modern" cultures as Catholicism has not. As a further consequence, Western ego and everything else have been priced on the hypothesis of the "free market." (Locke's philosophy was elaborate and to some extent corrected by the German philosopher Immanuel Kant; Hegel followed Kant, and Marx followed Hegel). Northrop's thesis has important implications for the contemporary dilemma presented to the Free World by the USSR. One secret of the "arrogance" displayed by Communist Russia toward us has been its assurance that its philosophical foundations are more modern and hence superior. Communist success is due to the deliberate, swift and powerful application of a philosophy, Marx's, in human history. The Marxist theoreticians understand the traditional Western philosophy and can predict its reasoning in a given set of circumstances. But we need to acquire "a similar realism grounded in a philosophical, economic and political theory which defines what we stand for." Hopeful sources of such a realism already flowing together in America are: (1) a search for coherence like that which has led the University of Chicago's Hutchins and other educators to the "medieval synthesis" of St. Thomas, in which the Lockean error was unknown; (2) a corrected, pragmatic view of precisely what is implied in scientific method, such as that developed by William James, John Dewey, and Alfred North Whitehead; and (3) a realization, such as Latin American cultures have, of the profound reality of sensuous, emotional and artistic experience.

Roman Catholic thought, honoring reason and concrete reality, may achieve in time a new synthesis of philosophy based on the new science. This periodic need for reconstructing philosophic theory is a peculiarity of Western culture. It has become so because the West has concentrated on one method of knowing the world: scientific theory followed by "verification." But all verification is partial at best. Northrop insists that mankind possesses riches and certainties to which the theorizing West has scarcely begun to wake up. But Einstein and recent philosophic thought indicate that the waking is at hand. In all knowledge there are not three terms, as Locke thought, but two: the aesthetic (*e.g.*, a book in your hand), and the theoretical (*e.g.*, your inferences about its writer). What Northrop calls the aesthetic component is what Western artists have sometimes called the sense of life; it is existence appreciated; it is what we know of life by seeing and feeling, by intuition, not by reasoning (existentialism?). To know this is the wisdom of the Orient; and in the great religions of the East, most purely in Buddhism, it has been cultivated through thousands of years as the ultimate reality. In the West, even artists were rarely content to render the sensuous world—the aesthetic component—for its own sake until 19th century Impressionism. Yet if all devotees of the theoretic component—Anglo-Americans in particular— can learn the religious value of direct experience, fanaticism and confusion would cease. This would mean, for one thing, that the arts would gain greater importance than the West has ever given them. East and West can get together by a "two-termed epistemic correlation" between the aesthetic and theoretical components of reality. The West, with its theoretical knowledge (of electromagnetic fields, atoms that are never seen, but "verified" by flashes, explosions, etc.) has much to offer to the East, once the error of the West's philosophic ways since Locke is corrected. The East, with its intuitive, contemplative knowledge of mother earth knows a lot that has no place in the West's scientific structure, and thereby finds the West's systematizing barren of much delight and wisdom. The need is to "correlate" the aesthetic and the theoretical into a philosophical ideal that will do for all civilization the world over.[55]

Problems of Leisure

One of the most significant consequences of the rapid development of scientific technology and the increase of productivity in industrial enterprises has been the process of uniform reduction of working hours. Even if we take into account the differences in this genesis among the various countries of the world, we can safely say that in Europe, the 45-hour week has become an established fact, while in the United States, this reduction has been approaching 36 hours, and the 7-hour day and the 5-day weeks has here become a commonplace.[56] If we recall that only about a 100 years ago the average work week in the U. S. was around 70 hours, we can appreciate the social range and rapidity of this development. In the course of one century, the working time in industry has halved to a certain extent; and this trend has been accelerated in the last few years, and there are all the indications that the working time is on the verge of being reduced even more rapidly in the years to come.

The basic problem here is how the "released" time can be usefully spent. It is a phenomenon which is in obvious contradiction to certain current ideologies, such as the heightened dignity of labor and the cult of efficiency, in which is reflected the Puritan element in the capitalistic doctrine, along with the much older conviction to the effect that leisure is a privilege, a condition of life reserved only to the upper classes. By and large, the ambiguity of the concept is contained in the very term used to designate it. The English "leisure" and the French "loisir," bring to the mind not so much a lapse of time as merely a quality and a style of life which traditionally accompanies social and economic privileges. In fact, it has been less than half a century since Veblen saw in "leisure," in the exclusive right to leisure, the characteristic feature of the class wielding economic and social power, distinguishing the upper class in the most unequivocal manner from the industrial working class, which at that time entirely lacked such a privilege.[57]

Today this condition of free disposal over time is tending especially in some of the most intensively industrialized societies to individualize one class, the industrial working class, which is emerging from a condition of having no leisure, while economic and social power is tending to be more and more concentrated in the hands of a minority of leaders—managers, scientists, technicians, etc.—who, if anything, are suffering from overwork. Basically, leisure for the working classes is one of the most striking results of the scientific applications of technology.

These trends are being interpreted within the framework of conflicting opinions. There are the optimists, the utopians, who see the future as a new earthly paradise; but there are also pessimists who stress that if man does not succeed in destroying himself with the atomic bomb, he will accomplish this anyhow with all the leisure time that he will have on his hands.

The working conditions of the workers have a direct influence on the way they spend their leisure. These conditions are, nearly universally, characterized by spasmatic activity, by lack of interest in the purpose of one's own work, by lack of any identification with the product of one's own daily labor, and by insufficient social participation in the life of the industrial enterprise (the factors which have been analyzed by a whole generation of sociologists and psychologists, from Durkheim to Elton Mayo, or by Friedmann and by Fromm, stressing especially the conditions of "alienation" and "deflection") .[58] It is only natural that the workers in the best of cases search in their leisure that satisfaction and self-realization which they cannot obtain in their work. The way in which leisure is spent is consequently determined by the fact that it would assume the character of typically compensatory behavior. In some respects, then, this type of leisure activities is conditioned by the influence of the negative factors produced by the working conditions which produce dissatisfaction and tend to warp the personality of the worker.

The negative aspects of this "leisure behavior" have not escaped the attention of the managers of many industrial concerns when organizing their employees' leisure and which aim precisely at compensating these elements of disharmony and dissatisfaction produced by working conditions.[59] Among the many we can mention such as: the monotony of factory work, the privations of young workers, the frustrated creative urge, the deterioration of family ties, etc.

Here is the concrete task of industrial leaders to find and seek the suitable means so that the process of industrialization can be switched from the negative aspects to more positive ones, changing it from a curse for humanity, a condition of deprivation, insecurity and isolation, to the promotion of creative leisure organizations which would enrich the cultural and "adult educational" aspects of the "non-working" hours. Such aims would promote working conditions which would help the workers to identify themselves, of their own free will, with their work and participate creatively in the company's activities, even in management, to increase their specialized and civic knowledge, adjust themselves culturally to their new responsibilities. There is a clear correlation between education and recreation.

Revolutionizing Human Relations in Industry

In more recent decades another Industrial Revolution is influencing U. S. industry: Human Relations in Industry, whose aim is to give the American worker a sense of usefulness and importance—and thus improve his work. Its goal is to make life more fun by making work more meaningful.[60]

Because of the non-human aspects of the technological behavior system, modern analysts of human relations have, until recently, ignored the human aspects of the production system. Whatever studies had been made focused on the technological behavior systems of work as being determinate of working behavior. The seeds of the contemporary change were sown by two great pioneers in this field, Frederick Winslow Taylor, a one-time day laborer, and Elton Mayo, an Australian immigrant, turned Harvard sociologist. Taylor, who died in 1915, was the father of scientific management; he increased industrial production by rationalizing it. Mayo, who died in 1949, was the father of industrial human relations; he increased production by humanizing it. It was Mayo's *The Human Problems of an Industrial Civilization* (1933) (and especially F. J. Roethlisberger's and W. J. Dickson's *Management and the Worker*, Harvard, 1939) which stressed that working conditions, if varied, did not produce expected changes in working behavior. This resulted in a very fruitful emphasis on the social aspects of work.

Taylor, while working at the Midvale (Pennsylvania) Steel Works in the 1880's, discovered that the workers, and not the bosses, determined the production rate. Having learned their jobs by rule of thumb, the workers wasted steps, motion and time. Using a stop-watch, Taylor discovered that he could determine the most efficient speed for every operation by breaking it into its component parts. Later he studied employees shoveling ore, coal, and the like for Bethlehem Steel, and found that the output varied widely according to the size of shovels. He experimented with a shorter shovel holding 30 lbs., instead of 34 and discovered that each man's daily tonnage rose—until a 21-lb. load was reached. Thereafter output fell. Taylor then set 21¼ lbs. as the ideal shovel—and the yard force was cut by two-thirds while daily loadings rose from 25 tons per man to 45. Taylor's pioneering in time and motion studies helped bring the mass-production era which helped workers to raise not only their output but their wages as well. But he also geared human operations to the precision of machines, making the managers think of workers as little more than machines that had to eat. The consequences were the fight for the utmost utilization of time, the "speed-up," where assemblies moved always a little faster than men's natural work pace.

In 1923 Elton Mayo was called into a Pennsylvania textile plant where the labor turnover was high in spite of various wage incentives. He discovered that the workers were unhappy, due to the machines depriving them of virtually all human contacts, and fell into melancholy and hypochondria. Mayo advised 4 daily rest periods and the hiring of a nurse who would listen to the workers' complaints. The changes were startling. Four years later, Western Electric at Hawthorne Works (near Chicago) experimented with two rooms of varied lighting, and could not figure out why both groups produced high. Mayo discovered that the answer lay in both groups being singled out for special attention.

Mayo's discoveries were hailed as a landmark in social science—although, actually, every worker could tell the manager that every human being wants to have his work appreciated, that the boss is interested in him and appreciates what he does. In the background was, therefore, the idea of a class warfare predicted by Karl Marx.

Mayo's new science began to penetrate. The legal change came with the 1937 U. S. Supreme Court Decision upholding the Wagner Act, forcing the management to learn to live with unions. The change was speeded up by World War II, which forced maximum output because of the patriotic necessity and brought in also housewives and other workers free of old suspicions and hostilities. Surveys after surveys exploded one of management's folklore—that workers wanted only more money. After 10 years of polling workers, Elmo Roper discovered that the four chief desires of workers are: (1) security ("the right to work continuously at reasonably good wages"); (2) a chance to advance; (3) treatment as human beings; and (4) dignity.[61]

Other studies followed. They pointed out especially that the robot nature of work thwarts the craving for prestige; the constant increase in mechanization is reducing the sense of personal pride and self-identification with the final product; the hope of advancement is constantly lost in the growing tendency to choose management material not from the bench but from young, college-trained technicians.

Conclusions

On October 4, 1957, the "Sputnik Day," the care and health of American Science suddenly became an important subject, and even more important than inflation, desegregation, foreign aid and full employment. Since then, a torrent of concerned and worried words has been flooding us, and much has been achieved: three American satellites wheel in orbit; a thousand American communities are rearranging school curricula to produce more scientists; the President has added to his staff an official science adviser; the Federal government has increased its educational programs and its support for education in the natural sciences (including mathematics), and Congress has appropriated special funds to promote science in the American educational system; educators throughout the country have been seeking ways to attract students to careers in science—and wondering how they can hold the interest of those they have.

As we have seen, science has been praised, as well as condemned. Whatever the case, the fact remains that science and technology are "what is" and there is no other way to handle its problem than to start with "what is" in order to accomplish the needed changes—as seen in the following chapters.

America has a history of the unique interplay of the three shaping forces of the modern Western world:[62] industrialism as a technology, capitalism as a way of organizing it, and democracy as a way of running both; from these comes an immense dynamic force which moves hand in hand with the great motif that men can find well-being and freedom here. Since technology is today inseparable from science, and is being used in the present struggle for world supremacy,

"Our development of more and better trained scientific, engineering, and other professional manpower is important to our national survival. It is fundamentally an educational problem, involving the whole organization and system of education in the United States. The problem can be attacked from many approaches. There is no single solution."[63]

Selected Bibliography

Irving Adler, *The Tools of Science: From Yardstick to Cyclotron* (New York: John Day, 1958). A simplified study of some of the tools important to man's scientific progress.

W. I. B. Beveridge, *The Art of Scientific Investigation* (New York: W. W. Norton, 1951). Illustrated with 16 portraits of outstanding scientists, it is flavored with numerous anecdotes of how discoveries have been made and by quotations from great discoverers.

Glenn O. Blough, Julius Schwartz & Albert J. Huggett, "Objectives in Elementary Science," *School and Society*, LXXXVI, 2133 (May 24, 1958), 244-250. Book review of *Elementary-School Science and How to Teach It.*

N. H. D. Bohr, *Atomic Physics and Human Knowledge* (New York: John Wiley, 1958). Essays concerned with atomic physics and its relationship to other fields of knowledge.

John Ely Burchard, Ed., *Mid-Century: The Social Implications of Scientific Progress* (New York: John Wiley, 1950), discussions of such problems as the present state and future promise of science; technology and the conservation of natural resources; the problem of specialization in modern education and in society generally; etc.

G. W. Corner, *Anatomist at Large: An Autobiography and Selected Essays* (New York: Basic Books, 1958). An autobiography and scientific and philosophic essays by one of the prominent biologists of the United States.

W. E. Dick, *Atomic Energy in Agriculture* (New York: Philosophical Library, 1957). How atomic energy has greatly expanded the field of radiation genetics.

Sir Arthur Eddington, *The Philosophy of Physical Science* (Ann Arbor, Mich.: University of Michigan Press, 1958). By one of the world's leading astrophysicists.

P. G. Frank, *Philosophy of Science: The Link Between Science and Philosophy* (Englewood Cliffs, N. J.: Prentice-Hall, 1957). An examination of the relationships between scientific and philosophical theories discussing such concepts as idealism, materialism, Newtonian relativity, etc.

R. G. Gabriel, *The Course of American Democratic Thought: An Intellectual History Since 1815* (New York: Ronald Press, 1940), Ch. 14, "The Science of Man," 161-172. An excellent survey of the application of scientific theories to human relations in American climate.

E. W. Hall, *Modern Science and Human Values* (Princeton, N. J.: D. Van Nostrand, 1956). An examination of the historical ideas in medieval theories of motion, economic thought, 17th century Hobbism and present-day Existentialism.

J. H. Hildenbrand, *Science in the Making* (New York: Columbia University Press, 1957). Interpretations of the rightful role of science study in American education.

Edward Hutchings, & others, eds., *Frontiers in Science: A Survey* (New York: Basic Books, 1958).

Sir James Jeans, *Physics and Philosophy* (Ann Arbor, Mich.: University of Michigan Press, 1958). An explanation of how the quantum theory and the new science have had an effect on modern philosophic thought so that the principles of physics, even though not understood by most laymen, have profoundly influenced basic thinking outside the field of science.

W. S. Jevons, *The Principles of Science: A Treatise on Logic and Scientific Method* (New York: Dover Publications, 1958). Republication of a treatise by a 19th century English logician and economist; with a new introduction.

Norman Landsell, *The Atom and the Energy Revolution* (Baltimore, Maryland: Penguin, 1958). The social and political impact of atomic energy developments; the demand for energy in the world, sources of power including atomic power, the sources of materials for atomic power, the way different countries and international bodies may exploit it.

Jean Lindsay, "Science in the Dark Ages," *History Today*, I (July, 1951), 25-34.

George F. Kennan, *Russia, the Atom and the West* (New York: Harper, 1958). Prompted world-wide discussions because of the startling military, political and economic suggestions.

D. C. McClelland, A. L. Baldwin, Uri Bronfenbrenner & F. L. Strodtbeck, *Talent and Society* (Princeton, N. J.: D. Van Nostrand, 1958). Explores new concepts of achievements and new methods of assessing the contribution of personality and achievement.

L. C. Pauling, *No More War* (New York: Dodd, Mead, 1958). The noted scientist paints a grim picture of atomic warfare and condemns atomic testing and fallout as serious dangers; he calls for cessation of testing and for international agreements for peace.

Peaceful Uses of Atomic Energy (Proceedings of the Second International Conference on the Peaceful Uses of Atomic Energy) (New York: United Nations, 1958).

Frederick Pollock, *Automation. A Study of Its Economic and Social Consequences* (New York: Praeger, 1958).

Harlow Shapley, *Of Stars and Men: the Human Response to an Ex-*

panding Universe (Boston: Beacon Press, 1958). An essay centering around the position of mankind "in the universe of physics and sensation," and considering the continuity in the series of living organisms, from inorganic organism to man, and the question of human destiny.

Charles Singer & others, eds., *A History of Technology:* Vol. 4: *The Industrial Revolution c. 1750 to c. 1850* (New York: Oxford, 1958).

R. G. H. Siu, *The Tao of Science: An Essay on Western Knowledge and Eastern Wisdom* (New York: John Wiley, 1958). A remarkable essay on the nature and limitations of Western science, on the wisdom of Eastern "no-knowledge," and on the advisability of their mutual supplementation in organized research.

J. W. Still, *Science and Education at the Crossroads: A View from the Laboratory* (Washington, D. C.: Public Affairs Press, 1958).

UNESCO Sourcebook for Science Teaching (1957). Handbook based on the work of numerous experts sent by Unesco to help towards the improvement of scientific education in the various countries.

C. R. Walker, *Towards the Automatic Factory: A Case Study of Men and Machines* (New Haven: Yale University Press, 1957). Study of the introduction of automation into an American iron works; takes account of psychological and sociological and other factors; the author is optimistic.

JOSEPH S. ROUCEK, Chairman, Departments of Political Science and Sociology, University of Bridgeport (Connecticut), was born in Czechoslovakia (1902), emigrated to the United States in 1921 and became an American citizen in 1927. A former pupil of Dr. Eduard Benes (who then taught at the Czechoslovak Commercial Academy and later became Foreign Minister and President of the Republic of Czechoslovakia), he received his B.A. degree from Occidental College, and his M.A. and then Ph.D. from New York University (1927, 1937). He started his teaching career as Professor of Social Sciences at the Centenary Junior College (Hackettstown, N. J.) (1929-1933), then as a staff member and Visiting Professor in various American, Canadian and South American Universities (Pennsylvania State University, 1933-1935; New York University, 1935-1939; Hofstra College, 1939-1948; Kent State University, 1940; San Francisco State College, 1941, 1942, 1944, 1945, 1946; College of the Pacific, 1942; University of Wyoming, 1944; San Diego State College, 1945; Occidental College, 1948, 1949; University of British Columbia, 1951; University of Puerto Rico, 1952; Portland State College, 1954). He is the author (*The Balkan Politics,* Stanford University Press, 1948; *The Working of the Minorities Treaties Under the League of Nations,* 1928; etc.), co-author, editor and collaborator of some 100 books (*Our Racial and National Minorities,* Prentice-Hall, 1937, 1939; *One America,* Prentice-Hall, 1956; *Sociology: An Introduction,* Littlefield, Adams, 1951; *Comparative Education,* Dryden Press, 1956; *Social Control,* D. Van Nostrand, 1956; *Recent Trends in Sociology,* Philosophical Library, 1958; etc.). The pre-Communist governments awarded him the Knighthood of the Crown of Yugoslavia, and the Knighthood of the Star of Romania. He is also a member of the editorial boards of *The American Journal of Economics and Sociology; Social Science; Journal of Human Relations.*

NOTES FOR THE CASE FOR AND AGAINST SCIENCE AND SCIENTISM

1. A poll conducted for the National Association of Science Writers and New York University, cited in *Publishers' Weekly,* CLXXIV, 8 (August 25, 1958), 10; the Survey Research Center, University of Michigan, interviewed 1,919 people and the results were analyzed in *Science, the News and the Public* by Hillier Kireghbaum (New York University Press, 1958), and *The Public Impact of Science in the Mass Media,* by Robert C. Davis (Ann Arbor, Mich.: Survey Research Center, 1958).

2. Concerned with the communication of facts with science, the survey group asked where the public gets its chief news about science. Sources mentioned included: newspapers, 64%; TV, 41%; magazines, 34%; radio, 13%. One noteworthy finding was that 42% of the people interviewed wanted newspapers to print more medical news, 28% would like to see more space given to other scientific news. Science news interest was especially marked in the West. To get more science into the newspapers, 19% would be willing to give up some society news; 17% some sports news; 10%, comics. Books played a minor part in this survey, but it was shown that only 47% of those questioned read at least one book in 1957, and only 20% read five or more books, mostly fiction; mention of science books was negligible.

3. John J. O'Neill, "Newest Conception of Science, a Chain Reaction in Knowledge," New York *Herald Tribune* (June 1, 1957).

4. James B. Conant, *On Understanding Science: An Historical Approach* (New Haven: Yale University Press, 1947).

5. P. H. Phenix, *Philosophy of Education* (New York: Holt, 1958), Ch. 17, "Knowledge," 297-320, & Ch. 18, "Science," 321-340.

6. James B. Conant, *Science and Common Sense* (New Haven: Yale, 1951).

7. From this point of view, modern science, which began in about 1600, had little effect on industry or the practical arts for its first 200 years. Indeed Dr. Conant contends that it was not until the launching of the electrical or dye-stuffs industries in about 1870 that science became of real significance to industry.

8. Jessie Bernard, "Can Science Transcend Culture?" *The Scientific Monthly,* LXXI, 4 (October, 1950), 268-273. John U. Nef, *Cultural Foundations of Industrial Civilization* (New York: Cambridge University Press, 1958), traces the development of science and culture and their

influence on industrialism and the several bases of civilization which contributed to our material development.

9. Bernard, *op. cit.*, 270.

10. Howard Becker, "Science, Culture, and Society," *Philosophy of Science*, XIX, 4 (October, 1952), 273-287; Becker, *Through Values to Social Interpretation* (Durham: Duke University Press, 1950).

11. Henryk Mehlberg, *The Reach of Science* (Toronto University Press, 1958), a technical study of the philosophy of science, its limitations and its methods of problem solving; James Stokley, *Science Remakes Our World* (New York: Ives Washburn, Jr., 1946); Herald Holton, Ed., Science and the Modern Mind (Boston: Beacon, 1958), Fred Reinfeld, *Miracle Drugs and the New Age of Medicine* (New York: Sterling Publ. Co., 1958); etc.

12. James Cavanaugh, "Facts About Lenses," *Camera 35*, II, 4 (1958), 279 ff.; Joseph S. Roucek, "A Sociological Comment on the Box-Camera Fan," *Southwestern Social Science Quarterly*, XXXVIII, 19 (June, 1957), 51-54.

13. For a bitter evaluation of this school of thinking, see: Pitirim A. Sorokin, *Fads and Foibles in Modern Sociology and Related Sciences* (Chicago: Regnery, 1956).

14. F. S. C. Northrop, *The Logic of Sciences and the Humanities* (New York: Macmillan, 1947).

15. Bertrand Russell, "The Science to Save Us From Science," New York *Times Magazine,* (March 19, 1950), 9 ff.

16. For instance, on September 14, 1958, 73 eminent scientists from East and West met in the third "Pugwash conference" to begin discussions about the dangers of atomic war; significantly, Premier Nikita S. Khruschev of the USSR sent a message of greetings to the congress, hoping that scientists would join efforts in the struggle for the prevention of atomic war. On September 16, 1958, Dr. James K. Pollock (Professor of Political Science at the University of Michigan) opened the fourth world congress of the International Political Science Association in Rome, and warned against the "political naïveté" of the space age's scientists since "foreign and military policy is hardly the task of the nuclear physicist."

17. P. W. Bridgman, "How Far Can Scientific Method Determine the Ends for Which Scientific Discoveries are Used?" *Social Science*, XXII, 3 (July, 1947), 206-212. The view that "science is ethically neutral" is also propounded by G. A. Lundberg, "Can Science Validate Ethics?" *Bulletin of the American Association of University Professors*, XXXVI, 2 (Summer, 1950), 262-275.

18. See such works as: John Dewey, *Logic, the Theory of Inquiry* (New York: Holt, 1930), one of the best comprehensive discussions of the theory of science; George A. Lundberg, *Social Research* (New York: Longmans, Green, 1942), discusses different techniques of research in social science; Robert S. Lind, *Knowledge for What?* (Princeton University Press, 1939), urges that social science should aid in the solution of social problems; Henri Poincaré, *The Foundation of Science* (Lancaster,

Pa.: The Science Press, 1914), greatly influenced more recent works; G. A. Lundberg, *Can Science Save Us?* (New York: Longmans, Green, 1947), promise and limitations of sociology as a natural science applied to the regulation of human affairs; Lyman Bryson & Louis Finkelstein, Eds., *Science, Philosophy and Religion* (New York: Harper, 1943); J. R. Baker, *The Scientific Life* (New York: Macmillan, 1943); Paul W. Taylor, "Social Science and Ethical Relativism," *The Journal of Philosophy,* LV, 1 (January 2, 1958), 32-34; Lawrence K. Frank, "Research for What?" *The Journal of Social Issues,* Supplement Series, No. 10 (1957); etc. Notice also that the social usefulness of science is the dominant theme of Marxian thinking; see: Joseph S. Roucek, "Fate of Sociology in the Soviet Union," *Ukrainian Quarterly,* XIII, 3 (September, 1957), 234-242.

19. This whole problem has produced quite a literature in the field known as the Sociology of Knowledge; see: Leo P. Chall, "The Sociology of Knowledge," 286-304, in Joseph S. Roucek, Ed., *Contemporary Sociology* (New York: Philosophical Library, 1958).

20. For the difference between the ideological and scientific reasoning, see: Joseph S. Roucek, *Social Control* (Princeton, N. J.: D. Van Nostrand Co., 1956), Ch. XII, "Ideologies," 185-204.

21. John Ely Burchard, Ed., *Mid-Century: The Social Implications of Scientific Progress* (New York: Wiley, 1950); Bertrand Russell, *The Impact of Science on Society* (New York: Columbia University Press, 1951); R. N. Anshen & others, *Science and Man* (New York: Harcourt, Brace, 1942); S. A. Nock, "The Scientist and Ethics," *Ethics,* LIV, 1 (July, 1943), 14-28; Sir William C. Dampier, *A History of Science and Its Relation With Philosophy and Religion* (New York: Macmillan, 1946); P. P. Wiener, Ed., *Readings in Philosophy of Science: Introduction to the Foundations and Cultural Aspects of the Sciences* (New York: Scribners', 1953); Philipp Frank, *Philosophy of Science: The Link Between Science and Philosophy* (Englewood Cliffs, N. J.: Prentice-Hall, 1957); A. Wolf, *A History of Science, Technology, and Philosophy in the Sixteenth and Seventeenth Centuries* (London: G. Allen & Unwin, 1950); E. A. Heath, Ed., *Scientific Thought in the Twentieth Century* (New York: F. Ungar, 1953); etc.

22. F. E. Hartung, "Science as An Institution," *Philosophy of Science,* XVIII, 1 (January, 1951), 35-54.

23. E. P. D. Wightman, *The Growth of Scientific Ideas* (New Haven: Yale, 1951); P. P. Wiener & Aaron Noland, Eds., *Roots of Scientific Thought: A Cultural Perspective* (New York: Basic Books, 1958); George Sarton, *A History of Ancient Science: Through the Golden Age of Greece* (Cambridge: Harvard University Press, 1952); George Sarton, *A Guide to the History of Science* (Waltham, Mass.: Chronica Botonica, 1954); etc.

24. James G. Frazer, *The Golden Bough* (New York: The Macmillan Co., 1943), 48-52, 711-712.

25. *Ibid.*

26. Hartung, *op. cit.*, 38.

27. R. J. Turgot, *On the Progress of the Human Mind,* trans. by Mc-Quilkan DeGrande (Hanover, N. H.: The Sociological Press, 1929), 6.

28. H. S. Williams, *A History of Science* (New York: Harper, 1904), 2 vols., I: Chapter VII, "Greek Science in the Early Attic Period," 140-178, V, "Post-Socratic Science at Athens," 179-188, VI, "Greek Science of the Alexandria or Hellenistic Period," 189-254; M. R. Cohen & I. E. Drakkin, *A Source Book in Greek Science* (New York: McGraw-Hill, 1948); Sir William C. Dampier, *A History of Science and Its Relation with Philosophy and Religion* (New York: The Macmillan Co., 1946), I, "Science in the Ancient World," 1-64; S. F. Mason, *Main Currents of Science Thought* (New York: Henry Schuman, 1953), Part One, "Ancient Science," 5-52.

29. H. Butterfield, *The Origins of Modern Science, 1300-1800* (New York: Macmillan, 1956).

30. D. J. B. Hawkins, *A Sketch of Medieval Philosophy* (New York: Sheed & Ward, 1947); Bertrand Russell, *A History of Western Philosophy* (New York: Simon & Schuster, 1945), 428-490; J. L. La Monte, *The World of the Middle Ages* (New York: Appleton-Century-Crofts, 1949), Book I.; Lynn Thorndike, *History of Medieval Europe* (Boston: Houghton, Mifflin, 1928); etc.

31. A. R. Hall, *The Scientific Revolution, 1500-1800* (London: Longmans, 1954); Herbert Butterfield, *The Origins of Modern Science, 1300-1800* (New York: The Macmillan Co., 1951); Charles Singer, E. J. Holymard, A. R. Hall, & T. I. Williams, Eds., *A History of Technology* (New York: Oxford, 1958), III: *From the Renaissance to the Industrial Revolution, c. 1500 to c. 1750;* W. P. D. Wightman, *The Growth of Scientific Ideas* (New Haven: Yale, 1951); A. R. Hall, *The Scientific Revolution, 1500-1800: The Formation of the Modern Scientific Attitude* (Boston: Beacon, 1956); etc.

32. Roller, *op. cit.;* Lynn Thorndike, *A History of Magic and Experimental Science* (New York: Columbia University Press, 1958), VII & VIII: *The Seventeenth Century.*

33. Mitchell Wilson, *American Science and Invention: A Pictorial History* (New York: Simon and Schuster, 1954); Brooke Hindle, *The Pursuit of Science in Revolutionary America, 1735-1789* (Chapel Hill, N. C.: University of North Carolina, 1956); F. D. Curtis & George G. Mallinson, *Science in Daily Life* (Boston: Ginn, 1958); etc.

34. L. L. Snyder, *The Age of Reason* (Princeton: D. Van Nostrand, 1955); Stuart Hampshire, *The Age of Reason* (New York: The New American Library, 1956); Crane Brinton, *The Shaping of the Modern Mind* (New York: The New American Library, 1953), Ch. 3, 83-111; Herbert Butterfield, *The Origins of Science* (London: Bell, 1949); A. N. Whitehead, *Science and the Modern World* (New York: The Macmillan Co., 1925); Preserved Smith, *A History of Modern Culture* (New York: Holt, 1930-34), I, Chapters 1, 2, 3.; etc.

35. Quoted in Réné Fulop-Miller, *Leaders, Dreamers and Rebels: An Account of the Great Mass Movements in History and of the Wish-Dreamers that Inspired Them* (New York: The Viking Press, trans., 1935), 269.

36. Fulop-Miller, *op. cit.*, Ch. VI, "The Dream of the Great Magic," section 2, "Machinery Has Done It," 267-276.

37. Roger Soltau, *French Political Thought in the Nineteenth Century* (New Haven: Yale, 1931); H. W. Laidler, *History of Socialist Thought* (New York: T. Y. Crowell, 1927), Chs. VII-XII; J. O. Hertzler, *History of Utopian Thought* (New York: The Macmillan Co., 1923); M. Beer, *History of British Socialism* (London: Bell, 1919); Thomas Kirkup, *History of Socialism* (London: G. Black, 1913); Lewis Mumford, *The Story of Utopias* (New York: Boni & Liverright, 1922).

38. Fulop-Miller, *op. cit.*, Ch. VI, "The Dream of the Great Magic," 252-333.

39. Joseph Stalin, "Economic Problems of Socialism in the USSR," supplement to *Soviet Weekly* (October 30, 1952), ii.

40. Robert C. Tucker, "Marxism—is it Religion?" *Ethics*, LXVIII, 2 (January, 1958), 125-30.

41. But Marxism must be credited with making our understanding of ideologies possible by its attack on the theories and practices of the bourgeois society. "A clear ideology," Engels concurred with Marx, is "the evaluation of the reality not from the reality itself, but from imagination."

42. William Barrett, *Irrational Man: A Study in Existential Philosophy* (New York: Doubleday & Co., 1958).

43. Waldo Frank, *The Rediscovery of Man: A Memoir and a Methodology of Modern Life* (New York: George Braziller, 1958).

44. Karl Jaspers, "Is Science Evil?" *Commentary*, IV, 3 (March, 1950), 229-233, claims, for instance, that "In our present situation the task is to attain to that true science which knows what it knows at the same time that it knows what it cannot know. This science shows us the ways to the truth that are the indispensable precondition of every other truth . . ."

45. Editorial, "Science Since 1900," New York *Times*, January 7, 1950.

46. See: Dexter Masters & Katherine Way, Eds., *One World or None* (New York: McGraw-Hill, 1946); of particular interest is the concluding statement, "Survival Is at Stake" by The Federation of American (Atomic) Scientists, and the article, "There is No Defense" by Louis N. Ridenour, who developed new types of radar at the Massachusetts Institute of Technology.

47. Pitirim A. Sorokin, "Physicalist and Mechanistic School," 1127-1176, in Joseph S. Roucek, Ed., *Contemporary Sociology* (New York: Philosophical Library, 1958); Sorokin, *The Crisis of Our Age; The Social and Cultural Outlook* (New York: Dutton, 1957); S. M. Rosen & Laura Rosen, *Technology and Society* (New York: The Macmillan Co., 1941);

Stuart Chase, *The Proper Study of Mankind: An Inquiry into the Science of Human Relations* (New York: Harper, 1956); C. R. Walker & R. H. Guest, *The Man on the Assembly Line* (Cambridge: Harvard University Press, 1952); Jacques Maritain, *Freedom in the Modern World* (London: Sheed & Ward, 1935); George Orwell, *Nineteen Eighty-Four* (New York: Harcourt, Brace, 1949); Karl Mannheim, *Man and Society in an Age of Reconstruction* (New York: Harcourt, Brace, 1941); *Public Reaction to the Atomic Bomb and World Affairs: A Nation-Wide Survey of Attitudes and Information* (Ithaca, N. Y.: Cornell University Press, 1947); Hans Speier, *German Rearmament and Atomic War* (Evanston, Ill.: Row, Peterson, 1957); Herbert Agar, *A Declaration of Faith* (Boston: Houghton, Mifflin, 1952); Bernard Brodie, Ed., *The Absolute Weapon: Atomic Power and World Order* (New York: Harcourt, Brace, 1946); John Ely Burchard, Ed., *Mid-Century: The Social Implications of Scientific Progress* (Cambridge, Mass.: Technology Press, 1950); Norman Cousins, *Who Speaks for Man?* (New York: The Macmillan Co., 1953); Walter Isard & Vincent Whitney, *Atomic Power* (New York: McGraw-Hill, 1955); Anatol Rapoport, *Science and the Goals of Man: A Study in Semantic Orientation* (New York: Harper, 1950); Theodor Rosebury, *Peace or Pestilence?* (New York: McGraw-Hill, 1949); W. M. Smith, *This Atomic Age and the Word of God* (Boston: Wilde, 1948); George Thomson, *The Foreseeable Future* (New York: Cambridge University Press, 1955); etc.

48. Anthony Standen, *Science is a Sacred Cow* (New York: Dutton, 1950).

49. Stuart Chase, reviewing Standen's book, New York *Herald Tribune*, April 23, 1950.

50. Similar attacks on science have been offered by: Martin Gardner, *In the Name of Science* (New York: Putnam, 1952); D. W. Hering, *Foibles and Fallacies of Science* (New York: D. Van Nostrand, 1924); Joseph Jastrow, Ed., *The Story of Human Errors* (New York: Appleton-Century, 1936); D. S. Jordan, *The Higher Foolishness* (Indianapolis, Ind.: Bobbs-Merrill, 1927).

51. In the field of social sciences, the arguments against "scientism" are summarized in A. H. Hobbs, *Social Problems and Scientism* (Harrisburg, Pa.: The Stackpole Sons, 1953) and *The Claims of Sociology* (Harrisburg, Pa.: The Stackpole Sons, 1951); L. Bloomfield, "Linguistic Aspects of Science," *International Encyclopedia of Unified Science* (University of Chicago Press), 1939, I, no. 4; M. R. Cohen & E. Nagel, *An Introduction to Logic and Scientific Method* (New York: Harcourt, Brace, 1934); Anatole Rapaport, *Science and the Goals of Men, A Study in Semantic Orientation* (New York: Harper, 1950); Bergen Evans, *The Natural History of Nonsense* (New York: Knopf, 1946, 1958); G. A. Lundberg, *Can Science Save Us?* (New York: Longmans, Green, 1947); Lundberg, *Foundations of Sociology* (New York: The Macmillan Co.,

1939); Karl Pearson, *The Grammar of Science* (London: Walter Scott, 1892); etc.

52. W. C. Dampier, *A History of Science and Its Relations with Philosophy* (New York: Macmillan, 1949); R. N. Anshen, Ed., *Science and Man* (New York: Harcourt, Brace, 1942); Anatol Rapaport, *Science and the Goals of Man, A Study in Semantic Orientation* (New York: Harper, 1950); Lyman Bryson, *Science and Freedom* (New York: Columbia University Press, 1947); Philipp Frank, *Philosophy of Science: The Link Between Science and Philosophy* (Englewood Cliffs, N. J.: Prentice-Hall, 1957).

53. Hans Reichenbach, *The Rise of Scientific Philosophy* (Berkeley, Calif.: University of California Press, 1951). This philosophical trend is, however, only one of the many philosophical interpretations of science.

54. F. C. S. Northrop, *The Meeting of East and West: An Inquiry Concerning World Understanding* (New York: The Macmillan Co., 1946).

55. The relationship between the scientific and aesthetic aspects of culture patterns has been of interest to many scientists and thinkers; for a brief but a succinct introduction to this field, see: Harry Elmer Barnes, *Historical Sociology: Its Origins and Development* (New York: Philosophical Library, 1948); also: Barnes, *History of Historical Writing* (Norman, Okla.: University of Oklahoma Press, 1937); Barnes, Ed., *An Introduction to the History of Sociology* (University of Chicago Press, 1947); J. B. Bury, *The Idea of Progress* (New York: The Macmillan Co., 1932); F. Stuart Chapin, *Cultural Change* (New York: Century, 1928); Pitirim Sorokin, *Contemporary Sociological Theories* (New York: Harper, 1928); W. D. Wallis, *Culture and Progress* (New York: McGraw-Hill, 1930).

56. F. Orlandini, *Problems of Creative Leisure: Organization of Creative Leisure in an Industrial Enterprise and In Its Vicinity"* (The Green Meadow Foundation, Six International Conference, 1957, mimeographed).

57. Thorstein B. Veblen, *The Theory of the Leisure Class* (New York: The Macmillan Co., 1899).

58. Emile Durkheim, *On the Division of Labor in Society*, trans., George Simpson, New York: The Macmillan Co., 1933, first published, 1893); Elton Mayo, *The Human Problems in An Industrial Civilization* (New York: The Macmillan Co., 1933); Georges Friedmann, *Industrial Society*, ed. with an introduction by H. L. Sheppard (Glencoe, Ill.: The Free Press, 1955); Eric Fromm, *Escape from Freedom* (New York: Farrar & Rinehart, 1941), and *Man For Himself* (New York: Rinehart & Co., 1947). Actually, observation of division of labor and its analysis can be traced back to the Greek philosophers. Adam Smith, *The Wealth of Nations* (1776) was quite impressed with technological division of labor, and used the example of common pin manufacture to suggest the reasons

for specialization and division of labor. For a brief summary of some pertinent literature, see: Louis Kriesberg, "Industrial Sociology, 1945-55," in H. L. Zetterberg, Ed., *Sociology in the United States of America* (Paris: UNESCO, 1956).

59. J. M. Anderson, *Industrial Recreation* (New York: McGraw-Hill, 1955), 14.

60. Robert Dubin, *The World of Work: Industrial Society and Human Relations* (Englewood Cliffs, N. J.: Prentice-Hall, 1958, Ch. 4, "Behavior Systems," 61-76, & Ch. 11, "Automation and Human Relations," 191-212; F. W. Taylor, *Principles of Scientific Management* (New York: Harper, 1911); E. K. Berrian, *Comments and Cases on Human Relations* (New York: Harper, 1951); G. U. Cleeton, "The Human Factor in Industry," *The Annals* of The American Academy of Political and Social Science, 274 (March, 1951); G. H. Duggin & F. R. Eastwood, *Planning Industrial Recreation* (Lafayette, Ind.: Purdue University, 1941); T. A. Ryan, *Work and Effort* (New York: Ronald, 1947); Ver Shlakman, "Business and the Salaried Worker, *Science and Society*, XV (Spring, 1951), & "Status and Ideology of Office Workers," *Ibid.*, XVI (Winter, 1951-52); H. C. Smith, *Music in Relation to Employee Attitudes, Piece-Work Production, and Industrial Accidents* (Stanford University Press, 1947); Harry Tipper, *Human Factors in Industry* (New York: Ronald, 1922); A. B. Waring, *First Report on the Joint Committee on Human Relations in Industry* (London: Her Majesty's Stationery Office, 1954); E. P. Cathcart, *The Human Factors in Industry* (London: Oxford, 1928); George Katona, *Psychological Analysis of Economic Behavior* (New York: McGraw-Hill, 1951); T. M. Ling, Ed., *Mental Health and Human Relations in Industry* (London: H. K. Lewis, 1954); G. Marot, *Creative Impulse in Industry* (New York: Dutton, 1918); W. C. Menninger, *Social Change and Scientific Progress* (Cambridge: Mass. Institute of Technology, 1951); R. F. Tredgold, *Human Relations in Modern Industry* (London: Duckworth, 1949); Elton Mayo, *The Human Problems in an Industrial Civilization* (New York: The Macmillan Co., 1933); N. A., "Human Relations: A New Art Brings Revolution to Industry," *Time* (April 14, 1952), 96-7.

61. Elmo Roper, "Discrimination in Industry: Extravagant Injustice," *Industrial and Labor Relations Review*, V (July, 1952).

62. Max Lerner, *America As a Civilization: Life and Thought in the United States Today* (New York: Simon & Schuster, 1958).

63. Charles A. Quattlebaum, *Development of Scientific, Engineering, and other Professional Manpower* (A Report Prepared in The Legislative Reference Service of The Library of Congress, Washington, D. C.: Government Printing Office, April, 1957), 3.

NATIONAL WELFARE AND SCIENTIFIC EDUCATION

WILLIAM W. COOLEY
Harvard University

Science educators in America today have the responsibility both to prepare the scientists needed for understanding and controlling the environment in our increasingly technical society, and to communicate to laymen an understanding of science and scientists. In this chapter, some of the broad problems concerned with the education of scientists and their relationship to the national welfare will be considered. This is not to imply that the responsibility which science educators have to laymen is unrelated to our national welfare. Certainly science cannot survive in a society in which only the practicing scientists have an understanding of the scientific process, its limitations and basic accomplishments. However, the preparation of scientists is perhaps more directly related.

Before proceeding further it must be made clear that "national welfare" is used here in its broadest sense. Too often discussions of this sort quickly move to the problem of national defense and the need to "keep up with the Russians." This is unfortunate both for scientist recruitment purposes and for educating the public (upon whom support for the scientific endeavor ultimately depends) about the possible benefits of science. The tendency has been to sell science for its contributions to hot and cold war efforts, when its greatest potential assets are in serving mankind, not destroying it. However, the intended purpose of this chapter is not to evaluate and compare the possible fruits of the research laboratory, but rather to discuss trends in scientific education as it is related to national welfare. We shall briefly consider the expanding role of science and scientists in our society and then describe some of the attempts being made to improve the efficiency of educating more and better scientists.

Scientific Manpower

Although it is not appropriate for a discussion of this sort to make value judgments regarding what is in the best interest of the national welfare, it perhaps can be demonstrated that the supply of scientific manpower has been and will continue to be a major concern in the

United States and an important factor in its rapid growth and increased standard of living.

To show the concern with which educators, scientists, statesmen, economists, psychologists and other responsible citizens view the problems of supplying America's scientific manpower, one need only point to the tremendous volume of literature on the subject which has pyramided in the last ten years. One study[1] listed 25 pages of *selected* bibliography of references related to the problem of supplying scientific talent. In fact, entire volumes have been devoted to presenting annotated bibliographies of the literature in this area.[2]

The enormous growth of science and technology in the United States is obvious to every citizen aware of what is going on around him, but the available statistics are even more impressive.

1) Employment in science and technology is increasing faster than the total labor force.[3]

Year	Worker per scientist and engineer
1870	1090 to 1
1900	300 to 1
1950	60 to 1

2) The federal government's investment in scientific research and development has been rapidly increasing.[4]

Year	Estimated amount in Millions of dollars
1900	5
1910	15
1920	40
1930	69
1940	97
1945	1606
1950	1143
1955	2020

3) The federal government is playing a larger role in the support of research and development.[5]

Year	Millions Total	% Gov't	% Industry	% Non-profit
1941	$ 900	41	57	2
1945	1,520	70	28	2
1949	2,610	59	38	3
1952	3,750	60	38	2

A number of studies using various indices as criteria (such as the number of doctorates granted in the major fields of science) [6] show that scientific activity has about doubled every 10 years.

These and similar statistics indicate that our society is becoming more and more dependent upon an expanding scientific frontier, as it formerly was upon an expanding geographic frontier. On this basis it is perhaps safe to contend that the national welfare is related to our scientific personnel resources.

Moving from description to prediction of manpower supply and demand is a big step, however, Dael Wolfle, one of the most informed and prolific writers on the problem, pointed out at a recent conference[7] on science and education, that the manpower prediction problem is harassed by a few uncertainties, one of which is that no one knows *how* to predict scientific manpower supply and demand! One reason why the task is so difficult is that previously determined economic principles dealing with raw materials and manufactured goods cannot be applied. Manpower prediction has been under the scrutiny of the social scientist for such a short period that he has not had sufficient time to develop dependable procedures, nor to discover many consistently observed relationships.

Predictions involving scientific manpower are especially difficult to make since they involve forecasting (or making assumptions about) such eventualities as peace, continued "cold" war or "hot" war; an expanding economy or a depression; major breakthroughs which may lead to whole new industries; and the economic policies of the political party in control. However, ignoring possible minor fluctuations of short duration, almost all[8] studies using over-all trends as the basis of their prediction foresee a shortage of scientific manpower unless preventative measures are taken. For example, one report[9] claimed that our supply of new scientists and engineers will be about 100,000 below the number needed by 1967, and this figure was based on the assumption that supply and demand are now in balance.

In their recent report, the education panel of the Rockefeller Brothers Fund felt that our current and projected shortage of scientific talent was sufficiently severe to warrant special attention, the only field of study being thus singled out. The report cited two basic ideas which should be kept firmly in mind when considering the problem of educating scientists:[10]

1. The crisis in our science education is not an invention of the newspapers, or scientists, or the Pentagon. It is a real crisis.
2. The U.S.S.R. is not the "Cause" of the crisis. The cause of the crisis is our breath-taking movement into a new technological era.

From a defense standpoint alone we cannot afford to be caught short of the most important resource in any conflict—highly skilled human beings. But certainly as important as defense considerations is the problem of maintaining a sufficient corps of research scientists who can carry on man's continuing struggle to further understand his environment and bring more and more of it under his control. The view of man as a fragile but intelligent creature in a Universe which cares not whether he flourishes or perishes makes the solution of the scientific manpower problem seem even more imperative.

The responsibility of supplying an ever increasing number of men and women with a sound scientific background rests primarily on the science educator. However, in their concern over the problem, many leaders in government and industry have made suggestions toward its solution. Their proposals have generally been along the following lines:

(1) Encourage potential scientists to take the high school courses which would make it possible for them to continue in science. (An effort to increase the quantity of scientists).

(2) Make available to promising students a high school science program which will improve his preparation for college science work and will give him a realistic idea of what scientists do. (An effort to improve the quality of scientists.)

The focus has been on the high school program as the key point in the career development of scientists, although what goes on in elementary school and post-high school science programs is also very important. Such careers as those in science generally involve an early commitment, or at least require an educational program which will make it possible for an individual to continue in a direction which could lead to a scientific career if he so desires, and the student's first major educational decisions (and related vocational decisions) are made in high school.

In the two approaches mentioned above it is necessary to be able to decide *whom* to encourage and *whom* to have participate in whatever special programs are offered. The next question which needs to be examined is: What is known about identifying potential scientists?

Identifying Potential Scientists

In order to discuss potential scientists it is first necessary to define what one means by "scientist," an extremely difficult task. An appropriate definition (for one concerned with science education) is "anyone occupying a vocational position which required at least a four year college major in one of the natural sciences or in engineering." Although most classification systems do not consider an engineer to be a scientist, such a definition is useful here because our purpose is to talk about all the people who might benefit from special pre-college programs in the sciences. Students who continue scientific studies in college and then enter into vocations which necessitate that training, can be considered as potential scientists in high school and, had they been identified as such, they could have benefited from special science work and improved guidance procedures.

Partially for these reasons, many writers today are calling for early identification of the future scientists, but few realize the limitations of our knowledge in this area. It will be profitable to consider briefly the nature of current research on the problem to see the extent to which future choice of and success in a scientific career can be predicted, and how much this might be improved in the near future.

To identify anything one must know the differentiating characteristics, at least the necessary if not the sufficient conditions for membership in a given set. If potential scientists can, in fact, be identified, they must differ in certain respects from non-potential scientists. To identify these distinguishing characteristics, research workers have utilized several different approaches: 1) comparison of college students who plan on becoming scientists with those not intending to go into science; 2) comparison of practicing scientists with non-scientists; and 3) comparison of previously determined antecedent characteristics of men who have become scientists with those who have gone into other fields (longitudinal studies). Although the longitudinal approach is the most valid because it infers neither the eventual occupation nor the early characteristics of adults, few such studies have been made because of the time and expense involved. In general the procedure has been to determine characteristics (test score data, questionnaire information, personality traits, etc.) in which the scientists differ from the non-scientists more than could be expected to occur by chance. Then the significantly different traits are summarized into a portrait of the scientist or potential scientist. The problem is that these are neither necessary nor sufficient conditions for becoming a scientist, but merely group tendencies. It is not possible to use this information in the classification of new individuals. It is interesting but not functional to know, for instance, that the scientist *tends* to be the first-born child of a middle-class family, the son of a professional man, who often feels lonely, shy and aloof from his classmates. That the scientist was always above average in general intelligence in school presents us with a necessary condition, but unfortunately defines 50% of the population rather than the 0.5% or less who actually do practice a scientific vocation.

One reason why the results of most previous research have little direct application to the problem of identifying potential scientists is that the attributes included in the lists were based on studies using tests of significance for individual items. Significant differences between scientists and non-scientists are the result of *some* of the individuals forcing the scientist mean away from the non-scientist mean. The deviates would not necessarily be the same ones for each characteristic; hence a composite based on separate tests of significance for each trait is not meaningful.

At present, then, the situation seems to be this: previous research has identified a number of ways in which scientists as a group differ from non-scientists but these differences are not directly useful for identification purposes since they either describe only a few scientists or include many more non-scientists than scientists. Guidance personnel have no way of utilizing such information since they have no way of estimating the various combinations of traits which indicate potential success as a scientist for a given individual.

This is not to say that the previous efforts have been in vain. They have shown that group differences do exist and we now have a foundation upon which future research can be based. But it does not seem that a continued search for separate group differences will be fruitful.

One promising new approach involves the application of multivariate methods of analysis. These procedures examine *in combination* the known information about individuals and make it possible to determine the permissible tolerances in the various combinations of abilities and personalities exhibited by people going into the various sciences.

Because of the overlap in attributes between occupational groups we still cannot expect to pin-point the potential scientists, but multivariate procedures should be a much more efficient way to define the potential scientist pool at the various stages of development. The task now seems to be one of setting up longitudinal studies wherein the attributes (which previous research has shown to be differentiating) of children are measured and the educational and career development observed until they enter into vocations and begin to achieve some job stability. The resulting information, after subjecting it to a procedure such as multiple discriminant analysis, will then enable us to conclude whether a new individual can be considered to be like those criterion groups who have gone into the various scientific occupations. Initial attempts along these lines indicate the promise this procedure can have for guidance purposes.[11]

The application of more highly developed methods of statistical analysis is not going to solve the identification problem alone, however. The multivariate models will be worthless if the data gathering instruments are unreliable or unrelated to the phenomena we are trying to predict.

The measurement of mental abilities has been developed to the extent that quite useful information is available. It is possible, for instance, to predict whether or not a person has the ability to survive the schooling required to become a scientist, certainly a necessary condition of being a potential scientist. But as was pointed out earlier, this includes many times the number of people who become scientists.

To further narrow down this potential scientist pool the inclusion of personality information seems to be a promising approach. It is perhaps reasonable to assume that not all people can be expected to maintain a sustained interest in natural phenomena, things, precision, order, and other aspects of the scientist's daily activities. The problem, however, is one of developing suitable instruments for measuring such "intangibles" as need for achievement, introversion, egocentricity, perseverance, tolerance for ambiguity, postponement of gratification, or whatever combinations of significant personality variates may be differentiating. After such instruments are developed and standardized through longitudinal investigations it will perhaps be possible to give more assistance to the student who comes forward with the question: Am I like people who go into, and are successful in, [a particular scientific occupation]?

The Increased Role of Government

As was pointed out earlier in this chapter, the government has been steadily increasing its support of scientific research and development. Along with this it is beginning to assume more and more of the responsibility of making sure qualified students are able to continue their scientific education, thus maintaining a sufficient pool of research workers for their various activities. This situation has made it necessary for federal agencies to support research which attempts to learn more about identifying and nurturing human talent. Government assistance now makes it possible to conduct large scale research projects of a longitudinal nature, which appears to be the most satisfactory way of investigating the development of specialized abilities, and which could probably not be done without federal aid.[12]

A conclusion at which one soon arrives when considering our present and expected future ability to predict continued pursuit of and success in a scientific career is that any plans for high school programs which involve *selecting* certain students for advanced work in science, with the expectation that these students will be the future scientists, is highly unreasonable. Such proposals not only assume a predictive procedure impossible to validate, but also are probably inconsistent with the basic philosophy of schools in a democratic society. No demand for scientists can be so extreme as to justify such a means for procuring them. Still, programs for "picking out" the potential scientists at early ages and placing them in special classes or even special schools continue to be proposed and in some cases implemented. The problem of what types of special pre-college science programs are being promoted and which ones do seem justified will be discussed next.

Providing for Potential Scientists

It is possible to describe what science educators are saying to each other in the literature about pre-college education of scientists, but what is actually happening in the 40,000 autonomous school districts of the United States is very hard to determine with any precision. For instance, the literature regarding gifted students has been so voluminous that there are now bibliographies of bibliographies on the subject. However, several surveys of actual school practices seem to indicate that incredibly few school systems have programs for identifying, encouraging, or making special provisions for their more intellectually able students. One report[13] shows that as late as 1954, only about 5% of the schools gave any sort of formal attention to gifted students.

Although general trends are discernable only with great difficulty, several studies have determined the common elements of the science programs for high schools which have been successful in producing a disproportionately large number of potential scientists, the criteria for selecting schools usually being the "production" of winners in the national Science Talent Search (STS). One such study listed the following points in common for seven high schools:[14]

1. The student is identified early and given guidance and direction. (Usually by 9th grade, using a high general intelligence cutoff.)
2. He is encouraged to take advanced courses in science.
3. He is engaged in individual research projects.
4. Books and scientific apparatus are available to him.
5. There is an opportunity for scientific exploration through extra-curricular activities.
6. He has the opportunity to work with highly competent, enthusiastic instructors.

These six points pretty well define what a school must do in order to have a sizable number of STS winners. That these are also the desirable and optimum conditions for producing successful scientists has yet to be demonstrated. It is possible that such programs tend to discourage students below that high IQ cutoff, or that high ability students are "pushed" into science when personality considerations might have indicated that law or business would have been a more satisfactory vocation.

The main objection seems to be with the first step usually involved in such special science programs, the use of tests of mental ability as the main if not the sole criterion for determining an individual's suitability for special attention in science. The reason, of course, is that not all high IQ people make suitable scientists and not all scientists have high IQ's. The assumption that one must be in the upper one percent on any group intelligence test in order to become a scientist is still often made, in spite of the fact that several studies have found this claim untenable.[15]

The Rockefeller Report[16] points out that although tests are most effective in measuring academic aptitude and achievement, "no single test should become a basis for important decisions" and that other kinds of data besides test scores should be used in an effective program of identifying the more able students. This seems reasonable, but the fact remains that we still do not know how to predict success in science on the basis of several variates, or at least complete and reliable data are not available upon which such prediction equations can be based.

Thus the current dilemma of science educators seems to be this: they want to provide special pre-college training for students going into science, yet there seems to be no sure way of deciding who should participate in such programs. Most schools with special programs, the STS, summer programs for science students, etc., use tests primarily measuring verbal-mathematical ability as the main criterion, a procedure which has only slight justification and which may be doing harm. However, the programs have to be limited somehow, and tests of mental ability seem to be the most reliable instruments for predicting future success in school work.

Programs of Self-Identification

One possible solution is to design programs which do not involve selection by examination. One of the most well-known high school programs in science education is the program at Forest Hills, New York initiated by Brandwein during World War II. (The details are described elsewhere and need not be gone into here.) [17] The future scientist "track" in their three-track science program is open to all students. It is limited in size only by being a challenging program and involving an amount of time and effort that only sincerely interested students are willing to expend. Such a process of self-identification involves more than just the student's general intelligence. His interests, motivation, and other personality characteristics are involved in his continuation along this third track. Another important feature is that the student is able to move into and out of the program if he so desires. (The three separate tracks, non-college, pre-college, and future scientist, begin at 10th grade.)

Of course, there are also difficulties with such a program. It is feasible in only the largest high schools, and half of our youth are in schools with an enrollment of less than 400. It also might be argued that a program of self-identification is open to question as much as one involving some process of selection. It implies that the immature adolescent is more able to evaluate his fitness for and interest in science than are psychometricians.

On the other hand, the fact that the student is given an opportunity to experience "sciencing" (to "try science on for size"), and that the process considers more than his measurable mathematical-verbal ability, seems to make this procedure more justifiable than one where the top 25 students on an aptitude test are selected for special treatment. It should also be pointed out that the student is not entirely alone in making his decision to participate in the pre-scientist program. Experienced teachers, whose judgments have been sharpened by years of observing the career development of former students, can also be of some help.

But all is not lost if the school is too small to have a three-track or even a two-track program. One promising activity of the special programs does not require a formal class situation, namely, individual research projects. Several studies of scientists have pointed out the importance of having the experience of "finding out for oneself" early in the career.[18] The student having such an experience, who temperamentally is predisposed to scientific research, will very likely commit himself to science once having experienced the thrill of personal discovery. He may then be less likely to be discouraged by much of the concentrated effort usually involved in becoming a scientist.

One difficulty is that it takes a highly competent science teacher to initiate an effective program of student research. The teacher must be able to guide the student beyond the usual high school texts and make available (or assist in obtaining) the references and needed materials for this type of program. The teacher is at the heart of the problem of implementing any program for assisting potential scientists. The intricate multivariate models for identification, the clever administrative arrangements for grouping, etc., and the research to determine how students might be more adequately prepared to realistically undertake a career in science are all in vain if in the science classroom stands an uninterested and ill-prepared science teacher. Fitzpatrick[19] has gone so far as to claim that *all* we need is an efficient corps of secondary school science teachers. "They will resolve the curriculum problems, and they will give us the scientific manpower that is essential to our desires, our comfort, and our security." If one is not willing to admit that more and better science teachers is a sufficient condition for solving the scientific manpower problem, it certainly cannot be denied that it is a necessary condition.

The Challenge

Supply and demand studies indicate that a program of action is needed. They also show the increased dependence of the federal government upon a continually increasing supply of scientists and engineers.

To increase the quality and quantity of scientists, leaders in government and industry have emphasized the importance of early identification and special pre-college attention to potential scientists for both guidance and training. In practice the difficulty has been an inability to identify potential scientists, an uncertainty as to what needs to be done differently in training future scientists, and lack of a corps of science teachers sufficiently trained to carry their students beyond the usual high school texts.

Therefore, the science educator's challenge seems to require a general attack on three fronts:

1. Research to learn more about career development and how we can aid able students to make realistic educational and vocational decisions.

2. Design and implement programs based on the evidence of sound research which will improve the quality of future scientists.

3. Improve the quality of science teachers through use of increased compensation and through training programs for future and in-service science teachers.

Industry and the federal government are now sufficiently aroused to support such programs of action. Upon the extent to which science educators meet this challenge, our national welfare depends.

Selected Bibliography

D. M. Blank and G. J. Stigler, *The Demand and Supply of Scientific*

Personnel (New York: National Bureau of Economics Research, 1957). A thorough analysis of the problem of predicting manpower supply and demand. Not merely tables and predictions.

Paul F. Brandwein, *The Gifted Student as Future Scientist* (New York: Harcourt, Brace, 1955). Describes the high school science program at Forest Hills, New York. His discussion of identifying and providing for gifted students in pre-college work is already a classic in this field.

Kenneth E. Brown and Philip G. Johnson, *Education for the Talented in Mathematics and Science,* Bulletin No. 15 (Washington: U. S. Office of Education, 1952). This small pamphlet is primarily a summary of the opinions expressed at a conference (Fall, 1952) of the Cooperative Committee on the Teaching of Science and Mathematics of the A.A.A.S.

Charles C. Cole, *Encouraging Scientific Talent* (New York: College Entrance Examination Board, 1956). This work is primarily concerned with the loss of talent from high school to college and what might be done about it.

William W. Cooley, "Attributes of potential scientists," *Harvard Educational Review,* XXVIII (Winter, 1958), 1-18. Considers the problem of identifying potential scientists, briefly organizes known distinguishing characteristics, and makes suggestions for further research.

Garford G. Gordon, *Providing for Outstanding Science and Mathematics Students* S. Calif. Ed. Monographs #16 (University of Southern California Press, 1955). Identifies science methods used in 142 high schools which are considered successful in producing future scientists.

Nelson B. Henry (ed), *Education for the Gifted* The Fifty-seventh Yearbook of the National Society for the Study of Education. (Chicago, Ill., 1958). The most recent comprehensive discussion of the general problem of identifying and providing for gifted students.

National Manpower Council, *The Utilization of Scientific and Professional Manpower* (New York: Columbia University Press, 1954).

Rockefeller Brothers Fund, Panel V. *The Pursuit of Excellence: Education and the Future of America* (New York: Doubleday and Co., 1958). The most thoughtful consideration of the general problems in American education to come out of committee work in a long time.

Anne Roe, *The Making of a Scientist* (New York: Dodd, Mead & Co., 1952). This clinical psychologist considers the family background and the results of projective testing of 64 eminent scientists in an attempt to find general relationships.

Scientific Personnel Resources. National Science Foundation, (Washington: NSF, 1955). A summary of data on supply, utilization and training of scientists and engineers.

Donald E. Super and P. B. Bachrach, *Scientific Careers and Vocational Development* (New York: Teachers College, Columbia U., Bureau of Publications, 1957). The first systematic review and criticism in this area of research.

C. W. Taylor (ed.), *Research Conference on the Identification of*

Creative Scientific Talent (Salt Lake City: University of Utah Press, 1956). Especially interesting are the discussions following each of the papers.

L. M. Terman, "Scientists and non-scientists in a group of 800 gifted men" Psychological Monographs, LXVIII (1954). A further chi-square analysis of the data collected in his gifted student longitudinal study.

Thomas Alva Edison Foundation Institute, *Strengthening Science Education for Youth and Industry* (New York: N. Y. University Press, 1957). Proceedings of the seventh Thomas Alva Edison Foundation Institute.

D. Wolfe, *American Resources of Specialized Talent* (New York: Harper & Brothers, 1954). Especially useful as a single source book for survey data in this area of research.

Marguerite W. Zapoleon, *The Identification of Those with Talent for Science and Engineering* (Washington: The President's Committee on Scientists and Engineers, 1956). Considers also the problem of guiding those identified in elementary and secondary schools.

WILLIAM COOLEY is currently an Instructor and Research Associate at Harvard University's Graduate School of Education. He is a graduate of Lawrence College, where he majored in chemistry, and has taught secondary school chemistry and physics in Wisconsin and in Hawaii.

His graduate study has been in the fields of science education and measurement, both at the University of Minnesota where he received the Master of Arts degree and at Harvard University where he recently received the Doctor of Education degree. His thesis investigation was concerned with the problem of the career development of scientists.

At present Dr. Cooley is engaged in a five year study of the process of becoming a scientist in which procedures for improving the guidance and training of future scientists are being evaluated. This project is sponsored by the U. S. Office of Education and its expected date of completion is June, 1963.

NOTES FOR NATIONAL WELFARE AND SCIENTIFIC EDUCATION

1. Cole, Charles C., Jr., *Encouraging Scientific Talent* (New York: College Entrance Examination Board, 1956), 230-255.

2. Eller, Mable H. and Weiner, Jack. *Scientific Personnel, a Bibliography.* (Washington: Office of Naval Research, 1950).

Research and Development Board. *Scientific Manpower 1940-1950, A Selected and Annotated Bibliography* (Washington: U. S. Dept. of Defense, 1951).

Ibid. Supplement, 1950-1951.

3. National Committee for the Development of Scientists and Engineers. *Trends in the Employment and Training of Scientists and Engineers.* (Washington: National Science Foundation, 1956) p. 4.

4. National Science Foundation. *Scientific Personnel Resources* (Washington: NSF, 1955), 7.

5. Office of the Secretary of Defense. *The Growth of Scientific Research and Development* (Washington: Dept. of Defense, 1953), 11.

6. Office of Scientific Personnel, National Academy of Sciences—National Research Council. "Production of U. S. Scientists," *Science,* CXXVII (March 28, 1958), 682-686.

7. Held at Harvard University Summer School, July, 1958.

8. Charles C. Cole, *op. cit.,* 33.

9. The B. F. Goodrich Company. *A Study of the Scientific Manpower Problem of the U. S.* (Akron, Ohio: 1956).

10. Panel Report V, Rockefeller Brothers Fund, *The Pursuit of Excellence: Education and the Future of America* (New York: Doubleday & Co., 1958), 27.

11. Cooley, William W. *The Application of a Developmental Rationale and Methods of Multivariate Analysis to the Study of Potential Scientists* (Unpublished Ed. D. Thesis, Harvard University, 1958).

12. The following four research projects, all initiated in 1958, are examples of such projects being conducted with government support. Listed are the principal investigator, title of project, and the institution at which the research is being carried out: William W. Cooley, *Career development of scientists—An overlapping longitudinal study* (Graduate School of Education, Harvard University); F. B. Davis, & G. S. Lesser, *The identification of gifted elementary school children with exceptional scientific talent* (Hunter College); John C. Flanagan, *Identifying, Developing and Using Human Talents* (University of Pittsburgh); John L. Holland, *The prediction of academic achievement subsequent life achievement, and creativity by means of non-intellectual factors* (National Merit Scholarship Corporation).

13. United States Department of Commerce, Bureau of the Census. *Statistical Abstract of the U. S. 1954,* p. 140.

14. Witty, Paul and Bloom, S. W. "Conserving ability in the sciences." *Exceptional Children,* XXII (Oct. 1955), p. 10.

15. Cox, C. M. et al. *The Early Mental Traits of 300 Geniuses* (Stanford University Press, 1926).

Strauss, Samuel. "Looking backward on future scientists," *The Science Teacher* XXIV (Dec. 1957), 385-387.

16. Panel Report V, *op. cit.* p. 29.

17. Brandwein, Paul F. *The Gifted Student as Future Scientist* (New York: Harcourt, Brace and Company, 1955).

18. Roe, Anne. "A psychologist examines 64 eminent scientists," *Scientific American,* CLXXXVII (Nov. 1952), 21-25.

19. Fitzpatrick, F. L. "Scientific manpower: the problem and its solution" *Science Education,* XXXIX (March, 1955), 97-102.

Background

HISTORY OF SCIENCE EDUCATION

JOHN H. WOODBURN
The Johns Hopkins University

Problems of Science Education and Their Origin

Science education had its origin when that first scientist, no longer willing to allow his discoveries and inventions to remain his private treasure, chose to share with his fellowmen the fruits of his labors. The teaching of science thus becomes inextricably entangled with the evolving enterprise of science. This entanglement creates many of the problems of science teaching and the efforts to solve these problems become the threads of continuity which form the history of science education.

Reviewed briefly, these problems are:

(a) Discoveries and inventions follow a mature scientist's investigation of a sequence of increasingly complex ideas whereas immature students of widely variable abilities and degrees of dedication are expected to grasp the end-products of these discoveries and inventions during the brief class periods of primary or secondary schools.

(b) The creation of a scientific discovery or invention hinges only on the elusiveness or complexness of an amoral phenomenon, but the teaching of it may make a sharp and sometimes controversial impact on the beliefs and values of people and on the way economic, political, and social problems are solved.

(c) Science is both product and process but scientists remember more clearly what they did than how or why they did it. Descriptions of the product of a scientific episode can be corrected in retrospect, but the passage of time renders an authentic portrayal of the process ever more improbable. Furthermore, a "principle of indeterminacy" may operate against the "scientific" clarification of "the scientific method."

(d) The continuously expanding volume of science renders ever more difficult the selection of that content which holds greatest promise of most efficiently transmitting to students a genuine comprehension of the total enterprise of science.

(e) In many cases, scientific principles are presented to students in a cultural context that is not at all consonant with the culture in which the investigations occurred whereby these principles were derived and interpreted. Whole world pictures or patterns of explanation may have changed.

Science Education and the Total Educational System

Chronologically, many facets of science education are interwoven with the evolution of the total educational system. Where, when, and under what circumstances various science subjects appeared in school curricula have been capably reviewed by several authors.[1,2,3] Frank, for example, begins with the mid-eighteenth century science as developed in the academies and describes it as "almost wholly book science and dealt with . . . by a sort of catalogue method."[4] By the middle of the nineteenth century, chemistry was well established but without laboratory. Lecture demonstrations, however, supplemented the textbook instruction. Natural philosophy (physics) was taught from books "with the topics cataloged one after another, beginning each with a problem, passing on to the explanation and closing with a statement of the principle involved."[5] Prior to 1860, the natural history included much of present day biology and "nearly all writers of texts accepted the doctrine of special creation and attempted to fit all plants and animals into the general scheme, like bits of glass in a mozaic."[6]

Downing, although dedicated to the point of view that science is a relatively recent innovation in the curriculum, begins his review with a reference to the *Academia Secretorium Naturae* which was founded in Naples in 1560. The scientific awakening of the sixteenth and seventeenth centuries was reflected in the origin of the English Royal Society (1662), the old French Academy of Sciences (1666), the Royal Academy of Science at Berlin (1700), and the American Philosophical Society (1743). Contemporary with the growth of these institutions, natural philosophy was introduced into the academies which "sprang up in protest against the classical education prevalent in the Latin schools and against the theological training in preparatory schools and universities."[7]

Establishing the Pattern of Science Education

In general, the pattern of science as a school subject was clearly established some time prior to 1763. Strong witness to this is the textbook, *A Course of Experimental Philosophy*, by J. T. Desaguliers[8] which was published at London in that year. Disregarding the changes in format and printing which have occurred in the past two centuries, Desaguliers' textbook is spectacularly similar to modern textbooks in its basic approach to science education.

Desaguliers parades an array of scientific principles (the 1034 pages of his two-volume text were adequate to include nearly all the principles of science known at that time) before the reader of his text. A few brief passages will portray his approach.

I shew'd, after explaining the first Law of Motion, how far it would serve to make us acquainted with the Motion of the heavenly Bodies, by shewing in what manner Gravity and the projectile Force keep those Bodies in their Orbits; but it required the understanding of the second Law to conceive rightly how they move in Ellipses that have the central Body in one of the *Foci,* and why their Velocities have successively accelerated and retarded.

But before I proceed to consider this, I must explain some astronomical terms, and shew what is meant by saying, that *the Planets and Comets in respect to the Sun describe Area's about it proportionable to the Times; as likewise the Satelittes in respect to their Primary Planets.* And this is a Truth known and own'd by all modern Astronomers, however they differ in accounting for the Causes of the celestial Motions.[9]

After having announced his basic principles (The Laws of Motion) and defining his terms, Desaguliers proceeds to show how these principles can explain natural phenomena or be applied to the technology of the day. For example,

Now let us suppose the Body T not to be in the Center of the Orbit, as the Earth is not in the Center of the Moon's Orbit, But to be distant from it the whole Length C T (*Fig* 19.) if the Moon or the revolving Body be observ'd at L and *L,* and so found to have gone through the Arc L *L* in one Day's Time; . . .[10]

Desaguliers extends the explanation by adding four corollaries which explain "why the revolving Body moves faster at the *Perihelion,* . . . the more excentrick (that is, the longer) the Ellipse is, the greater is the Difference of Velocity at the *Perihelion* and *Aphelion,* . . . why a Planet, tho' it be much more strongly attracted in its *Perihelion* than its *Aphelion* will not be drawn into the Sun;" . . . and "why those Planets, which are nearest to the Sun, perform their Revolutions in shorter Time than those which are farther off; . . ."[11]

Desaguliers acknowledges Kepler's contributions to the topic but introduces the process of science as he distinguishes these from Sir Isaac Newton's work because Kepler "gues'd that the Cause might be a Gravitation towards the Sun; but he did not demonstrate it."[12] As with modern texts, figures are included to clarify the mathematical relationships.

Each of Desaguliers' chapters (lectures) is followed by a chapter of annotations which extend the lecture and suggest how its contents may be further clarified. Student exercises are set up as scholia. These scholia suggest additional experiments ("For the sake of the Curious we shall mention a few more in the notes.") [13] and detailed directions are included for the design of essential demonstration equipment. For example,

To shew experimentally how a Planet accelerates its Motion as it approaches to the Sun, and retards, as it recedes from the Sun; fix the whirling Table (in the Condition represented by the 11th *Figure* of *Plate 24*) to its Frame, that it may be easily turn'd round in its horizontal Situation . . . .[14]

Criticism of Science Education

Desaguliers' textbook suggests an inference that is humbling almost to the point of humiliating. Whereas the scientific enterprise has made great strides forward in the maturity of its philosophy and in the efficiency of its tactics and strategy, it has provided a minimum of insight upon which improvement in science education could be hypothesized. The inference becomes all the more disturbing because of the amount of time, energy, and other resources which have been appropriated for the criticism of science education. Research tends more toward determining such things as at what age the various facets of science instruction should be offered, how much of the day's program should be devoted to science, how its objectives should be worded, how the learning of science might be intrinsically motivated, and how new content can be added. Actual comprehension and appreciation of those forces which give spirit and structure to science education and thus hold promise of redirecting its future must be sought in an analysis of events and circumstances of quite long standing—events and circumstances inextricably interwoven with the total science enterprise.

Production of Teaching Aids

The sophistication of the scientist in contrast to the naïveté of the student has urged the scientist and science teacher to invent teaching aids. The early writings of science contain many paradoxes, metaphors, similes, or dialogues for the benefit of the student. Similarly, especially in the early days of the Royal Society, many scientists prepared sometimes spectacularly clever demonstrations to insure the clarification of their discoveries and inventions. Good examples appear in Joseph Priestley's *Experiments and Observations on Different Kinds of Air*[15] and Faraday's *Experimental Researches in Electricity*.[16] With the gaps between scientists and the students becoming increasingly wide, the tapering off of this practice has deprived the science teaching profession of an essential contribution. There are more extensive repertoires of lecture demonstrations for those scientific principles which were developed prior to the 19th century than for those of more recent elaboration. Twentieth century scientific discoveries and inventions tend to evolve by way of mathematical relationships and to be expressed in the language of mathematics. Unless the newly derived principles can also be expressed in the language of common experience, their teaching must await attainment of sophistication in mathematics by would-be students.

Cultural Impact of Science Education

Returning to the second of the problems listed earlier in this chapter, the close relationship between science and science education brings another problem into focus. Whereas a scientist's investigation can be exceedingly frustrated by the elusiveness of the principle or relationship he seeks, unlike the teacher of science, he is not subjected to the impact of his discoveries or inventions on the religious, social, political, and economic beliefs and practices of a society until he becomes a teacher of science. The history of science education includes several episodes revealing how, at the time of their origin, many discoveries and inventions made sharp impacts on the beliefs and behavior of society. Sometimes these discoveries and inventions threatened a change in the status of those people within the society whose status was tied to acceptance of their authoritarian role.

Approaches to the Methods of Science

Clarification of the methods of science has been attacked from several angles. Students of logic have taken long and dedicated looks at situations in which scientists have been working. Much more recently, (See the September, 1958 issue of *Scientific American*) methods of the scientist, especially creativity, have attracted the attention of psychologists and physiologists.

Restoring appropriate aspects of the history of science to science education holds promise of solving more of the profession's problems. While at Harvard, President James B. Conant[17] developed his case-history method of teaching science. By simple announcement, Conant would bring his students up to a specific episode in the evolution of mankind's attempts to observe, describe, and interpret a natural phenomenon. He would then dwell on a selected episode in sufficient depth to transmit to his students the process or processes of science by means of which that particular episode was accomplished. Again by simple announcement, the students would be brought up to the current or modern state of the art or investigation of that same phenomenon. It can be argued that the process or processes of science thus revealed by detailed analysis of representative episodes can then be identified by the student in other episodes of science.

Eric Rogers[18] has developed what he would call the block-and-gap approach to the teaching of science. Again, Rogers would argue that complete coverage of one of the disciplines of science and, even more so, all the disciplines of science is impossible within a finite period of time. Forced to make some selection of either breadth or depth of content, Rogers chooses to select certain blocks of content, identify clearly the principles of science which were operating within the block of material, and hope that his students can use these principles to bridge the gaps created where equally significant blocks of subject matter apparently had to be omitted in order to release adequate time to really "cover" the blocks of subject matter which were selected.

Some efforts have been made (see Philip Frank) [19] to restore enough of the philosophy of science to the science curriculum to allow students to see how man's interpretations of natural events and circumstances are influenced by the world picture or philosophy prevailing at the time. By tracing the teachings related to one phenomenon, assuming that the phenomenon has been constant throughout mankind, it can be shown that man's interpretations of the phenomenon have evolved along with evolving world pictures. Paying attention to the philosophical implications of the content of science not only holds promise of solving some of the problems of science education, but raises very interesting questions. For example, does the evolution of an explanation of a natural phenomenon recapitulate the whole range of maturing thought patterns? Does this hypothesis apply equally to those natural phenomena which did not attract the attention of science until most recent times? By corollary, does the thought pattern that is predominant in the scientific theory or explanation serve as a clue to the probable longevity or usefulness of the theory?

The mid-point of the twentieth century is marked by intense interest in science education. People in many walks of life are becoming sincerely concerned with probable relationships between facets of national, domestic, and personal welfare and the success or failure of the nation's science teachers. Out of this concern will come trends forming the future of science education—trends to be discussed in many of the chapters which follow. Threads of continuity, however, will continue to have their origins deep within the evolving enterprise of science.

Selected Bibliography

R. Will Burnett, *Teaching Science in the Secondary School* (New York: Rinehart & Co., 1957). Chapter 3 presents the historical basis of conventional practices in the teaching of science.

James B. Conant, *Education in a Divided World* (Cambridge: Harvard University Press, 1949). A frank plea "to stop, look, and see" America's schools in operation and "consider the relation of the teacher's work to the future of this nation."

Frederick Eby, *The Development of Modern Education in Theory, Organization, and Practice* (New York: Prentice-Hall, 1952). This and the listing which follows provide a good perspective of how the development of science education has been entangled with the development of major educational systems.

Frederick Eby & Charles F. Arrowood, *The History and Philosophy of Education, Ancient and Medieval* (New York: Prentice-Hall, 1940).

Earl J. McGrath, *Science in General Education* (Dubuque, Iowa: Wm. C. Brown Co., 1948). A compendium of ideas and activities being explored midway in the twentieth century.

Paul Monroe, *Founding of the American Public School System* (New York: Macmillan Co., 1940). A History of education in the United States from the early settlements to the close of the Civil War period.

Roger D. Rusk, *How to Teach Physics* (Philadelphia: J. B. Lippincott, 1923). The first three chapters touch on the development of physics as a school subject.

JOHN H. WOODBURN, Assistant Director of Master of Arts in Teaching Program and Lecturer in Education, John Hopkins University (1957–), received his B.A. degree from Marietta College (1935), his M.A. from Ohio State University (1941), and his Ph.D. from Michigan State University (1952); Assistant Professor and member of the Board of Examiners, Michigan State University (1947-1949); Assistant Professor of Science, Illinois State Normal University (1949-1953); Assistant Executive Secretary, National Science Teachers Association (1953-1957); Science Services Consultant, U. S. Office of Education (1957).

NOTES FOR HISTORY OF SCIENCE EDUCATION

1. J. O. Frank, *How to Teach General Science* (Philadelphia, P. Blakiston's Sons, 1926), 1-23.

2. Elliot Rowland Downing, *Teaching Science in the Schools* (Chicago University Press, 1929), 1-23.

3. George W. Hunter, *Science Teaching at Junior and Senior High School Levels* (New York, American Book Co., 1934), 15-54.

4. Frank, *op. cit.*, 7.

5. *Ibid.*, 8.

6. *Ibid.*, 9.

7. Downing, *op. cit.*, 2.

8. J. T. Desaguliers, *A Course of Experimental Philosophy*, 3rd ed. corrected (London, 1763).

9. *Ibid.*, 356.

10. *Ibid.*, 357.

11. *Ibid.*, 358-59.

12. *Ibid.*, 359.

13. *Ibid.*, 408.

14. *Ibid.*, 408.

15. Joseph Priestley, *Experiments and Observations on Different Kinds of Air* (Birmingham, England, 1790).

16. Michael Faraday, *Experimental Researches in Electricity* (London, 1844).

17. James B. Conant, *Growth of the Experimental Sciences; An Experiment in General Education; Progress Report on the Use of Case*

Method in Teaching the Principles of the Tactics and Strategy of Science (Cambridge: Harvard University Press, 1949).

18. Eric M. Rogers, "Samples Versus Survey in Physics Courses for Liberal Arts Students," *American Journal of Physics*, XIV (November, 1946), 384-6.

19. Philip Frank, *Modern Science and Its Philosophy* (Cambridge: Harvard University Press, 1949).

RELIGION AND SCIENTIFIC EDUCATION

MEHDI NAKOSTEEN
University of Colorado

Secularism, Religion and Science

The abolition of sectarian religious instruction in our public schools does not necessarily imply that the secular curriculum should be devoid of information about religion in general or about Christianity and other American minority faiths in particular. To avoid basic information about religion in such areas of secular discipline as social studies, humanities, and the arts is not only undesirable, since it would lead to religious illiteracy, but also impossible. The descriptive survey of religion in the area of social studies is inevitable in such fields, for example, as history in which religious information is indispensable to an understanding of the development of Western civilization.[1] Similar involvements with religious studies may be noted in courses in economics, political science, sociology and psychology. So also in the humanities where the disciplines include expressions of all modes of human feeling, thought and action.[2] In the arts too, insofar as they express religious sentiments and beliefs, religious information and interpretation would be unavoidable.

It is, however, in science courses where information about religion may lead to controversies and conflicts between religious convictions and scientific conclusions. Therefore, in no other area of public school education the hazards of irritating the religious sensitivities of both students and teachers can be more evident than in science classes, particularly in such science courses as biology, physics, astronomy, anthropology and geology. Such topics as the age of the earth and the planetary system, the evolution of life, the nature of the universe, moral principles, human nature and destiny, evil and tragedy, birth and death, Biblical authority, theological creeds and dogmas,[3] etc., may arise as a result of scientific presentations, and demand both a hearing and a response. *To state and evaluate conflicting positions of science teachers regarding such religious questions is the main consideration of this chapter. We shall begin with a statement of the nature of religious experience and the scope of scientific discipline, including a brief analysis of the causes of historic conflicts between them. We shall conclude with a proposed program of religious education for science teachers.*

Religious Experience and Scientific Discipline

A. *Religious Experience:* Any attempt at a universally acceptable definition of religion is bound to lead to difficulties. Unlike the scientific discipline, on whose methods and aims there is, by and large, universal agreement, religious experience and the media through which it is channeled and expressed are too varied, regional and personal to allow for a dispassionate definition. True, common to all organized religions are the organized church, a system of creeds, a code of good life, some belief in a divine power or its equivalent, and some prescribed method of communication with this power. These common denominators take, nonetheless, unique forms and expressions, not only in various religions, but in factions and denominations within a given faith like Christianity, Buddhism or Islam.[4]

Religion is personal and private as it is also public and institutional. It may spring from the surface or depth of human emotions. It may have an intellectual and philosophic basis. It may express itself in mystical terms which relate the individual to his concept of God or the order of things. There is an ethical side to religion formulated in moral codes prescribing a mode of social relations among men. Religion may be born in a sense of awe and ignorance, in fear and hope, in yearning for immortality, in fancy and escape from the tensions of immediate realities, in escape from evil, in feelings of inadequacy, or in curiosity about the unknown.[5] Religion is ritualistic or simple, primitive or highly developed, reactionary or progressive, traditional or adaptive. The religious focus of worship may be one, two, or many gods; its ethical codes may range from a crude sense of justice to sublime concepts of love; its attitudes toward one's neighbor may be regional or universal. The concepts of God, soul, man, world, hereafter, sin and its consequences or punishments, family and its meaning and relationships, prayer and worship, differ in value and implementation in various faiths, and within divisions of the same faith. Yet, varied as religious institutions, creeds, codes of conduct and experiences may be, at bottom all religious experiences are man's allegiance or adjustments to the mystery of life[6] and the enigma of the unknown environment which surrounds him,[7] including the loyalties or responses which necessarily develop from this allegiance or adjustment, as well as the moral values or ethical principles which evolve from these loyalties or responses, and translate themselves into the conduct of life.[8]

B. *Scientific Discipline:* When we look at the net conclusions of any of the sciences we see them as "systematized and objective bodies of knowledge." However, it is not so much the conclusions as the way they have been arrived at and the attitude held toward them that characterize the disciplines we call science.[9] These disciplines may be stated in terms of two basic principles. *First* is the empirical principle, or the principle of direct and precise observation of or experimentation with an aspect of the knowable world, over a long period of time and under controlled conditions; the suggestion or development of a hypothesis of explanation tentatively validated only to the extent it can explain or justify the observed facts or experimental results; the deduction of what ought to happen if the proposed hypothesis is valid; and the final verification of the validity of such predictions by repeated tests. *Second* is the tangible or quantitative principle, which limits empirical observation to only the aspects of reality which are mathematically measurable and exactly determinable. By this twin method science attempts to discover the general laws that govern the behavior of observed realities, and by such discoveries hopes to predetermine and sometimes predict and control the course of future events.[10]

Science therefore refrains from the study of intangible spiritual and esthetic values unless they can be reduced to concrete or laboratory situations, as attempted by sociology and psychology.

Areas of Conflict Between Religion and Science

The fundamental conflicts between religion and science have developed historically in areas where the subject matter and methods of religious thought have clashed with the contents of scientific inquiry and with the scientific method. Authoritarian faiths and dogmas have always attempted and, until recent centuries, succeeded in undermining scientific inquiry and condemning scientific conclusions—save for a few notable exceptions when thought was relatively unhampered by faith.

Of these infrequent examples of freedom of inquiry in atmospheres of religious liberalism, two references are noteworthy. One is the example of the classical Grecian culture when Ionian science developed as early as the Sixth Century B. C., unrestricted by religious dogma—when no systematic creed or organized religious institution succeeded in overpowering the freedom of scientific inquiry and philosophic speculation, an intellectual freedom which continued more or less until the post-Aristotelian and post-Alexandrian religions restricted this freedom and undermined the fertile and creative Greek genius.[11]

Another is the development of science and philosophy in Islam under the Umayyads when Hellenic sciences and speculations flourished and advanced through Mohammedan scholarship in an atmosphere of religious liberalism and relatively free intellectual activity—a trend which continued to the 12th Century when Islamic culture fell prey to religious intolerance and fanaticism, and when theological dogmas and creeds replaced the original intellectual fertility of the Mussulman and kept it sterile until the present century.[12]

Why this conflict, and how can it be resolved? In one sense, the apparent clash between religious tenets and scientific findings is not so much between faith and knowledge as between our natural impulse to believe and our intellectual urge to inquire. For, the human mind performs two apparently conflicting, but ultimately complementary and supplementary functions. One function is that of fact-finding—exploring for the sake of discovery. When trained properly as an explorer and fact-finder, the mind carries on its investigations dispassionately, without emotional motivations or evaluative preferences. The other is that of counseling, persuasion and education. When posed as a counselor, the tendency of the mind is to deal with certainties and assurances, with dogmas and absolutes, with ready-made answers and categorical formulas of action. The indecisiveness of the scientific discipline which characterizes the evidence-seeking nature of the mind, gives way in the mind's pathfinding behavior to authoritarianism and sure-handed leadership, dictation and fanaticism. In this dual function the mind seeks, on one hand, to liberate itself from the tyranny of finalities, and to secure for itself on the other hand the comforts of absolutes. It thus thrives on both doubt and certainty.[13]

Two social institutions have historically symbolized these counseling and pathfinding functions of the mind, and have preserved for better or for worse the substance and continuity of selected aspects of human culture. These two are the institutions of state and religion, with the family playing an intermediary and interdependent role. Together, these institutions, with characteristics and functions already noted, have acted as instruments of stability and social integration, have kept the impulse of change in check, and have at their best saved society from disintegration and chaos.

Here then is an apparent conflict between science and religion which is of concern to the teacher and the public alike. While pure science is unconcerned with the moral consequences of its discoveries, institutional religion has historically endeavored to save men from the possible chaos resulting from ever-renewed and ever-changing scientific findings. While men of science have guarded against the dogmatism of tradition, churchmen have assumed the role of guardians of tradition and have favored the dogmas that have sustained this guardianship.[14]

The nerve-center of this conflict is sustained by two fears. Religion fears that the endeavors of science may upset or weaken the established forces of social solidarity, add to man's anxieties, frustrations and insecurities, and so reduce his sense of belonging. Science fears that the formalizing and categorizing tendencies of religion and the resulting fixed patterns of thought and behavior would lead to intellectual sterility and block the path of investigation and progress.

How can this conflict of aims and functions be resolved? The answer seems to be not so much in a division of labor, which would leave science to investigate and religion to indoctrinate, but in a new attitude of mutual endeavor, and support—an attitude which would enable men of faith to keep abreast of the new dimensions of our physical environment that scientific inquiry reveals, and encourage men of science to employ their disciplines and knowledge to the service and welfare of mankind. We need scientists with a religious passion for human well-being, and religionists with a devotion to truth characteristic of the scientific spirit.

Modern science, of course, does not claim that its methods are the only methods of knowledge, or that its conclusions are the only body of truth discoverable in human experience. It realizes the vast, deep and variable areas of esthetic, religious and moral experience, and the human truths discovered through intuitive reflection. It is not anti-religious, but anti-dogma and anti-superstition. It has no quarrels with faith in its high and noble and pure sense. Both the science teacher and the students of science should make a distinction between religious dogmatism and aggressive scientism on the one hand, and progressive religion and limited scientific discipline on the other.[15]

It is the categorical imperatives of materialism in its absolute inter- pretations and presentations of the physical world, such as characterized the sciences of the 17th, 18th and 19th Centuries, and of the naturalism of the same periods in its finalistic explanation of all life in terms of physical and chemical forces that stand in sharp conflict against religious experience and belief.[16] It is when science claims that its empirical and quantitative principles and conclusions are the only valid principles and the only dependable conclusions, that they run into difficulties with reli- gious, esthetic and moral experience.[17]

A suggested hope for the coexistence of religion and science, and for their mutual respect and interdependence may therefore be stated as follows: Modern scientific endeavor begins as a dedication to the objec- tive observation and understanding of the tangible world, and results in the application of its findings to human welfare. Religion also moves by two forces, devotion on one hand to whatever is considered the moving power underlying reality, and on the other good will to a chosen group or to all minkind. Thus, intellectual dedication and practical welfare, which are the two moving forces of the scientific endeavor, find their equivalent in religion in the feelings of devotion and the sentiments of good-will. Here, therefore, religion and science find two common de- nominators (but not necessarily two meeting points), though they take different shapes and may lead to different conclusions and applications in human experience. Science must allow for and respect whatever within man gives rise to the arts, the philosophies, and the cosmic-social exper- iences we call religion. Religion should also recognize that scientific dis- cipline and information cannot be encouraged in an environment of indoctrination or intolerace; that the chief characteristic of the scientific spirit is its capacity for "self-correction," which cannot be maintained in an atmosphere of religious conformity. Scientific achievement should be judged not only in terms of its objectivity and excellence, but also in view of its contributions to human good. Religious endeavors should in turn be judged not alone in terms of its lofty cosmic experiences, but also in terms of its efforts toward social progress, including progress in science.[18]

Position of Science Teachers on Religious Issues

Secular science instruction is bound to run into religious issues and problems arising out of the sectarian backgrounds of students and scientific materials and procedures. Typical of such religious questions are the following: Has man evolved from lower forms of life? Is the theory of evolution consistent with faith? Are there universal moral principles? Is this concept consistent with the pragmatic-scientific view of morality based upon human need, and derived from human experience? Is there life after death? Are we born in sin? Is predestination valid? Does man have a soul? Is the Bible infallible? Is there a personal God? Is the concept consistent with a materialistic-mechanistic interpretation of the universe? Can such concepts as resurrection, redemption and miracle be reconciled with the scientific outlook? Are there a heaven and hell? Are prayers answered? Is this consistent with the concept of law and order in the natural order of things? Is there a supernatural order of reality? Is there a Satan? Are there angels and evil or good spirits? Is Trinity scientifically valid? Can God break, violate or alter the laws of nature? Did Jesus ascend physically to heaven? Is Jesus returning physically to earth? Are the dead going to rise from their graves? Are non-Christian religions equally true religions? Is it necessary to join a church in order to live a religious life? Is revelation possible? Is religious indoctrination justified? Could there be a scientific faith? Can one live a scientific and religious life simultaneously? Are all scientists atheists?[19]

The science teacher must decide on the adoption of a point of view as to the manner of approaching or side-tracking these questions. And this point of view should include answers to some pertinent questions, among them: What are the legal and traditional issues involved in discussing religious questions in secular atmospheres? What limitations does scientific discipline impose upon the science teacher in discussing religious questions? What limitations do the religious backgrounds of students impose upon objective treatment of religious questions in science classes? Under what circumstances can the science teacher discuss religious questions? To what extent and in what areas should he discuss religious questions? To what extent and in what areas should science teachers refrain from discussion of religious questions? What training and other qualifications are necessary for effective discussion of religious questions by science teachers? Should discussion of religious questions be confined to teachers of social studies and humanities? Should science teachers who are by conviction totally against religion be permitted to discuss religious questions in science classes? Should science teachers and teachers in humanities and social studies acquire special techniques in dealing with religious questions in their classes?

Two conflicting points of view seem to guide the response of science teachers to these religious problems:

1. The majority of science teachers ordinarily *avoid* discussions of religious values for the following reasons, each of which is shared by at least some science teachers: (a) They feel it might interfere with scientific disciplines. (b) They feel inadequate to talk fairly on religious subjects for which they have no specialized training. (c) They feel a secular responsibility to avoid controversial discussions of religion. (d) They recognize basic conflicts between scientific conclusions and religious convictions. (e) They feel the students are in most cases religious illiterates and are therefore ill prepared to discuss religious matters in the context of religious experience. (f) They feel that religious experience and scientific inquiry are distinct and irreconcilable. (g) They feel that any discussion of religion should allot equal time and attention to all denominational variations, and indeed to all other non-Christian faiths and creeds, and that the very enormity and complexity of their specialized area should discourage conscientious science teachers from involvements in religious discussions of such wide range and variety. (h) They feel that in most cases the students may be even more familiar with certain details of religious life than science teachers who may have been out of touch with religious topics or indifferent to them because of their own time-consuming specialization. (i) They feel that questions of religion belong more adequately to social sciences and the humanities rather than physical sciences. (j) They feel that it is infinitely safer and wiser to avoid discussions of faith than to deal with personal biases in a discipline that is based on objectivity and dispassionate observation. (k) They feel that in a free society, with its traditions of religious freedom as well as freedom of open investigation and criticism, the interests of both religion and science may best be served by the science teacher's indifference to religious discussions and his contribution to religious enlightenment by the mere presentation of scientific evidence. (l) Some even feel that the aim of science teachers should be to reduce or remove the religious outlook and overcome it by scientific outlook, thus setting science in contrast to religion and in preference to it.[20]

2. In contrast to this group are those who variously feel: (a) that it is the responsibility of secular science teachers to help fill the gap between religious convictions and scientific conclusions; (b) that to do so would require special religious education for the science teacher to deal adequately with conflicts between faith and reason; (c) that science teachers have a responsibility to answer religious questions whenever they are raised by students without necessarily encouraging religious discussions, or involving students in religious debates; (d) that the science teacher should, whenever justified, avoid discussion of religious topics if he feels

inadequate or if he feels it is not the proper time and place for the discussion of a particular religious topic; (e) that the science teacher should take special courses in philosophy and religion in order to be able to converse intelligently on religious issues; (f) and that teachers of science should discuss religious questions in the open, and with the same frankness with which they approach scientific questions.

3. Maybe the problem can be resolved by a new philosophic attitude toward the two levels of scientific and religious experience. In the best scientific spirit, the science teacher should realize that religious convictions and experiences are highly personal and variable, and largely based upon authority and tradition and cannot be examined by the "quantitative" and "repeatable" methods of science. Yet, the experiences and beliefs should be considered by him as "valid" for those who hold them, and "true" on a *private* basis, and thus respected not as scientifically validated experiences or convictions, but as individually honored levels of belief and modes of life.

At the outset this approach sounds like asking science teachers to adopt two conflicting methods of dealing with the unknown: (1) the concrete, quantitative and repeatable methods of scientific inquiry, and (2) the personal and authoritative ways of religious intuition and indoctrination. But this apparent conflict may be resolved by the exposure of the science student to the contents and methods of science, and the ultimate and inevitable influence they are bound to exercise on him in preparing him to re-examine the validity of his religious values, and the consistency and grounds of his beliefs. Under no circumstances should the science teacher sacrifice, modify or betray scientific disciplines and conclusions in order to justify a religious bias, or compromise with a religious view.

It is the cult of "scientism," pointed out in the previous section, that the science teacher should distinguish from the modest and confined efforts of the scientific disciplines in discussions of religious problems in science classes. The science teacher should also be frank in distinguishing between a mature level of religious experience on one hand and religious creedism and moral dogmatism on the other. While he should indicate that there should be room in human experience for a religious as well as a scientific outlook of things, he should not cater to or defend religious absolutism and intolerance. While the science teacher realizes that his methods are still imperfect and his conclusions open to further inspection, and subject to change based upon new evidence, she should not endorse religious finalism, obsolete theology, or views inconsistent with its findings or interfering with its objective disciplines. He should point out to his classes the difference between authority in science, and authoritarianism in certain levels of religion. He should emphasize that authority in science is another term for concensus of scientific opinion, based upon certain established facts and theories, and acceptable only so long as they are not disputed and improved upon by the weight of new facts and experiments. Authoritarianism, on the other hand, whether in religion, morals or politics, considers all answers as final, all knowledge as absolute, and all further inquiry as useless. It allows of no doubt, tolerates no dissension, accepts no margin of error. It is, therefore, impossible for the science teacher to attempt to resolve the basic conflict between authority in science and authoritarianism in religion. On this point the responsible science teacher should stand his ground and tolerate no compromise.

To conclude: All scientific questions should be treated on strictly scientific levels, and discussions of religious questions should be at times avoided if the teacher feels inadequate to deal with them; if such discussions may interfere with the basic disciplines or purposes of the course; if they appear to disturb or annoy the student; if the questions belong more adequately to other school disciplines such as social studies or humanities; if the religious issues are of such a nature as to make them more properly the responsibility of the church or the home; if religious questions are asked merely to embarrass or annoy the teacher, if they are asked merely to amuse other students, or get a rise out of them; if they have no relevance to the subject under treatment; if the student who raises religious questions is not willing to reason out the answer; or if such questions are asked because the student does not consider the teacher adequate to answer the question, and does so in order to expose the teacher's unfamiliarity with the subject. With such reservations in mind, the apparent conflict between scientific disciplines and religious convictions seems to be fairly resolved without sacrifice on the part of the teacher or disappointment on the part of the student. In many cases religious questions are discussed by students with emotionalism and dogmatism. The science teacher should, by contrast, discuss scientific questions which may reflect upon religious views, or religious questions which may arise from discussion of scientific concepts with extreme patience, professional objectivity, pedagogical skill, respect for facts, respect for the student, and above all and always in the light of the basic objectives and disciplines of the course. It is as much of an obligation for the science teacher to modify, postpone, or even reject comment on religious questions which are vaguely stated or ill-motivated, as it is his responsibility to comment on them to the best of his ability and as long as the subject-matter may allow for digression into factual discussions of religion.

Science Teachers and Religious Education

In this brief statement of the religious problems confronting the science teacher, it may be well to point out two important factors contributing to religious enlightenment in an age of science. One is the responsibility of public schools in the factual study of world religions in general, and of Christianity and the Christian tradition in particular.

The other is the responsibility of our liberal arts colleges, including teachers' Colleges and schools of education, to provide the necessary religious education in both contents and techniques of dealing with religious issues for all our teachers, and most particularly for those involved in teaching social studies and the sciences. We shall close this chapter with a brief recommendation as to the role public schools and teachers training institutions should play in developing religious enlightenment in the context of a democratic education.[21]

A. *The Role of Secular Education.* If one of the basic roles of public education is to give all youngsters the academic opportunity for a critical appreciation of the American culture and certain aspects of the advanced cultures of the world, the responsibility for religious enlightenment as an important part of cultural education is obvious. In the words of Vivian T. Thayer, secular education should be dedicated to, "a fair hearing for all points of view, with special privilege accorded to none—a method that respects the claims of all contemporary faiths without exception, but senses an equal obligation to the future which it can discharge only by keeping open the channels of free communication and untrammeled inquiry. It is a method which has given birth to the American secular school and now relies upon that school for the education of Americans in a way of thinking and a way of life that may well constitute their most distinctive contribution to the creation of one world out of many cultures."[22]

There is no room in this chapter to discuss how such a dispassionate and factual presentation of religion should be worked out as part of the secular curriculum. Perhaps an appreciation of local and regional faiths and denominations should be the role of elementary instruction,[23] an understanding of religious institutions as a social and cultural force in America, that of secondary instruction, and a critical and comparative study of world religions that of the liberal arts discipline.[24]

The important fact to keep in mind is that whatever we do in this area on various levels of instruction should be done to the extent that a given community would allow—and should be set as an ideal in a culture committed to freedom of faith. Only such religious enlightenment can help students to appreciate, understand and criticize the heterogeneous religious beliefs and practices in our culture, and so arrive at an independent and personal religious belief of their own.

B. *The Role of Teachers' Training Institutions.* Perhaps in the last analysis the most basic problem in the proper treatment of religious and moral issues in science classes is the poor academic preparation of many science teachers in areas of philosophy, religion and ethics. Absence of adequate knowledge and perspective along these lines would force the science teacher to ignore religious and philosophic problems, expediently postpone such questions to undetermined "other times" or other courses, or attempt to deal with them from a position of relative ignorance.[25]

What should be the proper education of science teachers in areas of philosophy, religion and ethics? The question can be answered only in the light of one's philosophy of teacher education. I would therefore hazard an answer in the light of what I consider a fourfold qualification of a properly trained teacher.

(a) All teachers, including science teachers, should be critically familiar with the fundamental features of the American culture. This familiarity which is introduced in its simple forms in the elementary program and carried on further on the secondary level, is more particularly one of the basic aims of higher education. This familiarity and critical appreciation should include the areas of humanities, physical sciences, social sciences, the arts, as well as those of philosophy and religion, including the problems of philosophy, a critical survey of the Bible, a comparative study of religions, a survey of religious institutions and practices in the United States, an introduction to the philosophy and psychology of religious experience, and a survey of the history of the conflict between religion and science.

(b) The second basic qualification is mastery of subject matter in the area of scientific specialization. This can be accomplished only with a deep sense of professional devotion to one's chosen field of research, and a reverence for truth. Indeed this devotion and the resulting intuition into the nature of things would lead to a scientific-religious experience which some of the greatest men of science have considered equal in depth and penetration to the experiences of the mystics, the poets and the artists.

(c) The third requirement is the cultivation of techniques of presentation of the subject matter and the ability to arouse in each student a sense of exploration. The good teacher is a research worker and a research maker, developing untaught pupils into self-taught students. Applied to discussions of a religious nature, method implies wisdom and skill in handling such questions in a spirit of mutual concern and respect, and also in a way that would neither sacrifice the basic disciplines of science, nor contribute to religious confusion or ill-will.

(d) Last, and most important for all teachers, is the development of an educational point of view, a philosophic outlook that would give value and direction to various specialized disciplines, and would relate excellence in one subject or area to all subjects and to the all-round education of each student. This means that science teachers along with other teachers should include in their professional courses in education a survey of the history of Western education (including a survey of education in other cultures), and a course in philosophy of education aimed at a general evaluation of the basic theories of education, and at the development of a personal philosophic-educational outlook. For the science teacher dealing occasionally with religious topics, the presence of such a value point of reference would give strength and direction to occasional religious discussions, as it would give a moral-social perspective in the light of which such discussions may be carried on, not from a scientific point of view exclusively, but from the bigger outlook of man and his struggles, dreams and aspirations under the sun.

Selected Bibliography

Books:

American Association of Colleges for Teacher Education, *Focus on Religion in Teacher Education* (Oneonta, New York, 1955). Excellent and readable essays written by the faculty of the Western Michigan College, designed to express the views of each writer on the relation of his specialization to questions of a religious nature, covering the areas of social studies, humanities, various physical sciences, and psychology. The book contains a good bibliography organized according to subject-matter for further study.

John Baillie, *Natural Science and Spiritual Life* (New York: Charles Scribners Sons, 1952). A brief defense of the Christian faith, attempting to show areas of agreement and mutuality between science and faith, and drawing upon historical evidence to show that the scientific spirit has been stimulated at times by religious thought.

C. A. Coulson, *Science and Christian Belief* (The University of North Carolina Press, 1956). A lucid presentation of the place of science in a Christian philosophy. A distinguished physicist, Professor Coulson attempts to show that the models, concepts and laws of science arise fundamentally from profound religious feelings and activities sustaining scientific discipline.

Karl Heim, *Christian Faith and Natural Science* (New York: Harper and Brothers, 1953). A study of the encounter between 20th Century Physics and Christian existentialism.

Julian Huxley, *Religion Without Revelation* (New York: Harper and Brothers, 1957). An articulate effort by an eminent scientist to show that the scientific method can be applied to the realm of religion. The book is an impassioned plea for a new religion of humanism. Contains a short but useful bibliography.

William James, *The Varieties of Religious Experience* (New York: Longmans Green, 1902). Though written at the beginning of the 20th Century, the book is a classical study of the psychology of religious experience by one of the foremost philosophers of our time.

E. L. Mascall, *Christian Theology and Natural Science* (New York: The Ronald Press, 1956). A presentation of the regions of contact between theology and science, with a well presented attempt to show the baselessness of the conflict between them. The author, though a theologian by profession, shows a thorough grasp of modern scientific developments. The book contains a well chosen, extensive bibliography.

Joseph Needham, editor, *Science, Religion and Reality* (New York: George Braziller, Inc., 1955). The contributions to this volume are made by some of the "most distinguished scholars of this century" (Charles

Singer, Antonio Aliotta, Arthur S. Eddington, John W. Oman, William Brown, and Clement C. J. Webb), covering some of the great issues of science and religion, with an eloquent introductory philosophic essay by George Sarton.

Bertrand Russell, *Religion and Science* (New York: The Home University Library of Modern Knowledge, 1958). A searching presentation of the grounds of conflict between science and religion by an outstanding mathematician-philosopher of our time.

Arthur F. Smethurst, *Modern Science and Christian Beliefs* (New York: Abingdon Press, 1955). A study of the role of religion in a scientific age, showing areas of hostility between science and religion, and the way to a new conciliation between them.

R. C. Wallace, *Religion, Science, and the Modern World* (Toronto: The Ryerson Press, 1952). A brief but concise statement of the place of religion in a rapidly changing world, and in the light of sweeping advancements in areas of physical and biological sciences; with a well selected brief bibliography for further consultation.

Alfred North Whitehead, *Science and the Modern World* (New York: The Macmillan Company, 1954). A historical-philosophic overview and review of physical sciences, with a penetrating chapter on "Religion and Science," proposing conditions under which the historic conflicts between religion and science may be resolved.

Periodicals:

K. F. Mather, "When Science and Religion Meet," *Science Education,* XXX (March, 1946), 63-69. A good presentation of the basic aims of religion and science and their ultimate contribution to human welfare.

David Steinman, "Moral Armor for the Atomic Age," *Science Education,* XLII (March, 1958), 175-178. A good statement on how religion, art and science represent a trinity of the human search for the good, the beautiful and the true.

H. A. Webb, "Johnny, You Must Leave the Room," *Science Teacher,* XIX (Sept. 1952), 170-172. A statement on some restrictions imposed upon teachers of science and other disciplines because of religious beliefs.

MEHDI NAKOSTEEN, Associate Professor of Education (Comparative, History, and Philosophy of Education), University of Colorado. B.A., College of Wooster; M.A., Columbia University; Ph.D., Cornell University; one year of graduate study at Union Theological Seminary; one year of graduate study at Harvard University. Author of (1) *A Three-fold Philosophy of Education; Individual and Social Aims of Education,* and; *Theories of Education from Comenius to Dewey,* published in one volume, (Denver: Charles Mapes Publishing Co., 1943). (2) *The People and Culture of Iran* (Denver: The World Press, Inc., 1937). (3) *Religions of Iran* (Denver: Charles Mapes Publishing Co., 1943). (4) *Song of Uryan,* translation of the quatrains of Uryan, a contemporary of

Omar Khayyam, from Persian verse to English verse (Denver: General Printing Co., 1938). (5) "The Development of Atomic Theory Among the Greeks," *The Atom,* I (Fall, 1945), 11-14. (6) "The Islamic World," *Approaches to an Understanding of World Affairs,* (N.E.A.: Twenty-Fifth Yearbook of 1954, National Council for the Social Studies), 245-271.

NOTES FOR RELIGION AND SCIENTIFIC EDUCATION

1. Samuel I. Clark, "Religion and the Social Studies," *Focus on Religion in Teacher Education* (Oneonta, New York: A.A.C.T.E.), 1-18.

2. Frederick J. Rogers, "Religion in the Humanities," *Ibid.,* 19-23.

3. Bertrand Russell, *Religion and Science* (New York: Oxford University Press, 1956), 19-243.

4. Robert Ernest Hume, *The World's Living Religions* (New York: Charles Scribner's Sons, 1926), 1-17.

5. William James, *The Varieties of Religious Experience* (New York: The New American Library, 1958), 39-56.

Julian Huxley, *Man in the Modern World* (New York: The New American Library, 1956), 129-138.

6. Karl Heim, *Christian Faith and Natural Science* (New York: Harper and Brothers, 1953), 35-47.

7. Lecomte De Nouy, *Human Destiny* (New York: The New American Library, 1956), 15-21.

Karl Heim, *Christian Faith and Natural Science* (New York: Harper and Brothers, 1953), 71-108.

8. Irvin Edman, editor, *The Philosophy of Santayana* (New York: Random House, First Modern Library Edition, 1952), 144-216.

9. D. R. G. Owen, *Scientism, Man, and Religion* (Philadelphia: The Westminster Press), 18-20.

C. A. Coulson, *Science and Christian Belief* (Chapel Hill: The University of North Carolina Press, 1956), 29-63.

10. James B. Conant, *On Understanding Science* (New York: The New American Library, 1951), 20-40; Arthur S. Eddington, "The Domain of Physical Science" in *Science, Religion and Reality,* Joseph Needham, editor (New York: George Braziller, Inc., 1955), 193-225; George Sarton, "Introductory Essay," *Ibid.,* 3-22; E. L. Mascall, *Christian Theology and Natural Science* (New York: The Ronald Press, 1956), 47-86.

11. Bertrand Russell, *A History of Western Philosophy* (New York: Simon and Schuster, 1945), 6-7 and 22-23.

12. Will Durant, *The Age of Faith* (New York: Simon and Schuster, 1950), 235-344.

13. Karl Manheim, *Freedom, Power and Democratic Planning* (New York: Oxford University Press, 1950), 285-313; good bibliographical references, page 374.

14. Bertrand Russell, *Religion and Science* (New York: Oxford Uni-

versity Press, 1958), 7-19; Alfred North Whitehead, *Science and the Modern World* (New York: The Macmillan Co., 1954), 259-277.

15. Joseph Needham, editor, *Science, Religion and Reality* (New York: George Braziller, Inc., 1955), 4-22.

16. Hobbes reduced reality to matter, Gassendi to atom and Newton to simple and orderly laws of motion. La Mettrie described all human behavior in mechanistic terms, Holbach interpreted man within the order of nature, and Helvetius explained morality in terms of physical laws.

17. R. C. Wallace, *Religion, Science and the Modern World* (Toronto: The Ryerson Press, 1952), 1-16; E. L. Mascall, *Christian Theology and Natural Science* (New York: The Ronald Press, 1956), 7-32.

18. Robert F. Davidson, *Philosophies Men Live By* (New York: The Dryden Press, 1956), 421-425.

19. Bertrand Russell, *Religion and Science* (New York: Oxford University Press, 1958), 19-144.

20. W. C. Van Deventer, "Questions Concerning Religion in Science Classes," *Focus on Religion in Teacher Education* (Oneonta, N. Y.: A.A. C.T.E., 1955), 24-34; George Bradley, "Another Responsibility for the Science Teacher," *Ibid*, 35-39.

21. Bernard N. Meltzer, Harry R. Doby, and Philip M. Smith, editors, *Education in Society* (New York: Thomas Crowell Co., 1958), 454-459; Blaine E. Mercer and Edwin R. Carr, editors, *Education and the Social Order* (New York: Rinehart and Company, 1957), 484-488; George H. Williams, "Religion in the Public Schools of our Democracy," *Religious Education* LI (Sept.-Oct., 1956), 374-377; C. Winfield Scott and Clyde M. Hill, editors, *Public Education Under Criticism* (New York: Prentice Hall, 1954), 128-131 and 135-136.

22. Vivian T. Thayer, "The Secular Method in Education," *Education and the Social Order*, Blaine E. Mercer and Edwin R. Carr, editors, (New York: Rinehart and Co., 1957), 472.

23. Sara R. Swickard, "Teaching About Religion in the Elementary School," *Focus on Religion in Teacher Education* (Oneonta, N. Y.: A. A. C.T.E.), 45-47; Mate Graye Hunt, "Spiritual Values in Children's Literature," *Ibid.*, 48-54.

24. Charles I. Glicksberg, "The Religious Motif in Higher Education," *AAUP Bulletin*, 43, No. 3 (September, 1957), 449-457.

25. R. C. Wallace, *Religion, Science and the Modern World* (Toronto: The Ryerson Press, 1952), 33-48.

Formal Education Aspects

SCIENCE FOR GENERAL EDUCATION

Edwin F. Lange
Portland State College
(Portland, Oregon)

One of the major accomplishments of the present century has been the development of scientific curricula for the training of competent scientists in all of the branches of scientific study. This is evidenced by the tremendous strides made by American science during the present century and by the great number of Nobel prizes awarded to our scientists. On the other hand, the type of science training needed by the general populace remains a major problem in American education.

That some type of science education is desirable for all people in a democratic society has been expressed from several points of view.

In a democratic society, people are often called upon to pass judgment at the polls on problems which have a scientific basis. Experience has shown that scientific data is often overlooked as such issues are decided on emotional planes. The many bitter community struggles across the nation regarding the fluoridation of water is a case at hand. Today, problems of air and water pollution, conservation, mass inoculation and vaccination, space exploration, and radioactive fallout all have their roots in scientific study. Without some understanding of the sciences, it is difficult for the citizen to exercise intelligent judgment on these and other matters. He then becomes easy prey for emotionalism and mass hysteria.

The same type of reasoning might be applied to many vocations, though not in themselves scientific, they often come in contact with scientific problems. Such people as journalists, lawyers, and politicians often are required to render decisions which have a scientific bearing. It has been suggested that these people, possessing scientific facts and understanding scientific methods, will approach problems more realistically and intelligently.

As a part of a liberal education the study of science has value in satisfying the natural curiosity people have regarding how and why things react and behave as they do. In a scientific age considerable importance is attached to understanding oneself and the surrounding environment.

These ideas should constitute some of the major objectives for general education in the sciences.

Our times are characterized by considerable leisure time on the part of all the people. Never in the history of man have all the people of a nation pursued hobbies, do-it-yourself activities, sports, and self-satisfying projects as has our present populace. More people than ever before are finding relaxation and recreation in hobbies that are scientific in nature. Such activities as astronomy, nature study, rock and mineral collecting, photography, skin diving, and prospecting for radio-active minerals are attracting increasingly greater numbers of people. Not only do these activities prove satisfying, but often amateur scientists have made discoveries important to the history of science. Children too are caught in this trend. This is evidenced by the increased number of summer nature camps and the yearly production of scientific toys. Each year manufacturers produce new experimental scientific kits from which children receive satisfying experiences. Such toys and kits represent all of the sciences from astronomy to zoology.

As a branch of the school curriculum, science instruction is rather recent. Dating back to the 17th century the idea of gaining knowledge through experimentation was forcefully proposed by Francis Bacon and was implemented by the founding of the Royal Society of London and other scientific societies by people of many vocations and professions, who were interested in learning for the sake of learning. These societies had such objectives as the "improvement of natural knowledge" (Royal Society), and the "promoting of useful knowledge" (American Philosophical Society).

A hundred years ago the movement of scientific education received great impetus from the famous essay of England's great scholar, Herbert Spencer. In his essay, "What Knowledge is of Most Worth?", Spencer came to the conclusion that education in the sciences was of most value. He proposed that the purpose of education was the preparation for complete living. Then by analyzing the needs and activities of life and noting how the various studies contributed to these needs, he concluded that of all subjects, a knowledge of science was always the most useful preparation for living.

In the succeeding years, Thomas Huxley (1825-1895) skillfully and eloquently wove the sciences into a pattern of liberal education through his lectures and essays. Many excerpts could be quoted from his writings, but the following is representative of his views:[1]

"Considering progress only in the "intellectual and spiritual sphere," I find myself wholly unable to admit that either nations or individuals will really advance, if their common outfit draws nothing from the stores of the physical sciences. I should say that an army without weapons of precision and with no particular base of operation, might more hopefully enter upon a campaign on the Rhine than a man, devoid of knowledge of what physical science has done in the last century, upon a criticism of life."

As a result of these movements various forms of science teaching found their way into the curricula of elementary, secondary, and higher education of both Europe and America.

In this country an ardent champion of science education was Thomas Jefferson. He believed that a study of the sciences would contribute to the development of the nation as well as form a foundation for personal vocations and recreation. As a member of the Virginia legislature in 1779, he drafted a bill proposing "teaching sciences generally and in their highest degree" in three phases of public schools, elementary, college, and university grades. Again, after his retirement from the presidency of the United States, he was concerned with the development of a science curriculum for his new University of Virginia. Early in 1800 he wrote Joseph Priestley asking for suggestions for a course of science study for the University. Discussion centered on such courses as botany, chemistry, zoology, anatomy, natural philosophy, mathematics, and astronomy.

Up to the present century science for general education was not a major problem. Relatively few students continued their education beyond the elementary level. Many did not complete the elementary school. Life, too, was relatively simple. However, with the coming of the 20th century and its great scientific and technological developments, a large increase in enrollments in the high schools, colleges, and universities of the nation completely altered the educational pattern. The change has resulted in great technological specialization in vocations and professions. The change in the college curriculum is characterized by specialization, proliferation of courses, and often by fragmentation of knowledge. This has occurred in the traditional liberal arts colleges as well as in technical institutions. Areas of learning outside of the sciences have equally been affected by this trend. Many beginning college courses are more and more becoming professional courses rather than general courses. Such beginning courses often carry rigid prerequisites which general students are unable to meet. Educators across the nation have been greatly concerned with these trends, and many interesting experiments and counteracting developments have been proposed. Not only have the colleges and universities wrestled with the problem, but all phases of public education have been affected.

Elementary School Science and General Education

Perhaps, the greatest change that has occurred in the past twenty-five years in the elementary school curriculum has been the development of a comprehensive, integrated, course of science instruction, first begun in 1890, from the first to the eighth grade, replacing the traditionally limited and meagre nature study program and other offerings of the past century.

Such instruction in science fundamentals is entirely general as it is not the function of the elementary school to train for specialized vocations and professions. However, such programs are having significant effects on identifying outstanding scientific talent.

The elementary science program is materially aided by the availability of beautifully illustrated modern text books for each grade level. Text books have been amply implemented with a tremendous variety of supplementary books which in and of themselves create interest in the various branches of science. These books are well printed and illustrated, and stress scientific accuracy. Every branch of science is covered by these books which are doing much to increase the general science education of all the people.

A great impetus to science teaching on the elementary as well as the secondary level has been the rise and popularity of science fairs in many sections of the country. In areas where science fairs are successful, hundreds of students participate with exhibits and projects. Among the participants are large numbers who do not contemplate a career in science.

In recent years much attention has been given to developing activity programs in elementary science teaching. Experiment books are available for various phases of elementary science. Science kits, both commercial and locally developed, are to be found in many of the nation's modern schools. Almost all training programs for elementary teachers include work in the physical and biological sciences and often courses in the methods of science teaching.

Unfortunately, in spite of all of these commendable accomplishments which one might find in the better schools of the nation, hundreds of elementary schools almost completely ignore all forms of science teaching. Not all states have yet worked out a comprehensive eight year science program. In other schools, older teachers without training have shied away from this newer trend. Lack of leadership has prevented many schools from obtaining proper library and activity resources.

The elementary school, with a sound science program can have great effects on the general education of the entire populace. Teachers who conscientiously attempt to teach science, report that children in general are tremendously interested in all aspects of science. A good foundation at this level does much to encourage continued interest in the sciences. Studies have shown that many scientists first become interested in science during their elementary school days. A basic elementary program will not only contribute to an enlightened citizenship but will also help in providing an adequate supply of competent scientists for the future.

Secondary School Science for General Education

The high school science program has traditionally been looked upon as general in nature for all students. Changes in recent years have somewhat altered this view. Some forty or fifty years ago a large percentage of high school students completed courses in chemistry and physics. Today these courses draw little more than five percent of the high school population in any one year. On the other hand, two courses, general science and biology, unknown in the high school curriculum fifty years ago, attract from twenty to thirty percent of the annual high school population. In some schools all ninth graders study general science and in others all tenth graders complete a course in biology. Most high schools require the completion of one or more science courses for graduation.

This abrupt change in popularity of high school science courses was largely brought about by a changing high school population. Today practically all youths of high school age are to be found enrolled in the secondary schools of the nation. Fifty years ago the percentage of high-school age youth enrolled was small. It appears then that courses in general science and biology more adequately meet the needs of all for the purposes of general education than do such traditional courses as chemistry and physics.

Chemistry and physics are looked upon as college and professional preparatory courses in many schools. In other schools attempts have been made to generalize these courses so that they would appeal to a greater number of students. Unfortunately, a great many high school students are denied the privilege of choosing chemistry or physics because these have not yet been included in the curricula of hundreds of small high schools. Many small high schools offer these courses only in alternate years. Even when offered, instruction in the sciences is often very inadequate because of the lack of well trained teachers and facilities. Competent teachers in such small schools often are assigned several preparations so that time for science teaching is used up in other activities. About half of the nation's secondary science teachers have assignments other than the teaching of science. Such fragmentation in science teaching can only result in inferior work. The trend of consolidations of small schools into larger districts should materially aid this very serious problem.

On the other hand, large high schools generally offer satisfactory work in these areas. Experimental work has been done with advanced science courses and in some cases college credit has been allowed for outstanding high school work. The present emphasis on identifying and enriching the curriculum for gifted and talented youth will make this practice more common in the near future.

Among the recent trends in high school science is the introduction and experimentation with a variety of new courses. The main objective of these courses seems to be general education for juniors and seniors of high school age. Among the newer listing are such names as applied science, applied chemistry, physical science, earth science, plant science, aviation science, and photography. The large variety of titles suggests that high schools are seeking courses in generalized fields of science that will meet the needs of those not contemplating higher education.

The Colleges and Science for General Education

One of the major problems in higher learning today is that of the general education of both the professional and paradoxically, the traditional liberal arts student. Curricular proliferation and over specialization has already been referred to. Professional curricula have become so crowded with required courses that students find little time for general or broadening courses. The problem, applicable in many fields, is indicated by the summary of a study of enginering education.[2]

The basic findings of this study are:

1. That engineering educators throughout the country are in nearly unanimous agreement that their students would profit—as professional men, as citizens, and as individuals—from a fuller acquaintance with the resources of the humanities and social sciences;

2. That a sizable number of these same educators are honestly fearful that attempts to incorporate into already overcrowded curricula a substantial program of humanistic-social studies may either jeopardize the quality of technical education, or lead to superficiality in the treatment of the humanities and the social sciences;

Similar conclusions have been reached by educators in many other professional schools, both in scientific and non-scientific areas. In many cases the collegiate training period has been lengthened to include adequate time for general education as well as the inclusion of new technological developments. Today five year programs are becoming more common in certain areas as dentistry, pharmacy, architecture and others. In the future more educational programs can be expected to increase their training period, and some that have already lengthened theirs, may increase them again.

Traditionally, American colleges have recognized the need of science in the liberal arts curriculum. Almost all colleges and universities require the completion of one or more years of science for graduation. The courses meeting graduation requirements might be specified or might be left to the choice of the student. The professional and technical nature of many beginning college courses has already been mentioned. It is common practice today to find science departments offering beginning courses such as chemistry or physics on two or more academic levels, a rigorous approach for majors and professional students, and one or more less rigorous courses for non-science majors. The latter type is usually less mathematical and technical, and has sometimes been criticized as being little more than descriptive in nature. The completion of only one such beginning course in a four year college program contributes little to a broad outlook of the sciences in general and the scientific method.

The broad view was introduced into the college curriculum some thirty years ago by survey courses in everything from science, to literature, to social science. Courses in physical and biological science became common and often popular. Sometimes both areas were treated in a single course. Science survey courses soon became the object of a bitter controversy with strong supporters on each side. On the one hand they were heralded as providing a broad general background in the sciences, and on the other side they were criticized as being so broad that they lacked depth and were largely quite superficial. As a result of such strife, many science survey courses were abandoned.

To meet the needs of general education, new types of science courses have appeared on many campuses. While beginning college courses in such areas as biology, chemistry, geology, and physics are highly standardized and quite similar from campus to campus, the newer courses often differ widely in objectives and content. One of the interesting approaches to the problem is the case history method developed by Dr. James C. Conant, a former president and organic chemistry professor of Harvard University. Assuming the student has acquired some basic knowledge of the sciences in high school, his course deals with "Tactics and Strategy of Science" or "an understanding of science." The course centers around a historical case study of some phase of the evolution of science. It stresses the problems which accompany the advances in science, the importance of new techniques and their influence on scientific progress, and the rise and development of scientific thought. A two volume set of Harvard Case Histories in Experimental Science, designed particularly for non-science majors has recently been published.

Another approach is the development of many so-called block-and-gap courses. These counter the argument that certain courses lack depth. They are designed to investigate fully a few concepts and at the same time purposely omit others. They often cut across traditional discipline lines and may draw from several academic areas, integrating basic ideas and knowledge. Usually problems are selected which do not cover the large array of facts of all of the various sciences, or all of the details of a single one, but stress scientific unity and the relationships between the sciences and other academic studies. On the upper divisional level these courses are considered as a synthesis of all of the preceeding work, and emphasize the development of human values and philosophies. Courses of this type have many titles such as "The Impact of Science on Modern Life," "History and Philosophy of Science," "Science of Civilization," "History, Philosophy, and Social Importance of Science," "The Origin and Development of Scientific Concepts" and many others, all of which differ greatly from the stereotyped beginning college courses and from the survey approach.

Although courses of the type described are to be found in colleges and universities all across the nation, such programs are still largely experimental and are accompanied by their own peculiar problems which differ somewhat from campus to campus. As technological and scientific advances continue, more attention will be given to science courses which meet the requirements of general education.

General Agencies of Science Education

While formal efforts are being made by the schools of the country to provide adequate science courses for general education, other agencies are also contributing to this goal. In recent years the museums have taken on a new look. Museums of science, history, and industry have become vital instruments in the dissemination of scientific information. Exhibits, once gathering dust, are now constantly being changed and made attractive so that they appeal to a wide variety of interests. Service clubs and social groups are catered to in the form of scientific programs. Many museums foster children's activities on a wide scale, such as field trips, nature camps, scientific displays, and the publication of scientific literature. In many cities, museums are truly becoming cultural scientific centers of their communities.

The press, radio and television are today also making significant contributions in the field of general education.[3] The sciences have received major emphasis in this respect, primarily because of the many startling scientific advances that have been made during the past few decades.

Television, the newest medium of mass communication, has made tremendous advances by taking appealing science programs into the homes of the American people. Such programs have created tremendous interest and will continue to capture the attention of the viewing public. Schools, too are experimenting with television in order to make teaching more efficient. Two types of school-television cooperation is now under way. The one consists of closed circuit television programs designed to utilize outstanding teachers on a wider basis within a college or between several schools. Preliminary results of these experiments are encouraging and indicate that many of the sciences are applicable to this type of teaching. In the sciences such courses are implemented by discussion groups and laboratory work. Since the college enrollment in the coming years is expected to increase faster than the training of competent staff, studies of educational television are particularly timely and significant. Equally important are the experiments being carried out between schools and television stations in the sponsorship of telecourses, whereby all or most of the usual classroom work can be completed in the home.

The newspapers and magazines of the country have from early times been important agencies in educating the citizenship in different areas of learning, including science. Practically all major newspapers feature syndicated columns written by leading science writers who attempt to interpret modern scientific achievement for the layman. Science feature articles in magazine sections are interestingly written and often well illustrated. Some of the country's leading scientists, including Nobel prize winners, have contributed articles of considerable scientific complexity to leading periodicals and have had them well received by the reading public.

Conclusion

Science education is important for all people in a democratic society in a scientific age. All phases of public education should contribute to the goals of general education. Media of mass communication, the press, radio, and television, have in the past and will in the future continue to make important contributions in this area. The major problem yet to be solved concerns itself with the type or types of science education most beneficial for all of our people.

Selected Bibliography

Carnegie Foundation for Advanced Teaching, *Liberal Education* (from the 1955-56 annual report). This pamphlet is a summary of the discussion of the trustees of the foundation concerning their views of what constitutes a liberal education and how American educational institutions can contribute to such a program.

James B. Conant, *On Understanding Science* (New Haven: Yale, 1947). An explanation of the historical case study approach to the field of science for general education; *Harvard Case Histories in Experimental Science* (Cambridge, Mass.: Harvard, 1957). Each of the two volumes contains four case histories selected from the fields of the physical and biological sciences; the volumes are suitable for general reading or could be used as texts or references in the new types of college science courses.

Commission on Secondary School Curriculum, *Science in General Education* (New York: Appleton-Century-Crofts, 1938). Deals with the problems of science in American secondary schools.

David B. Hawk, "Specialization in American Higher Education and the General Education Movement," *The Journal of Educational Sociology,* XXVIII (September, 1954), 19-24.

Thomas H. Huxley, *Science and Education* (New York: P. F. Collier, 1902). A collection of addresses concerning the place of science in the various curricula of public schools.

Earl J. McGrath, *Science in General Education* (Dubuque, Iowa: Wm. C. Brown, 1948). Describes the development of the general education movement and discusses in detail the science courses designed for general education on twenty-two different American college and university campuses.

W. Hugh Stickler, "Senior Courses in General Education," *The Journal of Higher Education,* XXV (March, 1954), 139-146 & 171.

ERWIN F. LANGE, Professor of General Science, Portland State College (Oregon State System of Higher Education), received his B.A. degree from Williamette University (Salem, Oregon) in 1933, M.S. from the University of Oregon in 1936, and Ed.D. from the Oregon State College (Corvallis, Oregon) in 1951; after teaching science and mathematics in Oregon high schools for seven years, he became connected with the Oregon State System of Higher Education in 1946, and is currently teaching organic chemistry, history of science, and courses in science for general education.

NOTES FOR SCIENCE FOR GENERAL EDUCATION

1. Thomas Huxley, *Science and Education* (New York: P. F. Collier and Son, 1902), 127-128.
2. The American Society for Engineering Education (University of Illinois, Urbana, Ill.), *General Education in Engineering* (1956), vii.
3. See: John Pfeiffer, "Making Popular Science More Popular," *Science,* CXXVII (April 25, 1958), 955-957, discusses the problem of bringing science to the non-scientists through the media of the press; Lynn Poole, *Science Via Television* (Baltimore, Md.: Johns Hopkins University Press, 1950), the opening chapter summarizes the possibilities of disseminating scientific information by TV and a number of popular science program are described; M. W. Thistle, "Popularizing Science," *Science,* CXXVII (April 25, 1958), 951-955, under the heading of "Five Barriers" discusses the problems encountered by science writers in making the knowledge of scientists available to the public; Warren Weaver, *The Scientists Speak* (New York: Boni and Gaer, 1947), a collection of radio talks given by leaders of American science during the intermission of the New York Philharmonic Symphony's Sunday afternoon concerts.

PRE-PRIMARY AND ELEMENTARY EDUCATION

ELEMENTARY EDUCATION

HANOR A. WEBB

George Peabody College for Teachers

A Beginning

"Just as the twig is bent the tree's inclined."—POPE.[1]

I pressed my finger into the tiny palm of our son. His fist tightened with a surprising squeeze, for he was only one day old. His response was to a sensation—touch—and due to *instinct*.

Many weeks later we moved a small light back and forth in front of him; his eyes followed it. A few more weeks and his head turned as the light went by. This response to a sensation—light—was due to *interest*.

Several months having passed, he would turn and look at the right place when we asked, "Where is the light?" He could also find with his eyes the door, the kitty, Mama. Thus checking his sense of hearing by his sight, he showed us *understanding*.

Thus began our son's science lessons—if you will accept a definition of science according to its derivation—the Latin *scire*, "to know." We customarily broaden the term science to mean organized knowledge—and that came in due time as our son picked out flat pebbles from round ones on the creek bank for us to skip on the water. He was then about two years old.

Science is also tested knowledge, and at three our son proved for himself (ignoring "Don't touch!" warnings) that a bee would sting. Maturer science involves speculation and hypotheses, so the boy at five was bringing to me fossils at every hill-climb, asking, "Daddy, how do you know this thing was once alive?"

His science lessons continued—and still continue. Our son, now a geologist, will use his senses, his interest, his understanding, his discrimination, his interpretative imagination, to the end of his span as a scientist.[2]

At the age of six our son, with thousands of others, began to receive formal instruction in science. Whatever the activities in school were called, it was still science if the objectives were to observe and understand the environment. My assignment in the present volume is to discuss science education at the elementary school level, which includes ages six through twelve, and grades one through six.

If your interest in children younger than six continues—as mine does —there are excellent discussions in certain bulletins[3] and books.[4]

This chapter is not written for the specialist in educational research, who is as capable as I may be in locating the recent studies in the teaching of elementary science.[5, 6]

This chapter is not written for the timid teacher, who seeks countless specific items of experiments, plans, practices for her elementary class in the science experiences. A list of books and other publications, at the end of this chapter, furnish these necessities for her.

This chapter is written for those modest scholars who intend to read the entire book for values it may contain for those interested in the broad fields of science and of education.

Objectives of Elementary Science

"Be ready always to give . . . a reason for the hope that is in you." —*The Bible,* I Peter 4:15.

The principal was plainly frustrated. "I asked my six teachers—each on her own—to write five reasons why science should be taught in the elementary school. I have planned to print a little bulletin for our parents." He handed a folder to me. "Open it. Here are thirty reasons— all different. I'm glad I don't head a really large school with fifty teachers!"

A helpful list of objectives for an activity can be prepared if certain rules—as to brevity, pertinence, clarity—are followed. Near these lines appear *Nine Reasons Why*—a statement of objectives for teaching science in the elementary grades. The motto of its preparation might have been *E Pluribus Unum,* for out of many, many lists prepared by thoughtful, usually experienced, teachers over a period of several years, this one list was derived. Thereafter, several thousand teachers have read—perhaps studied—the list in mimeographed form. It has been published once.[7] No one, in person or by letter, has seriously questioned its ideas or wording.

Because we know that these nine reasons have been used as "Nine Commandments" by many conscientious teachers, we will follow them as the outline for further discussions of this chapter.

Nine Reasons Why

Science is taught in today's elementary schools:

1. *To give practice in simple observations*—as background for future investigation and understanding of the environment.

2. *To give practice in purposeful activity*—as background for future experimentation and constructive labor.

3. *To enlarge the vocabulary with the names of simple objects and processes*—as background for the future use of technical terms.

4. *To give experience in combining the factual and the emotional*—as background for future appreciation of natural law and beauty.

5. *To guide emotional responses away from the highly subjective*—as background for future sensible attitudes and behavior.

6. *To start habits of scientific thinking in simple matters*—as background for scientific thinking in important future decisions.

7. *To start building attitudes toward the social effects of science*—as background for future cooperation in community programs of health and welfare.

8. *To develop simple concepts such as cause and effect, balance of nature, and the like*—as background for future understanding of broad concepts like conservation of resources, the laws of learning, and even the sacredness of truth.

9. *To develop a simple reverence for nature*—as background for future appreciation of the wisdom and power of God.

First Reason: Simple Observations

"Oh foolish people, and without understanding; which have eyes, and see not; which have ears, and hear not." *The Bible*, Jeremiah 5:21.

Science is taught in today's elementary schools:

1. *To give practice in simple observations*—as background for future investigation and understanding of the environment.

"Are you detectives very, very smart?" asked the grade-school youngster as his class visited the Police Station.

"No, some of us are rather dumb," replied the Chief, "but every good detective sees things nobody else would notice."

The child enters the pre-primary or first grade with considerable experience in the use of his senses, and prompt associations of feelings—pleasure, pain—with them. Interest, which is one of the stronger human instincts, has developed in him a marked curiosity. (Many animal pets, and some of the wild creatures, are also noted for curiosity; it is a sign of above-average animal intelligence.) This curiosity should be sharpened by training in school, and the chief way is to give practice in accurate observation.

Counting adds accuracy to observation. The primitive savage counted "one, two, three, many" when the numbers of friends or enemies seemed important. Some children entering first grade can go surprisingly far in counting, while others barely start. The good teacher makes it her duty to discover these limits, and to enlarge each of them. She lets "many" come at some point, but it is much higher than three. Some teachers strive for ten—the number of fingers or toes. Such counts as petals on a flower, ducks on a pond, puppies at their natural feeding places, or—horrors!—flies on a banana peel, come within these limits.

Measuring adds to observation. The youngest grade child can handle a ruler, and soon learns to count inches. The question is not "big or little?" but "how big? how little?" Many teachers believe in the practice of guessing a measurement first, then measuring. This is like throwing at a target; more practice, more accuracy.

Color is a quality of accurate observation. First grade children already know the names of certain strong colors, and of several highly colored things. Recognition of colors by their names should be increased to ten or more. Thoughts as to whether a color is "pretty" (pleasing) or "ugly" (displeasing) are discussed under a later heading.

Structure is an important observation. Is a tree slender or spreading? Is a leaf wide or narrow? Does a flower bulge (like a rose) or open flat (like a daisy)? Are clouds thick or thin? Is a rock sharp or rounded? These aspects may be of great importance in explaining what is seen.

Texture is a part of accurate observation. Is a surface like sand or like glass, fuzzy, furry, or hairy? There is always a reason why a surface feels the way it does.

Observations involving sounds, odors, and tastes are subject to more accuracy than is usually given them.

An inexpensive lens, preferably a reading glass, is a great aid to observation.

The senses may be fooled, and this should be remembered, and tested by measurement.[8]

Successive observations are of value. Sprouting plants or growing chicks, for example, must be seen day by day to learn how they grow. Today's weather is more important when compared with yesterday's, or even the weather of an entire week.

Reports of observations are highly desirable. "See and tell" periods are standard practices in countless elementary classrooms. Before the youngest pupils can write, they can draw! Their art may seem crude, but let us call it "primitive," and let the youngster feel it to be "a poor thing, but mine own!"

Outdoor observations are the best for elementary grades, studying the world of nature. Next best are those of home equipment in service. Poorest are those set up with specialized equipment, never seen by the child in his ordinary experiences. This will be discussed further in the next section of this chapter.

Opportunity should not be missed in observation. It is true—and to a degree helpful—that many school systems have curriculum guides, and each grade follows almost daily assignments. Yet if a mother bird built a nest in a tree outside the window, and children can look directly into it as the eggs hatch, the tiny birds are fed, their pin feathers appear, and all other marvels of growing up take place, it would be a cruel and unwise teacher who would sternly draw the shade because the topic Birds is listed for another grade or season. There have been such teachers!

Trends in Observation. Common sense as well as child psychology tell us that observation fits the early grades; that it is the foundation of an alert understanding; that it is the first step in science from the lowest to the highest levels.

We see more and better suggestions for observation in the newer articles and guides. We are sure that in elementary classrooms, year by year, there are more things to see, feel, etc. and remember, and fewer things to remember only.

Mary's Little Lamb—Modern Version

Mary had a little lamb, it was her precious pet.
　It said "baa! baa!" to all the boys and girls that Mary met.
It followed her to school one day; this was against the rule!
　No teacher of the *older* way allowed a lamb in school.
But her *new* teacher took it in, and said, "How glad I am
　Today our lesson to begin about this darling lamb!"
"Teacher, why do you love it so?" the eager children cry.
　"*I* was a child not long ago," the teacher made reply.
　　　　　　　　　　　—ANONYMOUS (through modesty).

Second Reason: Purposeful Activity

"The human hand as well as the brain has played a part in the development of science. . . . The hand has not been sufficiently recognized in elementary education."—CRAIG.[9]

Science is taught in today's elementary schools:

2. *To give practice in purposeful activity*—as background for future experimentation and constructive labor.

The annoyed teacher scolded: "Too many of you students try to work without thinking!"

The irrepressible one replied: "And too many of you teachers try to think without working!"

A bit impudent—yet true. I predict a brilliant future for that boy.

A purposeful activity in the classroom, related to science, has usually been called an *experiment.* Many modern teachers prefer to call it an *experience.* Both words come from the Latin, "out of trial." A distinction may be that in an experiment you do not know what will happen, but intend to find out, while in an experience you know what will happen, yet to watch it again will be interesting.

Manuals of activities in science, suitable for the lower grades, are numerous indeed. Some of the later ones are listed at the end of this chapter.

The successful experience (or experiment) will have these qualities:

1. *It will be simple.* It will not be above the interest and understand-

ing of the average child in the grade. This implies simple equipment, chiefly from the home and neighborhood. Some familiar items may be purchased from local merchants.

The elementary teacher should avoid the use of high school science apparatus. Such equipment does not contribute to the pupils' own development, and gives them a false concept of science experimentation.

Some of the published lists of apparatus recommended for the elementary grades are formidable as to quantity and expense. These are not the better guides. Some lists give both simple and expensive items separately.[10,11]

2. *It will be originated by pupils.* The teacher's guidance, of course, is not ruled out. Wherever possible the children should feel, "We thought of it first!"

3. *It will be performed by pupils.* They learn to do by doing—the best plan of all. Experiences that require too much time, are too complex, or are dangerous, should not be done. An occasional introductory demonstration by the teacher, to arouse interest and show good handwork, may be quite helpful.

Too many hands should not be working at a single experiment. Time is also wasted if most of the class must watch a few setting up equipment. The better plan is to have several groups of proper size getting ready to demonstrate, each in turn, an experiment before the class.

4. *It will be interpreted by pupils.* Their discussions should go as far as their maturity permits. The teacher will guide, correct, and praise the pupils as they seek understanding.

5. *It will be recorded by pupils.* The smallest child can draw what he sees. As soon as writing skill develops, let it be put to use in "notes, the way the men of science do."

Teachers are often surprised and gratified at how many items parents will lend or give for use in teaching science to their children. Parents often give a science item or book to a child for his birthday, which he in turn gives to his teacher for class use. Toys that work on scientific principles are examples.

Enrichment material for science is abundant, even for the lower grade levels.[12]

Field trips as science activities are discussed briefly in a later section of this chapter.

Science Fairs. One of the most stimulating aspects in science education is the increasing popularity of science fairs—the student-built and operated demonstrations prepared for public showing and parental praise. The displays of the upper grades may be more numerous and complex (often showing talent for real scientific research), yet the lower grades offer their full quota of showmanship. These Fairs are one-school, community, city-wide, State, Regional, and National.[13]

Visual and Auditory Aids. The use of slides, films, records, involves special understanding at all levels. If a school has the facilities for these aids, someone on the staff will have the training to supervise their use.

Evaluation of Activities. Do the experiments work out? Are they interesting—even exciting—to the children? Do the children get as much understanding of the principles of science as would be expected of them at their age?

How well do they remember what happened, and why? Do they recall more accurately the things they have seen, or have read about? This can be tested by questioning a week—even a month—later. The tests will, of course, be oral in the lower grades.

Evaluation in numerical terms is hardly necessary for science at this level. The experienced teacher, or the observant teacher gaining experience, can tell "how things are going." The hope, "May I do better next year in science," is usually realized.

Trends in Activities. The spreading popularity of Science Fairs, in which the elementary pupils participate, is the most encouraging trend in science activities.

The high school science laboratory continues to lose its influence as the source of experiments and equipment for the elementary grades; the home and neighborhood seem ever more worthy of the child's investigations. This is as it should be, even though teachers trained in high school and college science classes must revise their attitudes toward "science laboratory work" in the lower grades.

It is unfortunate that the study of nature first-hand becomes difficult in a city environment. Only the poorer teachers, however, give up completely.

"Every child should have mud-pies, grasshoppers, water-bugs, tadpoles, frogs, mud-turtles, elderberries, wild strawberries, acorns, chestnuts, trees to climb, brooks to wade in, water-lilies, woodchucks, bats, bees, butterflies, various animals to pet, hay fields, pine-cones, rocks to roll, sand, snakes, huckleberries, and hornets; and any child who has been deprived of these has been deprived of the best part of his education."—BURBANK.[14]

Third Reason: a Growing Vocabulary

"All words are pegs to hang ideas on."—BEECHER.[15]

Science is taught in today's elementary schools:

3. *To enlarge the child's vocabulary with the names of simple objects and processes*—as background for future use of technical terms.

"I can't understand what you mean by 'denser medium,' " the boy complained.

"I just can't make it any plainer," the teacher said.

"I believe I can," remarked the principal, at the moment an observer. "Son, here is my watch. Is it running?"

The boy listened. "Yes, sir, it is running."

"Running in air, of course. Suppose I dipped it under water; would it run then?"

"Yes, I think it might," replied the boy.

The answer was not what he expected, but the principal was rarely at the end of his rope. "Suppose," he continued, "I sank the watch in a pail of molasses; would it run then?"

"Of course not!" The boy was emphatic.

"Well," the principal went on, "let's call air, water, or molasses by the name 'medium.' What is the difference between them?" The boy knew, of course. "Water is thicker than air, and molasses is thicker than water."

The principal clinched the matter. "Son, say 'denser' instead of 'thicker'—it is the better word, for one board might be thicker in inches than another, but not denser. Remember, each of these fluids is a 'medium.' Now, what does your teacher mean by 'a denser medium?'"

"I've got it, sir! A medium that is thicker—I mean denser—might stop a watch! And I can tell the teacher why, too!"

For want of a word, an idea was almost lost!

Words *can* be tricky! The English language is unfortunately full of identical words of wholly different meanings (a fluid medium, a statistical medium, a medium that talks with ghosts) ; of words pronounced alike but spelled differently (cent, scent, sent) ; of different words with similar meanings (peak, summit, tip, top, tip-top) ; of apparent positives and negatives that really mean the same (flammable, inflammable—compare sane and insane). Add to these the bizarre rules of spelling, plus the ruleless (also ruthless) pronunciation (cough, through, borough), and we have—my Chinese students have told me—almost the most difficult language in the world to learn after childhood.

Many of our adults learn new words in the spirit of a game. The cross-word puzzles are examples. To capture the interest of pupils, science teachers should bring up new words in the same zestful spirit.

Words are added to the child's vocabulary in three ways:

1. *By reading,* he comes upon new words for which he must gain comprehension. Pictures give meanings to words in the simplest readers. Further along the teacher devises ways. There is a striking spread in the reading ability of children in any grade.

2. *By writing* the spelling of new words is emphasized and—we hope—remembered. The difference between the best and the poorest spellers developed in the grades is astonishing!

3. *By speaking* the pronunciation of words is learned. The teacher's influence in this respect is paramount. Older children can find the correct pronunciation in the dictionary—but what has happened to diacritical marks? Are they no longer taught? Certain dictionaries make good use of phonetic spellings to guide pronunciation.

The abilities of the best and the poorest speakers in any class are far apart. The teacher must be alert to stimulate the taciturn, and repress the loquacious. "Everyone in his turn to speak" is a good rule, easily enforced after the class gets used to it.

Words, correctly spoken and written, understood when read, are concerns of all instruction, including science.

One of my students once remarked: "When I read an article on science education, the scientist in the author seems to write clearly, but the educator is hard to understand. Why is that so?" Why, indeed, do so many professional educators write so obscurely? The ostentatious loftiness of their language makes one wonder why the Doctor should wish to appear so pompous! "Oh why should the spirit of mortal be proud?" —WILLIAM KNOX (1789-1825), author, England, as the first line of a hymn.

Evolution of Vocabulary Growth. Do the children enjoy learning new, and exact, words for things and principles? Do they use these words in their conversations without self-consciousness? Are they obviously mastering more difficult reading matter? Do they go voluntarily to the dictionary and the encyclopedia? These are signs of a growing vocabulary.

Trends in Vocabulary Development. The grading of text books has become a science. While certain word lists may seem restrictive—bright pupils can learn far more words than these—yet they contribute to authors' lucidity at definite grade levels. Experienced teachers become experts in laying a foundation of clarity for future scientific writers and reporters.

"Reading maketh a full man; conference [speaking] a ready man; and writing an exact man."—BACON.[16]

Fourth Reason: Combining Facts and Emotions

"Nothing in Nature is unbeautiful."—TENNYSON.[17]

Science is taught in today's elementary schools:

4. *To give experience in combining the factual and the emotional*— as background for future appreciation of natural law and beauty.

A child cares for a well-loved pet or flower; does this give him pain, or pleasure? Children take part in classroom experiments in science; is this work, or play?

To a child, and to many adults, if an activity is fun it is play, and if it isn't fun it is work.

No subject in our schools has a greater power to give pleasure than natural science. Other subjects approach it, of course; literature, music, art, even mathematics, exert their charms on certain students. It can be argued—also disputed—that subjects such as these require a talent for their appreciation, while one and all can enjoy science.

What are the stimuli to pleasant activity in science? Some are these:

1. *Something new.* Newness arouses curiosity, which is the spice of interest, in turn the basis of pleasure.

2. *Something my own.* Children delight in making things work through their own efforts. "I'm the boss here!" brings emotions of proper pride—always a pleasant feeling.

3. *Follow the leader!* Children do not want to be on their own resources all the time.

I remember a keen cartoon in which a group of small ones, standing before their teacher, asked plaintively: "Teacher, *must* we do what we want to do today?"

The teacher, wiser, more experienced, surely has the duty of guidance, especially into new science topics. If a previous class greatly enjoyed a topic and its activities, she should lead the present class into the same "green pasture."

Goals of Appreciation. What are the goals of appreciation, to which science in the elementary grades can contribute? A basic statement may be expressed in the words of Lydia Child (1802-1880), authoress, United States, in *Autumnal Leaves.*

"Nature is beautiful, always beautiful! Every flake of snow is a perfect crystal, and they fall together as gracefully as if fairies in the air caught water drops and made them into artificial flowers to garland the wings of the wind."

Is beauty spoiled by knowledge of its cause? Ask yourself this question: Is the rainbow any less beautiful to you because you understand how the rays of clear sunlight, bent within each drop of rain or mist, separate according to their wave-lengths, by mathematical laws, into the colors that strike your eye and lift your spirit?

Appreciation is much a matter of temperament, and temperament in little children is readily modified by imitation. A teacher who truly appreciates the beauties in which natural science is so rich may be the chief reason why members of her class develop their own, and individual, appreciations.

Only as high appreciation can the dedication of many to science be explained. Earnest research workers have exclaimed, in all truthfulness, "Science is my life!" We must remember, too, that men have died for the sake of science.

We are told that no well-trained naturalist has ever taken his own life, although other specialists have occasionally selected suicide as "the way out." If this be true, then nature has the power to develop an inner calm that strengthens the will to live.

Evaluation of Appreciation. Do the children continue to find science a pleasing subject as the weeks go by? Does their interest remain high, even though the experiments they perform and the questions that arise become more complex? Do certain things of striking beauty—of shape, color, pattern, harmony—first arouse their awe and then their inquiry?

The degree to which facts arouse feelings is by no means as readily measured as certain other growths in science. If a teacher is convinced that "nature is beautiful," she can tell when others join her in that appreciation.

Do facts sometimes arouse unpleasant feelings? This will be discussed in the next section of this chapter.

Trends in Appreciation. Authors of articles concerning elementary science lay increasing stress on cultivation of the feelings rather than the memory. Indeed, a thing is better remembered when it was fun to find it out.

Authors of science readers for the grades continue to put color into their sentences to match the colors from the printing presses.

With such aids, teachers should catch the spirit of discovering beauty in each science lesson. This will be sowing good seed, for "a thing of beauty is a joy forever."—KEATS.[18]

"You will find poetry [i.e., appreciation] nowhere unless you bring some with you."—JOUBERT.[19]

Fifth Reason: Objective Attitudes

"It was a bold man that first ate an oyster."—SWIFT.[20]

Science is taught in today's elementary schools:

5. *To guide emotional responses away from the highly subjective and toward the objective*—as background for future sensible attitudes and desirable behavior.

Far back in the human story this surely happened:

A man, primitive in form, face, and thought, beset by that most primitive of feelings—hunger—stood at low tide above an oyster bed. The day before with curiosity—another primitive human emotion—he had opened the shell and saw the slimy creature within. Since other slimy things had proved distasteful, he threw the shell and contents back into the water.

Today he was really hungry. Should he again reject the soft mass within the shell because it reminded him of things unpleasant? Or should he eat one, to find out what it was really like?

He opened a shell, dug out the pulp, and plopped it into his mouth. Perhaps he was surprised at the bland taste, and how easily the slick lump slipped down his throat. Others followed; his hunger was appeased; he called his fellows.

This unknown hero demonstrated the power of *objective* over *subjective* thinking—considering what a thing really is, rather than what it reminds you of. Today, any gathering of humans, young or old, may be divided into two fundamental groups: (1) those who are objective in attitudes, and can eat raw oysters, and (2) those who are subjective in attitudes, and cannot bite, much less swallow, raw oysters.

A major duty of science is to develop objective attitudes, and the earlier this begins with children the more effective this will be.

The problem is not to rule out the subjective in the thinking of a child or an adult, but to subordinate the subjective to the objective.

Characteristics of the subjective are squeamishness ("I can't pick up this earthworm!"), prejudice ("I hate to see women hold office!"), false pride ("One Latin student is worth five science students!"), stubborn refusal to accept truth (this earth *is* flat—*The Bible* says so in Revelations 7:11).

Converse characteristics of the objective are discriminative boldness ("I'll pick up a snake with a round head, but not one with a flat head), broad-mindedness ("Let the best man win, even if he is a woman!"), tolerance ("It takes many kinds of people to make a world!"), a balanced viewpoint ("I'm not the first by which the new is tried, nor yet the last to lay the old aside!"), and a willingness to interpret only after study, "rightly dividing the word of truth"—*The Bible,* 2 Timothy 2:15.

The objective is likely to consist of the direct and factual, rather than the abstract—which you can hardly decide whether you like it or not. This is particularly true when children are taught, for they are not interested in reading for reading's sake, writing for writing's sake, are appallingly hostile to mathematics for mathematics' sake, and do not understand science for the sake of science. Let a real-life situation be presented in any of the above studies in which the aspect—reading, etc.—is required as a tool to "get the story," and the children are alert.

Imagination follows. An objective attitude is a guide, not an obstacle, to imagination. Scientists are men of imagination, but they test their inspirations by observation and experiment. Imagination in a child should never be stifled; it should be guided. Imagination may be the stimulus behind genius that puts a great scientist at work expanding the sum of human knowledge after the child has become a man.

The wise teacher can smilingly steer imagination away from fantasy, if any science is involved. She must agree with Beecher that "a mind without imagination is like an observatory without a telescope."[21] She should guide the childish imagination toward the development of abilities in higher reasoning rather than the images of uninhibited dreaming. Imagination, like the electric current, is a power that may go far, but requires a return path to reality.

How Questions Arise. Imagination is a likely stimulus to questions in a science class. Objective answers should be given. This matter could be discussed at length, but let a few simple rules be given:

Do not ignore a question entirely. Make some reply.

Do not chide a questioner with "you're too young" or "you don't need to know that." Give some answer on the proper level of understanding.

Do not give a false promise, "We will discuss this later." Remember the obligation, even if you must make a written note of it.

Do not give dogmatic answers. "Yes" and "no" are rarely adequate for understanding.

Treat the ridiculous as important. No child should ever feel embarrassed when asking a question. Suppress class laughter sternly in this matter, even if you laugh with the class at other times.

Meet the heckler's question seriously. That spoils his fun, and he will soon stop the practice.

Avoid "show-off" questions of your own that imply, "You are ignorant, but I am *so* wise!" This advice is particularly needed by many high school and college teachers.

A further study of questions appears in the next section of this chapter.

Evaluation of Objective Attitudes. Let progress be shown in practical ways. Do the children eat a more varied diet? Do they go to bed earlier? Have they lessened, or lost, their fears of harmless insects, spiders, worms, even reptiles (as lizards and salamanders) ? Do they examine, rather than recoil from, things that are slimy or sticky or greasy? Have they overcome fear of the dark, fear of lightning, fear of doctors? And—perhaps the supreme test—would they eat raw oysters?

Trends in Objective Attitudes. A greater frankness in discussing matters, once taboo, is evident. These are examples:

1. Sex is presented more successfully, because more objectively, in texts, manuals, classroom guides.

2. Apparent conflicts between *The Bible* and science—the frequent source of questions by children—are discussed with better understanding in both the religious and secular press.

3. The great moral issue—whether science shall be used chiefly for peace or war—is being understood at an earlier school age. Possibly a longer period of contemplation will bring more positive adult judgments in this vital matter, once the next generation has the full responsibility.

"Facts, when combined with ideas, constitute the greatest force in the world."—ACKERMAN.[22]

Sixth Reason: Scientific Thinking

"What is the hardest task in the world? To think."—EMERSON.[23]

Science is taught in today's elementary schools:

6. *To start habits of scientific thinking in simple matters*—as background for scientific thinking in future decisions of importance.

"Miss Alice, do rocks ever grow?"

In some classes this query would have been followed by a gale of contemptuous laughter. Miss Alice, however, had led her students, by her own attitude, to consider *every* question worth serious consideration. She asked, "What makes you think that rocks might grow?"

"In any pile of rocks, some are little and some are big—and little things usually grow bigger." Well observed, and truly stated.

Miss Alice might have given a short and correct answer at once, but that was not her way of developing the thinking of her class. So the children put a few little stones in a dish with soil, and watered them, and waited three weeks, and came to a wholly convincing answer to the question.

When a question is asked in class, a teacher may do one of three things:

1. *She may answer it at once.* This may be justified if the question and its answer are simple, relatively unimportant, or—as often happens— unrelated to the topic under discussion.

2. *She may say, "Look it up."* This is proper if the answer is on record as a fact, and is not subject to practical testing.

3. *She may ask questions* that will determine what the children know already, what they would like to know, and which may lead to plans for exploring the unknown. This will be done by experiments, or by reading, inquiry of parents, and by other avenues of information.

At times the answer to a question can not be found by any method. (Example: Why does a pig's tail curl?) It is well for a child to learn early that complete knowledge often eludes us.[24]

Chief Goal of Teaching. One goal—many believe the chief goal—of teaching is to develop the ability of critical thinking. This is often called scientific thinking, because it is the method by which men of science carry on their reasoning. The term "critical" implies reasoned judgment, which may either favor or oppose an explanation, a point of view, or a decision.

Critical (scientific) thinking starts with facts, then draws conclusions. Scientific imagination starts with valid conclusions, then dares to go beyond them. Each of these qualities of the human intellect may be planted as seed in the minds of children, and may develop through nurture.

Adult scientists have no monopoly on critical thinking. The thoughtful in any interest or vocation have acquired this habit of the mind in action, along with objective attitudes, poise, and emotional stability.

How may an elementary teacher develop critical thinking in the minds of her pupils?

First, she must resolutely decide that she will do less of the children's thinking for them. This requires patience and tolerance on her part, and is difficult for certain temperaments.

Second, she should develop her own concepts in three categories of certainty: some ideas to accept, some ideas to reject, some ideas to wait upon.

She will accept the results of experiment, especially if it has been tried again. Some teachers say: "Try everything three times. The first result is an incident; the second a coincidence; the third proves the rule."

She can accept the records of past occurrences, including accounts of others' experiments, in well prepared texts and reference books. Examples are dates and effects of Vesuvius' eruption, scientific explanation of volcanic action, reports of medical research.

She can accept, with considerable confidence, published information concerning matters beyond her powers of experiment, such as the data of astronomy, the cause of seasons, the paths of world winds. Since elementary testing is practical for many published results, as the effect of sunlight on plant growth, the use of fertilizers, and the like, she should plan first hand experiences for her class in such phenomena of nature.

She will reject magic or spirits as explanations for any happening in nature, no matter how marvelous it seems. She will reject guessing before testing (except for practice in estimating measurements) —yet imagination beyond factual explanation is a sort of educated guessing. If certain ideas are readily confirmed or disproved by consulting references, she will insist on this inquiry. She will reject the judgments—often expressed with vigor—of persons who can not be considered experts on the basis of their limited experience, or who are obviously addicted to subjective judgments.

She will wait for further information on many things. This is called "suspended judgment"—but if she uses the term in expanations to children, she must be sure they do not picture "a judge hanging by his neck on a gallows" as more than one youthful imagination has done.

Children will accept, as a matter of good sense, the idea that scientists readily change their explanations if new facts are discovered; that mankind as a whole only learns in the school of experience; that Graduation Day—when all "courses" offered to students of nature have been completed—will never come; that when some law of nature seems to have an exception, a new law is awaiting discovery.

The children will develop—simply at first, but with growing comprehension— a respect for those men and women who are expanding man's knowledge in all areas, including science; whose fervor is truly religious, whether they hold a creed or not, based on the admonition, "Prove all things; hold fast that which is good" (*The Bible,* I Thessalonians 5:21) .

Many educators who write and speak on the subject of critical thinking urge that young pupils should not be allowed to generalize from scanty evidence. If it be at all possible, test for the answer. It is entirely too easy to jump to conclusions—this is not good thinking.[25]

Many tests offer some negative, as well as positive, proofs. A magnet draws a nail, a paper clip. Try an aluminum roofing nail, a brass paper clip. A magnet pulls another magnet; devise an experiment to show that a magnet may also push.

A child's thinking may often be in error. These misconceptions may be evidence of an active mind which only needs guidance. Even adults are prone to mix fact and fancy!

Critical (scientific) thinking is an attitude rather than a set of rules. With such an attitude, a teacher may meet each item of simple science, one by one, and feel confident that tender minds under her care are developing in the better ways of thinking.[26]

Evaluation of Critical Thinking. Do most of the children take part in each discussion? Are they learning to stick to the point? Are they willing to go to considerable trouble to make experiments? Do they find people (parents, others) who can answer certain questions for them? Are they using school encyclopedias freely as aids to their thinking? Do they seek a number of sources of information, rather than find satisfaction with a few? Do they insist that any effect they observe be traced to its cause? Are they willing to eliminate their own, their parents', and even their pastor's prejudices in arriving at the truth? Do they honestly and constructively criticize their own thinking?

Trends in Scientific Thinking. The current news of science and technology is before children of every grade level. The smallest scholar has heard of the atomic bomb, Dr. Siple at the South Pole, the *Nautilus* under the North Pole, metal moons in orbit around our planet, plans—more than dreams now—to land men on the moon. They ask: "Who plans these things? And why?"

Adults of today know—and children hear them explain—that achievement in these masterpieces of thought and execution depends on education, not inspiration. The challenge of the Russian "sputniks" has made a direct demand on education in the United States to "catch up"—if there be deficits in effort or erratic aim in purpose of our own training programs. Science instruction is sure to be strengthened at every level, including the elementary schools.

"Thinking is the hardest work there is, which is the probable reason why so few engage in it."—FORD.[27]

Seventh Reason: Social Attitudes

"If rational men cooperated and used their scientific knowledge to the full, they could now secure the economic welfare of all."—RUSSELL.[28]

Science is taught in today's elementary schools:

7. *To start building attitudes toward the simple social effects of science*—as background for future cooperation in community programs of health and welfare.

A teacher of social studies wrote these words upon the board one morning: "New Golden Rule: Think about the peoples of other nations as you would have them think about you."

Science is of most value when it contributes to the peace and prosperity of all men everywhere. Since foundations must be built in each area of learning, the early social attitudes that science can give children are important.

Science has at times been considered a competitor, not a collaborator, in the elementary curriculum. "If science gets in, what subjects do we drop?" has often been asked. More often today we hear, from an elementary teacher or principal, "We really ought to teach more science."

After reading the hundreds of pages that discuss the place of science in elementary instruction, one finds a broad consensus of opinion that science should not stand alone at that level. It should be merged with other subjects in effective ways. The terms "correlated," "fused," "integrated," are used, with distinctions of meaning more appreciated by their authors than by readers. Not all science educators approve of these recommendations.

Reading and writing, the first tools of learning acquired by the child, are increased in interest if science be their theme. Mathematics comes soon, and resourceful teachers bring bits of science into the counting and measuring. As each of these "3-R's" develops, science can contribute even more helpfully.

As the language arts are presented, the child can talk about himself in respect to his health, his habits, his recreation. He can talk and write about how simple things work, little things grow, big things move, make noises, smell. He is learning to describe in words the messages of his senses.

When the child begins to study the wide, wide world, he enters the area of the social studies. It is to be hoped that in his history he learns more about inventions and explorations and less about wars and politics; that in geography he studies more of peoples and climates and crops than of borders and rivers and mountain ranges; that in simple civics he is informed of the measures for health and sanitation as well as police and fire protection and the local courts. Wherever worthwhile aspects of his social studies are presented, partnership with science can be a major aid to his interest and information.

There are risks to science in these mergers. It may so easily be left out. Too many authors of texts in the social studies ignore science, leaving to teachers the responsibility of suitable blending of the areas.

There are risks, too, in the effort to make science stand alone, or nearly alone, in its own period in the elementary school. Students must be kept busy, yet experiments may be too tedious, too quantitative, too complex. The topics discussed may seem to have no bearing on the child's own life, hence of little interest. Of course this is poor instruction in science, but it can happen.

A more common risk of the science period is mere "talking about science." This does not develop the scientific attitude; it may be the early training of dilettantes in science.

Integration of science with the social studies seems to have been successful because of the common aspects of living to which both contribute. Who could deny that sanitation of a city is both a scientific and a social problem? Who would try to separate the science and the economics involved in the control of pests that attack crops in the country? A list of topics of cooperation would be extensive indeed: this has been explored in parallel columns in a recent study.[29]

Conservation's Foundations. The contributions of science to society become more fundamental, and their effects more lasting, in the vital matter of conservation of our Nation's natural resources. The study of conservation should begin early in all of our schools. It does begin early in some thirty States, and hundreds of communities, where activities with real meaning are training children for the future wise use of soil, water, minerals, forests, crops, wildlife, energy.

The interests vary, of course. In Colorado, minerals are a major concern; in Georgia, soil and trees; on the Maryland coast, oysters; in California, water.

An entire chapter in the present volume would be inadequate to give all the available suggestions as to programs in conservation, based on science and the social studies, in each grade of the public schools. Recent publications which the teacher may own and use have been issued by the U. S. Office of Education[30] and the U. S. Department of Agriculture.[31]

Resources in the Community. The alert elementary teacher will use both places and people in the community to enrich her pupils' experiences. Field trips may begin with simple walks around the school grounds "to see what we can see." A school built in a crowded city area will be at a disadvantage here. The fortunate school is one located within walking distance of a natural city park. Trips to important industries (as a dairy) and services (as a fire station) may be arranged, but the grade level for a profitable visit must be considered, and the spirit of welcome assured.

Certain people in a community are quite willing to speak to children on their occupation, special experience, or hobby. One father spoke to a suburban school on "How I Train Dogs to Hunt"; an articulate mechanic was most interesting on "Things that Go Wrong with Cars." A mother told the children "How to Cook the Fish Your Daddy Catches" with much humor. Naturally such contacts with adults, especially parents, is excellent from the standpoint of public relations.[32, 33]

The most important social effects in the elementary grades are the understandings that children develop as they work together. Science is not the only subject that invites cooperation, but it is the most consistently effective for the purpose.

Evaluation of Science-Social Attitudes. Social growth aided by science must often be sensed rather than measured. The specific effects of a certain "cause" in today's classroom may not appear for many years. Attitudes are rather intangible anyhow, yet a teacher may ask herself questions such as these about her class, and use her own judgment for the answers:

Have the children increased their appreciation of science in their home life?

Do the children suggest scientific solutions for the community problems they study?

Have the children improved their ways of planning together, of working together, of free discussion?

Are the shy more confident, the aggressive more controlled?

Trends in Social Aspects of Science. We study science to live better—this is indeed the trend in the cooperation of science with other subjects in school. Comparisons prove this: comparisons of texts used decades ago, and now; comparisons of the older readers with the newer, rich in science; comparisons of elementary school curricula, of elementary classroom equipment, of training courses for elementary teachers.

Science has applied for a job in the lower grades, has promised to work in harmony with others (no pushing, shoving), and has been accepted. Under the proper supervision, its work is highly satisfactory.

"The sciences are of sociable disposition, and flourish best in the neighborhood of each other; nor is there any branch of learning but may be improved by assistance drawn from other arts."—BLACKSTONE.[34]

Eighth Reason: Simple Concepts

"The greatest compliment we can pay to truth is to use it."—EMERSON.[35]

Science is taught in today's elementary schools:

8. *To develop simple concepts such as cause and effect, balance of nature, cycles of nature, and the like*—as background for future understanding of broad concepts, as the conservation of resources, the laws of learning, and even the sacredness of truth.

A bright Cuban boy in my class had never seen snow, but the first fall was not far away. An eager seatmate explained to him: "Snow is very white, as white as sugar, and covers the ground."

A few days later the lad wished to read to me a letter to his parents in Cuba. He translated: "I will soon see snow. I am told that it is beautifully white, and tastes sweet."

This error in his concept of snow was soon corrected by experience.

What is a concept? It is the effect of percepts.

What is a percept? Something is before you, and you notice it. Your senses, one or more of them, give your mind impressions of the thing; these are percepts.

Your mind then puts the percepts together into a broader comprehension of the thing—all of the attributes you have observed—and you have a concept. You might call it a mental image of an object or an action.

The concept may be old, or new, to you. If old, you may have a name for it, and many associations with it. Every concept, however, was new to you at first.

How very many concepts that children receive are new to them! How could it be otherwise, because of their few years? Education's supreme responsibility is that most of the concepts received by children should be worth while, and helpful no matter how old they become. To express it differently, the concepts of value should become fixed and remembered, while worthless ones fade away after a few moments notice.

Concepts are subject to revision when additional percepts are received. Children change or abandon their concepts readily, it is really not a wrench to learn that fairies are not real, or that "Santa Claus is just a man, maybe Daddy." Adults, however, often hold their early concepts against all proof of error. Both physiology and psychology have explanations for this stubbornness.

Cause and Effect. The most basic concept, to which science contributes much, is that of cause and effect. It is well to be certain about some matters. Of course the child has already learned that if he does certain things good (pleasure) will follow; if he does other things evil (pain) is the result. He acquires a belief in the *certainty* of consequences. (One is tempted here to discuss an aspect of discipline.) In school his concept of cause and effect will be confirmed and widened, especially if there are experiences in science.

Unfortunately, a concept of cause and effect does not rule out superstition. Fairies and fiends alike may be suspicioned as existing "causes," with effects good or bad according to the temperament of the sprite. The sole antidote for superstition is a belief in natural law—a substitution that may "overcome evil with good" (*The Bible,* Romans 12:21).

A concept of cause and effect does not rule out a belief in the supernatural. This problem in the elementary school will be mentioned briefly in the next section, and is discussed more fully in the earlier chapter on Religion and Scientific Education.

How may we be more certain of the effects of causes? In simple matters the best way is to repeat the experiment. Scientists call this "duplicate determinations," and if two measurements do not agree, then a third, or even many, must be performed. Wisdom suggests that a complex explanation should be thought over twice or more, starting at the beginning.[36]

Units in Science. Shall units in science be small, medium, or large? Much discussion has taken place as to this matter.

Small units are likely to be of little value, isolated, more like the treatment a small encyclopedia gives to a topic. They may seem unimportant to a student because they are so short.

Large units may pass beyond the interest span of children, may bore them, may seem to strain natural relationships to fill out the assignment, may replace items more worth study, may give an impression of complexity that baffles children.

Units in science of medium size would seem to be a proper compromise. These should vary in length of time required for them, depending first on interest, next on availability of experimental and study material, and—fundamentally—on the worthwhileness of the topic to the child's life and mental development. A teacher is under no obligation to exhaust a topic in science.

What Is a Scientist? The concept of a scientist is important for children in the grades. Inquiries have shown bizarre ideas they have obtained from television, comic books, and other sources. The scientist is no magician—these are fellows who are clever with their hands and equipment. Anything called magic is an illusion, for the senses are fooled. Magic is never really what is seems to be.

Scientists should not be presented to children as heroes, but as keen observers, clear thinkers, diligent workers. Many important discoveries have been made by scientists whose names are known only to those who search the records of their work.

Rarely is a scientist a true genius—although many of them possess exceptional ability. Rarely is he a recluse—although many prefer laboratory work to recreation. He is not a "crazy scientist" even though his actions in the field seem strange, and the symbols he scribbles in his notes have the appearance of cabalistic signs.

Trends in Teaching Concepts. Good teachers have always preferred that pupils remember concepts rather than lists—to learn by reason, rather than rote. Studies as to science concepts readily acquired by children in each grade are numerous. Since new concepts always arise ("the atomic bomb, rockets shot into space," and others) for the attention of even the younger pupils, such studies will continue.

"You need not tell all the truth, unless to those who have a right to know it all. But let all you tell be the truth."—MANN.[37]

Ninth Reason: Reverence

"And all God's earth is holy ground."—MILLER.[38]

Science is taught in today's elementary schools:

9. *To develop a simple reverence for nature*—as background for future appreciation of the wisdom and power of God.

They met on Fathers' Night at the Parent-Teacher Association.

"Frankly, Miss Alice," said one father, "I am worried about the science you are teaching my son in the fifth grade."

Miss Alice smiled sweetly. "I am not too wise," she said, "and by the time I misunderstood science, and your son misunderstands me, you probably do have something to worry about."

What man could argue further with such a modest lady?

Man was a naturalist before he was a scientist. Having intelligence far above any beast, he sought explanations for all of his environment that seemed strange. Out of these he used an instinct for order, plus his imagination, and worked out systems of primitive religions, composed—from our viewpoint—of countless myths, superstitions, taboos, rituals. Most characteristic was his belief in spirits, both good and bad.

"Science is organized knowledge," wrote Herbert Spencer (1820-1903), philosopher, of England, in his book, *Education.* Organized science has been compelled to struggle against jumbled superstition to this very day. At times believers in science thereby became doubters of the current religious doctrines, and were punished. This punishment, though changed in nature and lessened in severity, has not been wholly abolished.

"Science without religion is lame; religion without science is blind." —EINSTEIN.[39] Thus the most eminent of modern scientists and philosophers spoke for the vast majority of modern men of science. They, more than any others, can sense the amazing complexity of nature's orderliness, logically impossible as the result of mere chance. Scientists may not adhere to creeds and dogmas, but many—perhaps most—of them are deeply, devoutly, religious.

I found need at one time to express my own beliefs founded on both facts and faith, and these words seemed fitting:

"What, then, is science? Science is our faith

That some day all things true may be revealed.

What is this faith? It is our firm belief

That through some Power may all men's woes be healed.

What is this Power? Although it may seem odd,

Men are not sure, yet seek. They call it God.

Tussles with Theology. Even the tiny tots have tussles with theology. For many years I invited elementary teachers in my classes to relate incidents in which some child had questioned a fact of science as being "against the Bible." These incidents occur to all teachers, and I have notes on scores of them.

One of the most frequent objections raised by a child is this: "I am *not* an animal!" Parents have sent notes, insisting, "Don't teach my child that he is an animal! Man is in the image of God!"

One teacher joked about it—at some risk—saying: "So, Johnny, you are not an animal. Well, there are only three kinds of things in the world—animal, vegetable, mineral. If you are a vegetable, are you a cabbage? Or perhaps a carrot? I'm sure you are not a mineral, for rocks sit still in their places." She heard no more from home about it; perhaps a parent saw the point.

Another teacher took the matter seriously. She told her class: "You and your parents know that humans have bones, and flesh, blood and organs just like our animal pets and the wild creatures of the fields and forests. We, and they, have brains, and can feel and think. But we believe that when an animal out in the woods dies, that is the last of it. When a human dies we know that the body which is like an animal never lives again, but we believe that every human has a spirit that lives on, and maybe forever. That is really enough difference between us and the lower animals to make us very thankful."

She could have added that the "image of God" is not in skeleton and organs, for it is written that "God is a spirit" (*The Bible,* John 4:24).

Another matter often raised is the date of Creation, set at 4004 B.C., which does not tally with references to mankind's life tens of thousands of years ago, nor to the millions of years since the dinosaurs, the coal age, the deposit of limestones, the flow of granites. The wise teacher looks up Bishop James Ussher (1581-1656) of Ireland, and his hobby of genealogies. She may also display a copy of *The Bible* that does not publish his dates atop the pages.

These are but a few of the questions, related to theology, that come suddenly in any grade. A certain technique has proved helpful at these times; it is called "the bouncing question." The principle is this—when you are asked a question of this type, ask another.

The bouncing question gives the teacher three advantages she needs:

First, it gives her time to think. An answer after a few moments' deliberation may be very different from one given instantly.

Second, she may learn something of the questioner's background of interest and information, and thus judge what, and how much, to tell him.

Third—and often crucial—she may learn whether the question is sincere, or planned by some adult to trap her. This is particularly likely to happen in some community where religious beliefs are distinctly conservative, and she could be highly embarrassed if her answer ("Yes, the world is really round!") is against certain local interpretations of the Scriptures.[40]

Let us give one example of the "bouncing question," using the most dangerous inquiry that comes with fair frequency in elementary classrooms.

"Miss Jennie, do you believe we came from monkeys?"

Miss Jennie bounces the question. "Why, do *you* believe it?"

"No, I don't!"

"Then why do you ask?"

"Well, Grandpa said he bet you believed it, because you teach science."

So this was the angle! Miss Jennie would be on guard.

"Have you ever known anyone whose Grandpa was a monkey? Not just a funny man, but a real monkey?" This argument would not be allowed to become too serious.

Then, several "bounces" later, this comment from Miss Jennie: "Monkeys interest me a lot. Haven't you noticed, in the zoo, some monkeys that look like your Grand"—she checked her sarcasm just in time—"like people we know? If you are really interested in monkeys, how they live what they eat, how they play, our encyclopedia has a fine article, with pictures, about them. Will you read this and tell us the most interesting facts about monkeys?"

The emotional was thus replaced with the intellectual; the crisis was averted.

This classroom incident—which actually happened—is typical of the way a wise teacher follows the veteran Apostle Paul's advice to a young teacher, Timothy: "But foolish and unlearned questions avoid, knowing that they do gender strifes" (*The Bible*, II Timothy 2:23). The implication is less that the questions are foolish, than that the questioners are unlearned.

More sound advice comes from a noted philosopher: "Truth is a good dog, but it should be careful of barking too close to the heels of error, lest it get its brains kicked out," wrote Oliver Wendell Holmes (1809-1894), physician, author, United States, in *Autocrat at the Breakfast Table*.

The teacher should treat a "theological" question in class much as any other question, except she should be aware of its emotional associations. She should avoid an opinion, which might be quoted and expanded, on conservative convictions held in the community. The "bouncing question" is one way out of trouble.

Science can indeed lay the foundations of sincere faith. If once the concept of a Creator is accepted, at least as a postulate, then "the heavens declare the glory of God; and the firmament sheweth his handiwork" (*The Bible,* Psalm 19:1). Only rarely is there objection to the mention of God in the elementary classroom, provided the teacher does not "preach."

It has not proved difficult for experienced teachers to make clear, even to younger pupils, the difference between facts and faith. Men base their beliefs on both. For some explanations the facts are clear, undeniable, and should be accepted as truth. Any proved law of nature is an example. For other explanations we devise theories and hypotheses, depending on the amount of valid evidence, believing that the natural laws will some day be discovered.

But for the postulates—causes that can only be assumed—we must have faith; faith in a better world for posterity through some triumphant philosophy; faith in a future life for the human personality; faith in a Supreme Being. Our minds feed on facts, our spirits on faith. Little children can accept this distinction, even if some extremely advanced thinkers can not.

Evaluation of Reverence. This growth in understanding may be observed, even if not measured. Have the children lost their pleasure in killing birds, butterflies, spiders, worms, due to an expanding sense of the sacredness of life? (Reasons for swatting flies, and picking potato beetles, still exist.)

Do the children regard dramatic aspects of nature, such as fine clouds, lightning, the rainbow, with awe yet with understanding? Have they put away all superstitions relating to these, and similar vivid displays?

Have the children sensed the truly amazing structures of a bud, and the perfect sequences of flower, fruit, seed, and seedling, as evidence of a consummate wisdom that is far beyond human planning? Does their science thus strengthen their belief in Omniscience?

Do they relate the mighty forces of the universe, as the heat of the sun, the momentum of the planets, the power inherent in the rotation of the galaxy (as they learn of these things) as aspects of Omnipotence?

Trends in Reverence. Theological quarrels with science grow less with each passing decade. Fifty years ago I "walked on eggs" as a young science teacher in an ultra-conservative community. Today the science teacher in that same community, in the big new high school, still needs to be careful. Science is also taught today in the town's elementary school. If it is in these teachers' hearts to do so, their science lessons may both train the minds and lift the spirits in that community.

"Nature is but a name for an effect whose cause is God."—COWPER.[41]

A Summary, in Brief

How may one summarize nine reasons why science is taught in elementary schools?

The answer may come through an analogy. A plant is nourished by separate fertilizers—salts of nitrogen, potassium, phosphorus, plus many other minerals, water, and carbon dioxide of the air. These are combined by nature's amazing skill into a growth of fruitfulness and beauty. An animal is fed by carbohydrates, fats, proteins, plus many minerals and oxygen of the air. These combine with incomprehensible precision to form a creature of strength and intelligence.

For teaching science, as for these natural processes, we may know what goes in, but the blending makes the product. The teacher's skill, like that of the farmer with useful plants, or the chef with tasty foods, joins nature in making the finished product—science-minded children.

Out of the "Nine Reasons Why" the following ideas are typical of things that were developed and should be remembered:

Science is not studied and learned in the classroom only.

Science is taught for the child's sake, and not for the sake of science.

Science can not fail to interest children, but these interests are modified by their home conditions, their community environs, their parents' occupations and temperaments. Teachers will have strong, but never sole, influence on their pupils' interests.

The goal of science teaching at the elementary level is primarily social development. We wish these children to learn to live in a scientific age, and—for many—to make a living in scientific activity. It is to be hoped that this chief purpose remains in the science studies of higher grades.

Both life and labor require adjustment. Men, no longer living like savages, must cease to think like savages. Science is a practical aid to such adjustment.

Finally, it is hard to imagine a man or woman without religion, although many are without a creed. Science is an ally of true religion. Science searches for truth, and finds much of it; faith should not hold fast to falsehood, but seek truth in its own realm. Scientists think more clearly on facts and faith than many others seem to do. Children do not become trained thinkers in either science or religion in a few years, but they can begin to understand both the physical and the spiritual in the elementary grades.

Published Aids to the Teaching of Science in Elementary Schools

Current prices may be obtained on inquiry.
Packet:

Tunis Baker, *Baker Science Packet,* Tunis Baker, 42 Carolin Road, Upper Montclair, N. J. Box of 153 science experiments on cards, convenient for filing, sorting, assignments.

Books:

Glenn O. Blough and Albert J. Huggett, *Elementary School Science and How To Teach It.* xii, 532 p. (New York: The Dryden Press, 1951). Alternating chapters on content and method.

Guy V. Bruce, *Science Teaching Today:* Vol. I, *Experiments with Water,* 1950, 46 p.; Vol. II, *Experiments with Air,* 1950, 44 p.; Vol. III, *Experiences with Fuels and Fire,* 1951, 31 p.; Vol. IV, *Experiences with Heat,* 1951, 39 p.; Vol. V, *Experiences with Magnetism and Electricity,* 1951, 72 p.; Vol. VI, *Experiences with Sound,* 1951, 44 p.; Vol. VII, *Experiences with Light and Color,* 1951, 47 p. (Washington: National Science Teachers Association, 1201 Sixteenth Street, N. W.) Most of these experiences may be readily adapted for use in elementary grades.

R. Will Burnett, *Teaching Science in the Elementary School,* xvi, 541 p. (New York: Rinehart & Co., 1953). Six chapters of philosophy, nine of experiences and explanations.

Anna Botsford Comstock, *Handbook of Nature Study,* xx, 937 p. (Ithaca, N. Y.: Comstock Publishing Co. 24th Edition, 10th Printing, 1951). The first, and still the most comprehensive, series of lessons on all nature—animals, plants, the earth, the sky. "The Science Bible" for elementary teachers.

Gerald S. Craig, *Science for the Elementary School Teacher,* new edition, xii, 894 p. (Boston: Ginn & Company, 1958). Seven chapters of philosophy, eighteen of lessons and activities. Excellent on grade-by-grade topics.

Julian Greenlee, *Better Teaching through Elementary Science,* xii, 204 p. (Dubuque, Iowa: William C. Brown & Company, 1954). Incidents in many classrooms brought together as if experiences of teachers in a single school.

Don Herbert, *Mr. Wizard's Science Secrets,* 264 p. (Chicago: Popular Mechanics Press, 1952). The popular television demonstrator presents more than 150 of his stunts for home performance.

Carleton J. Lynde, *Science Experiences with Home Equipment,* 244 p. (1949); . . . *with Inexpensive Equipment,* 280 p. (1950); . . . *with Ten-Cent Store Equipment,* 276 p. (1950) (New York: D. Van Nostrand Company). Precisely 200 demonstrations in each volume, with diagrams and explanations.

E. Laurence Palmer, *Fieldbook of Natural History,* xii, 664 p. (New York: McGraw-Hill Book Company, 1949). Extremely comprehensive volume on the skies, minerals, plants, animals, that are likely to be met in outdoor study. Features columns of description, with drawing or photo on top.

Recent Bulletins:

Glenn O. Blough, *It's Time for Better Elementary School Science,* 48 p. (Washington: The National Science Teachers Association, 1958). Report of some fifty conferees, consultants, and correspondents, chiefly in conference at Washington in May, 1958.

Maxine Dunfee and Julian Greenlee, *Elementary School Science: Theory and Practice.* x, 65 p. (Washington: Association for Supervision and Curriculum Development, 1957). Chiefly a compendium of practical ideas from 176 published sources, but exceedingly well selected and arranged by the authors.

Periodicals:

Articles or activities in elementary science are published in each—or nearly every—issue.

Cornell Rural School Leaflet, Ithaca, N. Y. One teachers' number, and three children's numbers, each school year.

Elementary School Science Bulletin, National Science Teachers Association, Washington 6, D. C. Eight issues during school year. Special plan for copies to each teacher in a school.

Grade Teacher. Darien, Conn. Monthly. Science pages in each issue.

Instructor. Dansville, N. Y. Monthly. Science pages in each issue.

Junior Natural History. American Museum of Natural History, New York, N. Y. Monthly during school year. Interesting, instructive.

The Science Teacher. National Science Teachers Association, Washington 6, D. C. Eight issues per year. Covers broad interests, including elementary science.

Lists:

Many books in science for young readers are published each year. Lists that classify and—to a certain extent—select these titles appear at intervals. Recent ones are these:

Paul E. Kambly and Eleanor E. Ahlers, "The Elementary School Science Library for 1956-1957," *School Science and Mathematics,* LVIII (June, 1958), 478-489. Fourteenth annual listing.

Verne M. Rockcastle and Eva L. Gordon, "Science Books for Children, 1950-1957," *Cornell Rural School Leaflet,* LI (September, 1957), 1-64. Previous lists, 1949, 1942, 1938.

Research, Recent:

Muriel Bueschlein, "Implication of the Findings of Recent Research in Elementary School Science Education," *School Science and Mathematics,* LVIII, 610-613, November, 1958.

". . . the most urgent need continues to be teacher preparation." There are other problems, of course.

Jacqueline Buck Mallinson, "Survey of Recent Research in Elementary-School Science Education," *School Science and Mathematics,* LVIII, 605-609, November, 1958.

Covering published literature for the period January, 1956-August, 1957. Chief items: programs, teacher education, grade placement of principles, efforts to vitalize, sources of current information, science writings for children.

HANOR A. WEBB, Professor of Chemistry and Science Education, Emeritus, George Peabody College for Teachers, Nashville (Tennessee), received his B.A. degree from the University of Nashville (1908), M.S. from the University of Chicago (1911), and Ph.D. from the George Peabody College for Teachers (1920); Professor of Psychology and Education, West Tennessee State Normal School (1912-1914); Professor of Chemistry and Biology in the same institution (1914-1917); Instructor in Chemistry, George Peabody College for Teachers, 1917-1920; Professor of Chemistry and Science Education (1920-1953), and Emeritus (1953). As the senior author, he authored two general science texts (1925, 1936) and has written numerous articles for periodicals and encyclopaedias, in addition to having been editor of *Current Science* (weekly, Columbus, 1927-1947), the science issues of *Education* (Boston, 1936-1953), The High School Science Library (annual, 1924-1944); and Keystone Visual Units in Science (slides, 1930-1933). He has also served as Chairman, Department of Science Instruction, National Education Association, 1926; Secretary, Nashville Section, American Chemical Society, 1934-1950; President, National Association for Research in Science Teaching, 1937; President, Tennessee Academy of Science, 1946; Secretary, National Science Teachers Association, 1946-1952.

NOTES FOR ELEMENTARY EDUCATION

1. Alexander Pope (1688-1744), poet, England, in *Epistle IV*.

2. The median age at death of scientists in the United States is about 60. This figure comes from the necrology reports of the weekly *Science* for the period July, 1957, to June, 1958. This is considerably less than the expectation at birth in 1955 of 69.5 years.

3. Such as Marion L. Faegre and Caroline A. Chandler, *Your Child from One to Six,* Children's Bureau Publication No. 30, 1945, Washington 25, D. C.

4. Arnold L. Gesell, *First Five Years of Life,* 1940, Harper; *How a Baby Grows—a Story in Pictures,* 1945, Harper; *Infant Development,* 1952; New York: Harper & Brothers.

5. Fifth Annual Review of Research in Science Teaching, *Science Education* XLI (December, 1957), 375-411.

6. Ellsworth S. Obourn, "Analysis of Research in the Teaching of Science, July, 1955-July, 1956, Bulletin 1958, No. 7, U. S. Office of Education, Washington 25, D. C.

7. Hanor A. Webb, "Nine Reasons Why," *The National Elementary Principal,* XXXIIII, (September, 1953), 22.

8. One of the simplest, among countless examples: Draw two parallel lines of the same length. At each end of one line draw two short lines flaring forward from that end. At each end of the other line draw two short lines flaring backward from the end. Even though the two longer lines are placed in matching positions, one below the other, the apparent difference in their lengths is surprising.

9. Gerald S. Craig (1893-   ), science educator, United States, in *Science for the Elementary School Teacher.*

10. Robert Stollberg, "Materials for Elementary School Science," *The National Elementary Principal,* 32d Yearbook Number, XXXIII (September, 1953) 265-256.

11. Herbert S. Zim, "Science for Children and Teachers," Bulletin 91, Association for Childhood Education International, Washington, D. C., pp. 13-24.

12. Henry Harap, "Free and Inexpensive Learning Materials," Division of Field Services and Surveys, George Peabody College for Teachers, Nashville 5, Tenn. $1.00. A book-sized bulletin, revised at frequent intervals, listing sources (chiefly industrial) and costs.

13. One of the larger community fairs is the Greater Saint Louis Science Fair; the Chairman of its Committee is Mr. Norman R. D. Jones, Professor of Biological Sciences, Harris Teachers College, Saint Louis, Mo. In one gigantic field house, 3404 student-built exhibits were displayed in 1958. Of these, 906 units were by fifth- and sixth-grade students, 177 by first-, second-, third-, and fourth-grade students—a total of 1083 displays from the elementary grades of the area's schools.

One year I served as a judge in this great Fair, and was amazed at the quality and variety of the students' work and interest.

14. Luther Burbank (1849-1926), horticulturist, "plant wizard" of California, made many talks to teachers. I have not been able to locate the time, place, or first publication of the address in which these words were spoken.

15. Henry Ward Beecher (1813-1887), clergyman, United States, in *Proverbs from Plymouth Pulpit.*

16. Francis Bacon (1561-1626), scientist and author, England, in *Essays: of Studies.*

17. Lord Alfred Tennyson (1809-1892), poet, England, in *The Lover's Tale.*

18. John Keats (1795-1821), poet, England, in the first line of *Endymion, Book i.*

19. Joseph Joubert (1754-1824), essayist, France, in *Pensées.*

20. Jonathan Swift (1667-1745), satirist, Ireland, in *Polite Conversation,* Dialogue ii.

21. Henry Ward Beecher (1813-1887), clergyman, United States in *Star Papers.*

22. Carl W. Ackerman (1890-   ), editor, United States, in an address, September 26, 1931.

23. Ralph Waldo Emerson (1803-1882), essayist, United States, in *Essays: Intellect.*

24. Hanor A. Webb, "One Thousand and Two Childish Questions," *School Science and Mathematics* XXXVIII (May, 1938), 504-510.

25. Fasten two apples or oranges to strings about two feet long

(Scotch tape will stick to the peel). Hang them two inches apart, swinging freely. Blow *vigorously* between them, in an effort to swing them apart by the blast of air. You'll be surprised!

26. Paul E. Blackwood, *How Children Learn to Think,* Bulletin 1951, No. 10, U. S. Office of Education, Washington, D. C.

27. Henry Ford (1863-1947), automobile manufacturer, inventor, philosopher, United States, in interview, February, 1929.

28. Bertrand Russell (1872-   ), sociologist, philosopher, England, in *The Conquest of Happiness.*

29. Cloy S. Hobson, "Science in the Common Learning Program," *The National Elementary Principal,* XXXIII (September, 1953), 155-160.

30. Effie G. Bathurst and Wilhelmina Hill, *Conservation Experiences for Children,* Bulletin 1957, No. 18, U. S. Office of Education, Washington, D. C.

31. Albert B. Foster and Adrian C. Fox, *Teaching Soil and Water Conservation—a Classroom and Field Guide.* Bulletin PA-341, 1957, Soil Conservation Service, U. S. Department of Agriculture, Washington, D. C.

32. Various authors, "Using Community Resources," *The National Elementary Principal,* XXXIII (September, 1953), 165-202.

33. June Grant Mulry, "Uses and Abuses of Community Resources," *Elementary School Science Bulletin,* No. 39 (September, 1958), 1.

34. Sir William Blackstone (1723-1780), jurist, England, in *Commentaries.*

35. Ralph Waldo Emerson (1803-1882), essayist, United States, in *Journals.*

36. Student: "I have weighed this thing five times."

Miss Alice: "Fine! You're a smart boy!"

Student: "And here are the five different answers."

37. Horace Mann (1796-1859), educator, United States, in *Lectures on Education.*

38. Joaquin Miller (1841-1913), poet, United States, in *Dawn at San Diego.*

39. Albert Einstein (1879-1955), mathematician, philosopher, Germany, United States, in *Out of My Later Years.*

40. At least once each decade since 1915 my students have been threatened, and in a few cases discharged, from their school positions for teaching the planet's roundness. The most recent instance was in 1951.

41. William Cowper (1731-1800), poet, England, in *The Task.*

MATHEMATICS

W. W. SAWYER
Wesleyan University (Conn.)

Emotional Importance of Early Years

Evidence suggests that, as a rule, children form a strong liking or a strong dislike for mathematics while still in Grade School, this attitude often lasting throughout life.

In our culture, the general attitude towards mathematics is one of fear. Children do not feel this fear, until they are infected with it by adults, or acquire it through bad teaching. We thus have a vicious circle situation. How can children be taught to calculate confidently by adults who lack confidence themselves?

There is no simple answer. The enjoyment of mathematics must spread out in ever widening circles from those places where it already exists. Good teachers of arithmetic must be given the widest opportunities for influencing young children; they may become arithmetic specialists. Improved materials help to arouse enthusiasm in pupils, as do good texts and printed material, once children can read fluently. Mathematical clubs in which several high schools participate are an important means of advance. The 9th graders of today are the teachers of tomorrow. *Training College is, as a rule, too late to produce any change of emotional attitude.* Our strategic aim should be to produce good elementary teachers *ten years hence.* This means, among other things, that we should not only try to create a liking for mathematics in those destined to become brilliant research workers, but in the student body as a whole.

Causes of Fear and Dislike

An important cause is the work going *too fast or too slow.* Slowness causes boredom. Undue speed makes the subject incomprehensible. As children differ enormously in rate of development, the *lock step*—the whole class advancing together—inevitably produces both evils.[1]

Abstractness is another cause. We are so used to numbers that we tend to assume young children understand what numbers mean—which is by no means the case. The child is thus, in effect, required to repeat a sentence in a language he does not understand—a thing few adults can do.

The beginnings of arithmetic are extremely concrete. What are three two's? We only have to put three pairs of objects together : : : to see the pattern for six. In principle, all questions in arithmetic could be answered by an experiment with actual objects. But there are important practical difficulties. If a child learns 4 + 5 by putting four objects and five objects together, and then counting, two unfortunate results follow. (1) The child forms the habits of answering such questions by counting, which is most undesirable. (2) Young children do not count reliably. The child may well get 8 or 10 as the answer, and thus have arrived, by his own experiment, at a wrong result. Catherine Stern[2] has worked out, in admirable detail, an answer to this difficulty. She uses a variety of devices. One of her important ideas is to replace counting by measurement. A child finds that a 4-inch bar and a 5-inch bar will exactly fill a space 9 inches long, and thus discovers 4 + 5 = 9. This involves no counting, and leaves no possibility for error. Handling the bars gives the child the feeling that 9 is a big, long affair, while 4 is much smaller. This *feel* for the size of numbers is most important. One finds children who say cheerfully 4 × 8 = 56, when it is clear that this answer is much too big. 8 being smaller than 10, four 8s must certainly be less than four tens. But 56 is bigger than 40.

Catherine Stern gives many valuable ways for imparting this sense of the relative size and the order of the numbers. She also gives a fascinating game for teaching the difference between 3 + 2 and 3 − 2. The children throw dice. A throw of +2 is rewarded by the child receiving two blocks; a throw of −2 means the loss of two blocks. In this way, emotional significance is given to the signs + and −

The Rhythm of Discovery

To discover anything, it is necessary (i) to see clearly what question has been asked, (ii) to be able to test for yourself whether you have found the correct answer. By contrast, rote teaching merely requires pupils to memorize a rule and answer quickly. Discovery proceeds slowly, by trial and error; but in the long run it is both quicker and more certain. This is increasingly recognized in schools, but it still needs to be stressed. William Viertel[3] found that several teachers taught, "2 divided by 0 is 0." Consider what this means. "16 divided by 2" asks, for example, if you had 16 apples, to how many children you could give 2 apples each. "2 divided by 0" then asks, if you had 2 apples, to how many children could you give none each? It is clear you could give nothing to as many children as you liked. Either "No answer" or "Infinity" would be an understandable answer. The answer, "zero," is just nonsense, taught authoritatively. Presumably other nonsense is taught with equal conviction.

Observation

It is rarely stressed that arithmetic requires observation. Children enjoy noticing things for themselves. It is well known that the 9 times table has remarkable features. One has only to write the multiples of 9 column-wise to see several features. But every table has its characteristic pattern.[4] In the 7 times table the final digits are 7, 4, 1, 8, 5, 2, 9, 6, 3, (these are the final digits of 7, 14, 21, 28 etc.). One notices the rhythm of the changes here—down 3, down 3, up 7; down 3, down 3, up 7; down 3, down 3. If one reads the final digits of the 7 times table backwards one gets those of the 3 times table. Such coincidences intrigue children.

9
18
27
36
45
54
63
72
81
90

Surprise

Children love surprises. Tricks give an excellent way of securing accuracy. Tell a class, "Think of a number. Add 3. Double. Subtract 4. Halve. Subtract the number you first thought of." Ask a few pupils what answer they got. At least half the class should get the correct answer, 1. The rest will start looking, to see where they made the mistakes that cheated them of the surprise. A great variety of such tricks and surprises exist.[5]

The Changing Direction of Arithmetic

Traditionally, the purpose of arithmetic is to make commercial calculations correctly. Calculating machines are rapidly making this use obsolete. More important, commercial mathematics is incredibly dull. The growing uses are in science and in higher mathematics. In both of these, it is not individual results that are important, but rather patterns made by numbers. The regularities in the 9 times table are a very elementary example of such pattern. Exercises in arithmetic can be made to foreshadow algebra, *without in any way calling for additional knowledge in the elementary school teacher.* The following is an example. "Find

(a) $2 \times 3 - 1 \times 4$
(b) $3 \times 4 - 2 \times 5$
(c) $4 \times 5 - 3 \times 6$
(d) $5 \times 6 - 4 \times 7$"

Here of course, multiplication precedes subtraction. (a) for instance gives $6 - 4 = 2$. (b), (c), (d) also give the answer 2. This may lead a child to ask, "Does this go on for ever?" The teacher replies, "Try it for yourself. Work out $9 \times 10 - 8 \times 11$ or $74 \times 75 - 73 \times 76$." A *research element* has now entered arithmetic. An apparent coincidence suggests a general law, and leads to further calculations. Children begin to look out for such coincidences.

Some simple experiments are possible, by which students can learn to spot scientific laws.[6] The mathematics involved may only be the 2 times table. Boys particularly are pleased to see mathematics used to understand *things:* so much mathematics seems to be about nothing at all.

Boredom

There is no doubt that arithmetic is dragged out far too long in the present syllabus. From Grade 5 on, the abler students are restless. Something can be done by drawing on the ideas sketched above. But it should be recognized that in Europe gifted students have often finished the whole of arithmetic by the age of nine, and are starting geometry and algebra. The students who are able, but not gifted, would certainly be deeply into algebra in their 11th year. Our students are just as good as European students. Naturally, they get bored; they become idle or delinquent. It is essential, if an accelerated class is not available, to let them read ahead on their own. It might be interesting to try how the the best 6th graders respond to a European algebra book written for eleven year olds.[7]

Our Changing Situation

We should think of our syllabus as constantly changing and growing. In 1930, a mathematician could teach in school or university; there was little else for him to do. Now suddenly, industry requires Ph.Ds in mathematics by the *tens of thousands*. We let our abler students idle away four years from grade 5 through 8, and then work madly as Ph.D. students to make up for lost time. This is certainly not an efficient approach. The pressure of urgent needs is bound to change it. We should be preparing our minds for that change.

Selected Bibliography

The Arithmetic Teacher, published monthly by the National Council of Teachers of Mathematics; contains valuable information on teaching methods, educational research, educational systems abroad, etc.

Tobias Dantzig, *Number, the Language of Science* (New York: Doubleday Anchor Books, 1954). A fascinating account of the development from primitive tribes upwards; the later part of the book goes beyond elementary work.

Margaret Drummond, *Gateways of Learning* (London University Press, 1931). Deals excellently with the emotional and intellectual development of young children in arithmetic lessons; stresses individual development and the dangers of lock step.

R. V. Heath, *Mathemagic* (New York: Dover, 1953). This paperback contains tricks, curious results, "Number Symphonies," etc. from which questions with surprising answers can be easily manufactured.

E. H. Lockwood & D. K. Down, *Algebra* (London University Press, 1955). A cheerful book; occasional references to British currency, cookery, games, etc. should not disturb student readers; in comparison with American texts, the thoroughness and speed of development will be noted.

"Math in Science," weekly feature in *Current Science and Aviation*. These articles are primarily intended for Grade 7 and upwards; several simple experiments are within the reach of Grade 5; one or two are usable even lower.

The Mathematics Student Journal (published by the National Council of Mathematics Teachers, 1201 16 St., N.W., Washington 6, D. C.), is distributed mainly to High Schools and Junior High Schools, but each issue contains two or three pages of arithmetical curiosities, tricks, puzzles, and likewise, certainly suitable for 5th and 6th grades.

F. Polya, *Induction and Analogy in Mathematics* (Princeton University Press, 1954). It will be an exceptional elementary teacher who might wish to buy this book. However, besides giving a few interesting arithmetical facts, the work shows mathematics as a subject that is discovered and grows and in which guesswork plays a part. Only a small part is elementary.

W. W. Sawyer, *Mathematician's Delight* (Baltimore: Penguin Books, 1958). Originally written for engineers in the armed forces; the first four chapters discuss the reasons for the general fear of mathematics, and the fifth chapter reviews arithmetics; elementary teachers may be able to use some ideas from Chapter 7; the rest is more advanced.

Catherine Stern, *Children Discover Arithmetic* (New York: Harper, 1949). The outstanding book on the beginnings of arithmetic; teachers will find many excellent ideas, worked out in great detail. The first half of the book seems especially valuable; in the latter part, where large numbers are involved, both teachers and pupils may prefer to diverge from Stern's apparatus.

W. W. SAWYER was born in England (1911) and educated in Highgate School and St. John's College, Cambridge. In undergraduate and graduate work he specialized on relativity and quantum theory. Taught in several British Universities; 1945-47, head of the Mathematics Department, Leicester College of Technology, where he experimented on the teaching of mathematics through physical apparatus to machinist apprentices, 1948-1950, Head, Mathematics Department, University College of Ghana (West Africa); 1951-1956, at Canterbury College, New Zealand; 1957, Associate Professor of Mathematics, University of Illinois; 1958,

Professor of Mathematics, Wesleyan University (Connecticut). Designed Aston's mass spectrograph with second order focusing; various papers on differential equations. Is the author of several volumes (*Mathematician's Delight, Prelude to Mathematics, Designing and Making, Mathematics in Theory and Practice,* etc.); editor of *The Mathematics Student Journal.*

NOTES FOR MATHEMATICS

1. Margaret Drummond, *Gateways of Learning* (London University Press, 1931), is an excellent book in connection with this problem.

2. Catherine Stern, *Children Discover Arithmetic* (New York: Harper, 1949).

3. William Viertel. "Letter to the Editor," in New York State *Mathematics Teachers Journal,* VIII (October, 1958), 10.

4. See, for instance, W. W. Sawyer, *Mathematician's Delight* (Baltimore: Penguin Books, 1958), Chapter 5.

5. Beginning November 1958, *The Mathematics Student Journal* (Washington: National Council of Mathematics Teachers) is publishing material of this type in each issue; R. V. Heath, *Mathemagic* (New York: Dover, 1953) contains some useful examples; G. Polya, Induction and Analogy in Mathematics (Princeton University Press, 1954), Chapter I contains some arithmetical examples.

6. See: "Math in Science," weekly feature in *Current Science and Aviation* (Columbus, Ohio: American Education Publications), XLIV, September 1958.

7. Lockwood and Down, *Algebra* (London University Press, 1955), for example.

SECONDARY EDUCATION

GENERAL PROBLEMS OF SCIENCE EDUCATION IN AMERICAN HIGH SCHOOLS

MELVIN RONALD KARPAS
Willimantic State Teachers College (Conn.)
AND
LEO E. KLOPFER
RHAM High School (Hebron, Conn.)

Science teaching in the high school occupies a unique position in any program of science education that may be proposed. Not only must the training of scientists, engineers, and technical personnel find its roots in the science classes of the secondary schools, but the vast majority of our future citizens, who will terminate their schooling at this level, must also be made aware of their responsibilities in a scientific society within the framework of high school science instruction. This dual role of high school science teaching has long been recognized by science educators and continues to be emphasized today in all thoughtful proposals for the improvement of secondary science instruction. Yet all proposals for the improvement of science instruction will most likely result in little more than meaningless chatter of the crowd unless they come to grips with several persistent obstacles which have hampered the development of fully adequate high school science programs throughout the years of this century.

Science reflects the values of the social order in which it is found. David Lilienthal points out that in the hands of depraved men it could work for destruction[1]: ". . . It is clear that science in evil hands can make us slaves—well-fed perhaps, but more pathetic for that fact. On the other hand it is plain that men can use science to further human freedom and the development of human personality. But how it shall be used will be determined by choices made by the people, choices that are genuine only when based on a knowledge of alternatives. And the means whereby the people exercise a real choice depends upon a sacred and inviolable process, the dissemination of knowledge."

This paper is concerned with some of these obstacles which have been most evident in years past and which still remain with us today. These problems must be faced squarely if the present challenge to science education is to be met with any lasting reply. We shall seek to explore the depth and scope of three of these problems, to assess their present importance, and to indicate some means of solution. For easy reference, we characterize the three problems to be discussed as: (1) The Wary Curriculum; (2) The Closed Purse; and (3) The Tired Teacher.

(1) *The Wary Curriculum*

It is difficult to make all-inclusive generalizations about the high school science curriculum being offered to students in schools throughout the United States. This is due partly to the great range of types and organization in the schools themselves and partly to the large variety of courses which are offered under the general heading of *science*. Yet, it is possible to picture the high school science curriculum, in broad terms, as a *wary curriculum*. By this we mean that the high school science curriculum today is, in general, lacking in inspiration and originality, but is instead beset by course offerings which tend to preserve the traditional and which reflect a subservience to a variety of special interests.

Diversity of Science Offerings. The variety of science courses in which American high school students are enrolled is almost staggering. Not only do we find the better-known courses of general science, biology, chemistry, physics, which have long been a part of the high school curriculum, but also common are courses in physical science, earth science, physiology, applied science, applied chemistry, applied physics, and related science. In a number of schools, we may also find courses in botany, advanced biology, biological techniques, aviation science, senior science, survey of science, basic science, advanced chemistry, qualitative analysis, advanced physics, and various vocation-related science courses. Clearly, this diversity of science courses shows that attempts are being made to meet the needs and interests of various kinds of young people, and to achieve the aims of general education in science as well as the preparation of specialists. It is questionable, however, that the pattern of courses which are offered in most schools represents the considered concensus as to the suitability of the program. Rather, the curriculum seems to be a result of circumstances.

The circumstances which seem to be primarily responsible for the diversity of science courses offered in the high schools are the desires of college-preparatory students, the needs of the large and varied high school population, and a response to influences, often transitory, from outside the schools.

Pressures and Desires of College-Bound Students. The influence of secondary school students preparing for college on determining the nature of high school science courses has long been profound. Although the colleges no longer prescribe the content of secondary school science courses, as was customary throughout the closing decades of the nineteenth century and the early decades of the twentieth, the desire for a systematic, formal course in the sciences on the part of college-bound students has resulted in the widespread offering of this type of course in biology, chemistry, and physics. The rationalization for this type of course is usually stated to be the need to prepare students for the science examinations of the College Entrance Examination Board. In actuality, however, only a very small percentage of the students enrolled in biology, chemistry, and physics courses take the CEEB examinations in the sciences, so that this argument can hardly be accepted at face value. A more plausible explanation for the great durability of the systematic "college-preparatory" courses is the unwillingness on the part of science teachers and school systems to break with established patterns. Thus, the situation exists that the content of many courses in chemistry, in physics, and, to a lesser extent, in biology, is little different today from what it was thirty or more years ago. If we consider the major advances which have been made in these fields during this time, we could scarcely look upon this as a healthy situation.

Varied Needs of High School Population. It is known that, throughout the years of this century, the enrollment of young people in the high schools of the country has increased steadily. During the period 1900 to 1954, there has been a 1200 per cent increase in the number of pupils enrolled in grades 9 to 12 in the public schools. As this new swell of pupils came into the high schools, it soon became evident that the science courses designed for those who were aspiring to college were far from suitable for the much larger number of boys and girls who held no such aspirations. To give these young people a background of science as a part of their general education, new courses and new types of courses were introduced. General science courses, whose content cut across the traditional divisions of science into biology, chemistry, and physics, were the first of the new offerings designed for the youngsters of varied objectives and needs now enrolled in the high school. Courses in general biology, whose intent was to present materials which were significant in the lives of the students rather than to give a systematic treatment of the subject, as in the college-preparatory course, were also introduced and are still a very common offering today. Like general science, the course in general biology is designed to serve the needs of a vast body of students with widely divergent gifts and potentialities. Consequently, both general science and general biology courses have been developed with a variety of different approaches and forms of organization, as was demanded by the broad group being served. Again, in numerous places in the nation, the local situation and the demands of the growing numbers of diverse students, brought about the introduction of such courses as applied science, applied chemistry, and applied physics, whose purpose was to emphasize the practical rather than the theoretical aspects of science. More recently, courses in physical science and earth science, which seek to present selected areas of science in a broad context, have been introduced and have found acceptance in many schools, in an effort to reach the large student audience.

Special Interest Pleaders. The third major cause of certain offerings of science courses in the high schools has been the response of educators to the influence of special interest groups, outside the schools. While the intentions of these special interest groups are usually good and their motivations noble, it is doubtful that they exhibit much concern for the total educational pattern of youngsters or that they are particularly versed in the broad aims of a high school science program. Sometimes the influence of the special interest groups is transitory, as was the case in the late 1930s when a rash of science courses in aviation science, aviation physics, and aviation biology appeared. More often these outside groups achieve more permanent placement of courses in science curricula in various schools of the country. This is seen in such common offerings as conservation, consumer science, and the vocation-oriented science courses. We are not here seeking to cast aspersions on the value of such courses in selected local situations, for it is probable that students may frequently have meaningful experiences within their framework, but we do look askance at the unfortunate circumstance that the structure of a science curriculum can be quite readily influenced by people and organizations who can claim little competence in appreciating what the foundations of a sound education in science are. The existence of the wary science curriculum permits such encroachments.

Our nation is faced today with the construction and implementation of a sound, all-inclusive curriculum in science. Already the voices of the critics and the ill-informed are heard loudly on all sides offering suggestions for improvements, revisions, and innovations. In the past, the existence of the wary science curriculum has not made it difficult for new courses to be added. Unless there is a change and science educators will stand firm on their greater capabilities in determining the science curriculum of the high school, we may again see a mushrooming of new science courses added to those already being taught in an effort to appease the demands coming from outside the schools.

Need for Creativity in Science Experiences. There is, moreover, another facet to the consideration of the wary curriculum. Perhaps a diversity of science courses in different places would be a most salutory situation if most of these courses would make a unique and creative contribution to the science experiences of boys and girls. Our hopes in this regard, however, are not justified. Most research studies and surveys indicate that the content and organization of the overwhelming majority of science courses are determined by the material presented in existing textbooks. In this respect, Dr. William H. Kilpatrick speaking at Wheelock College Founder's Day ceremonies in Boston, January 31, 1959, said that ". . . The worst feature of education from the Renaissance to 1900, and even till today, is its overemphasis on book learning

. . . our high school has too much of it, and the liberal arts college even more of it."[2]

By the nature of the publishing business, wherein a large volume of sales of a particular book must be assured to obtain a return on the considerable expense of producing it, the contents of a textbook generally is limited to those topics and practices which are widely accepted in courses being taught. Thus, a vicious circle is established in which the course content is based on what is found in an existing text and the content of new texts is determined by the materials taught in a large number of courses. The inevitable result is stagnation. New materials find their way most slowly into textbooks and courses and major changes in content or approach are rare indeed. Little wonder, then, that the teacher of a standard high school physics course in 1959 would find himself quite at home in a text written eighty years ago. The situation may not be quite so drastic as this in all science courses being taught today, but the moral still holds true.

Lack of Aggressive and Conscientious Leadership. It is difficult to introduce into the wary courses of the wary science curriculum new ideas in regard to content or approach. It is even more difficult to have such new ideas, however worthwhile they are acknowledged to be, gain widespread acceptance among science teachers. One reason for this is the old bug-a-boo of the College Board examinations, or some other standard form of statewide or citywide terminal test, that it is claimed students must be prepared to pass. While this reason may be dismissed, at least in part, as an excuse for the desire of science teachers to remain with the tried and familiar topics and to preserve the traditional approaches and methods, there is a more cogent reason for the difficulty with which new ideas in science teaching make any headway. This is the deplorable shortage of conscientious and aggressive leadership in the science education field. Too few colleges and universities have strong science education departments which could provide spokesmen and rallying-points for the science teachers in the schools. For many years, there was no one in the United States Office of Education specifically charged with the responsibility for secondary school science, and, until recently, there was only a handful of Science Consultants on the staffs of State Education Departments and of large metropolitan school systems in the entire nation. This lack of highly-placed leadership has been responsible, in large measure, not only for the abortive careers of numerous new ideas and new directions in science teaching which have been advocated in the past, but has also been a significant factor in producing the conglomeration of science course offerings and the stagnation in course content, the features of the wary curriculum, which we have already discussed.

Hopeful Signs of the Present. Happily, there are hopeful signs today, and it may well transpire soon that the wary science curriculum is a creature of the past. In drawing the broad picture of the wary curriculum, we have perforce ignored the much smaller and widely scattered original and creative contributions to high school science teaching which have almost always been present through the years of this century. We have not alluded to the steady growth of the associations of science teachers, especially in the post-war years, to a point where they command the attention of all educators and of the public. Also, we have not mentioned such publications as *A Program for Teaching Science,* (National Society for the Study of Education, 1931), *Science in General Education,* (Progressive Education Association, 1938), and *Science Education in American Schools,* (National Society for the Study of Education, 1947), in which science educators set forth thoughtful, constructive proposals for sound science curricula and well-fashioned science courses. These occurrences and activities, which have been carried on largely in the background in the past, provided the seeds and the climate for the growth of the hopeful signs we can discern today.

Within the past five years, a number of nationwide programs have been launched to study and improve science courses and science curricula. The most widely-heralded of these programs is that of the Physical Science Study Committee, (described in detail elsewhere in this volume), in the field of physics, but promising investigations are also under way in biology and chemistry. The National Science Teachers Association has established a Commission of Education in the Basic Sciences, whose responsibility it is to investigate the fundamental problems of science teaching and to stimulate and support those developments which will contribute to its improvement. While these broadly-conceived studies are of great interest and of immense value, it is much more significant that, at the present time, there is a virtual ferment of science curriculum reorganization in many high schools and school systems throughout the country. Science teachers themselves, often with the aid of practicing scientists, are developing the kinds of curricula and courses which they know are best suited for the boys and girls whom they teach. To be sure, the science programs which are being developed differ in detail, but almost all of them reflect the teachers' creative insight into the needs of the young people in today's world and the teachers' desire to teach the science important in the present day. A further most encouraging sign that may be perceived is the increasing number of school administrators who are coming forth in solid support of a soundly organized and consistent science program. The science curriculum seems to be wary no longer.

(2) *The Closed Purse*

It is almost inconceivable that any teacher in the schools would be asked to instruct his students without paper, without chalk, or without desks and chairs. Nevertheless, it has been all too common in the past, and it still remains a pervasive problem today, that money has not been provided to the science teacher for equally important equipment, supplies, and facilities. The all-too-prevalent *closed purse* when funds for public education are required, while affecting all areas of the high school program, has hit the science area particularly hard because of the special nature and large variety of supplies and equipment, sometimes costly, which are needed for adequate science instruction.

In almost any conclave of science teachers where the problems of high school science teaching are discussed, the problems related to supplies and equipment are most often mentioned. Probably next in frequency are the problems related to science rooms and facilities. Almost always this lack of equipment and facilities with which to work has not been due to a failure of the science teacher to request them from the school authorities, but rather the cause may be found in the failure of school leaders to recognize their importance and to be satisfied to get along with only the most minimal requirements. The school leaders in their turn have been hampered by the paucity of available funds, even if they had wanted to supply the needs of science teachers.

Inadequacy of Supplies, Equipment, Space and Facilities. The dearth of money being channeled toward the instructional needs of high school science has resulted in a situation where there are too few and too small science laboratories in most schools. Since laboratory work is considered an almost essential *modus operandi* by most science teachers, the loss to adequate science instruction because of poor laboratory facilities has been great. Students have little opportunity, and in too many cases no opportunity at all, to come into contact with and experience at first hand the excitement of the materials and tools that form the basis of all scientific work. Is objectivity in science possible without the freedom to observe? "No one can go far in science, if he is not free to see things as they are. . . . Men have not always and everywhere been free to become scientists. Creative productivity, not only in science but also in art and other fields, has varied greatly from culture to culture and from age to age in the same culture. From the periods and circumstances under which productivity has been high, it seems clear that an essential factor in any culture where creative activity flourishes is a high degree of independence and individual freedom . . ."[3]

The thrill of discovery through experiments is lost and the manipulative skills needed for further work in science remain undeveloped. It is true that laboratory work of some form is provided in most schools for chemistry and physics courses, but the absence of individual laboratory activities in biology and general science courses is quite general and all too common. At their best, these courses usually include a variety of teacher and pupil demonstration experiments, but the demonstration, while advantageous in certain instances and certainly more economical, is a poor substitute for direct contact with and manipulation of the materials under study. Science becomes a passive, ladeled-out experience for students who cannot work in the laboratory, while its true nature of creative, aggressive activity is seldom revealed to them.

Improvisation of Materials. Another result of the perennial closed purse when supplies and equipment for science are requested is that the conscientious teacher must bend his efforts to improvising the materials which he needs for instruction. Many have been the ingenious and inexpensive inventions which high school science teachers have devised to provide equipment for their students or to illustrate an important principle. Tin cans, jelly glasses, balloons, discarded medicine bottles, dead batteries, broken windows, milk cartons, bicycle wheels, roller skates, garden hoses, old rubber tires, bird cages, scraps of wire, string, and sealing wax, among many other items, have all been pressed into service in the science classroom. No doubt, many an aggressive teacher has enjoyed the construction of his own original brand of equipment and his students have benefited immensely from his efforts, but the aggregate time spent in such improvisation has often been very large. Had he been free of the obligation to improvise his teaching tools by being supplied with adequate and sufficient equipment and supplies, this time could usually have been much better invested in time to think, in time to keep abreast of the rapid developments in science, in time to evaluate his own teaching strengths and weaknesses, in time to develop new ideas and to create new techniques, new courses, and new curricula. What a waste of the unique talents and training which the science teacher has when he must consume his energies in scurrying about providing *substitutes* for his teaching needs which the closed purse has denied him!

Positive Trends of the Times. It is heartening to note on the present national scene some positive signs of the slow loosening of the purse-strings of the closed purse. The most encouraging indicator of this trend is the recognition by school boards in many communities of the need for providing adequate funds for science facilities and equipment. Most new high school buildings being erected contain provisions for good science classrooms, varied laboratories, and storerooms, in accordance with the recommendations of far-sighted architects and science educators. Money is being expended to provide for the necessary water, gas, electrical, and other special facilities that must serve a good science teaching center. Funds for the renovation of science classrooms in existing schools are also being provided by boards of education with increasing generosity. Funds for special science equipment are also being contributed to the schools by industrial concerns and foundations, e.g., the program of the Research Corporation in its pilot project in the state of Connecticut. Scientific apparatus makers report a sharp increase in sales volume of science equipment and supplies to the schools. Finally, late in 1958, the U.S. Congress voted to provide federal funds for the purchase of science equipment and the improvement of facilities in secondary schools by adopting the National Defense Education Act. Among other provisions that will benefit a number of selected areas of national education, this Act provides for the matching of federal funds, on a dollar-for-dollar basis, with all qualified funds expended by a state or local school system for laboratory equipment and facilities and other special equipment suitable for use in providing education in science, mathematics, and modern foreign languages. As soon as the necessary machinery can be established in the various states to meet the qualifications of the Act, high schools contemplating the equipping or expansion of science courses and laboratories may expect to plan for two times as large a program as the local resources would allow. While this present Act will surely not answer all the demands made upon the closed purse by the needs of science instruction, it is certainly a step in the right direction and it provides a pattern of action for alleviating the failings which the lack of money has imposed upon high school science teaching.

(3) The Tired Teacher

The success of any high school science program depends ultimately upon the classroom teacher who is carrying it out. What is the status of the high school science teacher today? Unfortunately, the general picture is not a heartening one. In a word, the high school science teacher of today is *tired!* He is tired because of the heavy demands he feels he must fulfill to provide effective science courses in a sound science program. He is tired because science teachers are in short supply. He is tired because he is often not adequately trained for his varied assignments.

Overburdened with Responsibilities, Duties, and Demands. Every teacher in the schools carries numerous responsibilities and performs a variety of duties which are not readily discerned in a casual observation of his classroom activity. A moment's reflection makes it clear that the teacher must carefully plan and fully prepare himself for every classroom lesson, but there are also the less obvious responsibilities of grading papers and examinations, keeping records, compiling reports, and supervising students while they are on the school grounds. The science teacher shares all these responsibilities and duties with his fellow teachers, but he also feels called upon to carry a large number of additional responsibilities to make his science teaching fully effective. The science teacher must devote considerable time and effort to preparing demonstration experiments and laboratory activities. His devotion to the experimental approach in science teaching requires that he must find and take time to see that inventorying, arranging, cleaning up, and putting away equipment and supplies is properly done. He must spend time in familiarizing himself with, evaluating, selecting, and ordering the equipment and supplies appropriate for use in the courses he teaches. When the need arises, he must improvise equipment that is lacking or he must devise experiments with substitute materials to drive home an important idea. He must be alert to the values of science teaching materials which are made available by commercial companies and he must consider their value and suitability for his particular courses. Moreover, he must keep abreast of significant developments in the rapidly-moving fields of science so that his courses may be up-to-date. This listing of the responsibilities that the conscientious science teacher undertakes need be extended no further to make the point that he must devote long hours to his job and is often physically very *tired!*

Shortage of Trained Science Personnel. A second aspect of the situation of the tired teacher is the fact of the present shortage of science teachers. Beginning in 1951, the number of new college graduates prepared and certified by state agencies to teach science began to decline. The low point was reached in June, 1954, when only about 41 per cent of the number of new science teachers prepared in 1950 were graduated, but there has been some recovery since that time. Still, the demand for new science teachers to replace those who leave teaching or retire and to take care of increasing high school enrollments far exceeds the supply of new teachers being prepared. In a study by the Research Division of the National Education Association, the national demand for new science teachers in September, 1958, was estimated at 6700, but the supply of college graduates of June, 1958, prepared to teach science was only approximately 5800. Moreover, experience in previous years has shown that only about 60 per cent of the June graduates prepared in science actually enter science teaching positions the following September. Thus, the number of teachers prepared and available to teach science in the high schools fell short of the number who could have been employed by at least 3200! This shortage of new science teachers to fill existing positions has been part of the national picture for at least seven years past and promises to continue to be true in the future, unless a much greater number of young people can be attracted to the profession and trained. It is evident that an intensive effort must be made to increase the number of high school and college students who are planning to prepare for science teaching. Encouragement must be given by present science teachers, by guidance counselors, by industry, and by college professors in science and science education. In the larger context, the development of increasing numbers of science teachers must be encouraged through the efforts of school boards and citizens' groups to improve teacher salaries and to improve the conditions of employment. Among the conditions of employment that deserve attention are the provision of scheduled time to plan for laboratory teaching, the possibility of adding non-teaching laboratory assistants to the science staff, and the arranging of courses for teachers to up-date or refresh their knowledge either during the year or in the summer. In the meantime, however, the shortage of qualified science teachers works an increased hardship on the science teacher already on the job. The science teacher shortage adds the burdens of heavy teaching schedules and large classes to his already tiring responsibilities.

Qualitative Aspects of Science Personnel. Thus far, we have looked at the quantitative aspect of the science teacher shortage, but it is also important to inquire into the quality of the science teachers in the nation's high schools. Though it is most difficult to judge the *quality* of a teacher because of the great complex of teacher characteristics and personality which are involved in such an assessment, we can make an approach to the problem of teacher quality by asking how well his collegiate preparation qualified him for the science teaching assignments he must fulfill. No recent, complete nationwide study of this question has been made, though status studies on the preparation of science teachers within a number of states are available. On the basis of such studies and other data, the Report of the Conference on Nation-Wide Problems of Science Teaching in the Secondary Schools drew the following generalizations:

"a) a considerable minority of science teachers (particularly in smaller schools) are not certified in science;

b) a significant portion of teachers who are certified in science have relatively meager backgrounds in this area;

c) a fairly large number of science teachers have their science training concentrated in one area, to the near exclusion of other areas;

d) many teachers certified in science spend all or part of their time teaching non-science classes or engaging in other types of educational activity;

e) teachers of General Science tend to have less thorough backgrounds than those who teach special science subjects;

f) there is a tremendous range both in the scientific and the professional educational backgrounds of the nation's science teachers;

g) a considerable percentage of science teachers have graduate degrees or have done graduate work in science and education;

h) in general, the science teachers in larger schools tend to be better prepared than those in smaller schools."[4]

It is evident from the above generalizations that, on the basis of collegiate preparation, the quality of science teachers in the country at large is not as outstanding as it might be. This does not imply, of course, that there are no good or well-prepared teachers, for there are many of them, and they are doing a superior job. However, ill-qualified science teachers are among us and, if we consider the meager supply and rising demand, ill-qualified science teachers are here to stay. What can be done to alleviate this situation and to minimize the disadvantages and dangers which accompany poor science teaching? What *is* being done?

Converted Teachers. One administrative solution which is, of necessity, finding widespread currency in the schools is the *conversion* of teachers from other subject areas, where the supply is plentiful, into either full-time or part-time science teachers. However, the application of such emergency measures is seldom desirable in the long-run and is of limited benefit. The *converted* teachers often have a meager knowledge of science and of science teaching techniques, but, more seriously, they *may lack an interest in science itself.* When the teacher lacks interest in science, we cannot expect his students to become excited about science or to have much opportunity to pursue the trail of discovery through stimulating laboratory activities. Teachers who are unsure of their ground tend to stick closely to their textbooks, so that only the poorest type of science teaching results. In any form of teacher *conversion,* it is imperative that steps are taken to insure that the newly-created science teachers are capable of imparting to their students the spirit of science and are properly trained in science subject matter and instructional techniques.

Programs to Help Science Teachers. Several other better-conceived programs to assist the ill-prepared science teacher have gained momentum in the past few years. Increasing numbers of foundations and industrial concerns are providing funds for fellowships at summer institutes for science teachers where they may refresh or learn anew the subject matter and techniques of instruction which they need. Among fellowship offerings dating back a dozen years or more are the programs of the General Electric Company and the Westinghouse Educational Foundation. The National Science Foundation, using Congressional appropriations, is providing substantial grants to a large number of colleges and universities for Summer Institute programs and Academic Year Institutes, where science teachers may obtain a full year of training. The American Association for the Advancement of Science, under a grant from the Carnegie Corporation, is now conducting the Science Teaching Improvement Program, which includes a major study on the use of science counselors, who can guide and present new ideas and information to science teachers. This study is being carried out at the Universities of Oregon, Nebraska, Texas, and Pennsylvania State University. As programs of this type are continued and expanded, it will surely follow that the quality of science teaching in our high schools will be greatly improved.

Education in science has moved from the wings to the center stage of education. One can no longer mediate educational problems in the United States on the value assumption that science may be principally left to the higher institutions of learning. The urgency, appropriateness, and relevance of education in science at all educational levels have become central educational concerns.

Why hasn't the socal order made science education of a high-level an imperative in its schools, when it so deeply depends on science-based activities outside of its schools? When millions of people, including teachers, students, governmental agencies, and social agencies interact intimately with our science-integrated and science-orientated culture, how does it happen that group pressure to place science education in the schools is so low?

Selected Bibliography

Welden N. Baker and Merle E. Brooks, *Background and Academic Preparation of the Teachers of Science in the High Schools of Kansas 1955-1956* (Kansas State Teachers College, Emporia, Kansas. The Emporia State Research Studies, Volume 6, Number 2, December, 1957). Surveys the status of high school science teachers in one state and is typical of similar surveys made in a number of other states and localities. In its pages may be found broad implications concerning the training and competence of secondary school science teachers throughout the nation.

Sam S. Blanc, "Guideposts in Science Education," *The Science Teacher*, XXXV (March 1958, National Science Teachers Association, Washington, D.C.), 82-83, 109-112. A concise report on a survey of trends in science education at the elementary and secondary levels based on an examination of materials published during the last 20 years.

Paul F. Brandwein, Fletcher G. Watson, and Paul E. Blackwood, *Teaching High School Science: A Book of Methods* (New York, Harcourt, Brace, 1958). Primarily a handbook for the use of present and prospective high school science teachers, this book contains many valuable discussions on the problems of science teaching, science courses, and science curricula. Also included is a section containing extensive reports of the actions taken in three different communities in an effort to improve science teaching.

Kenneth E. Brown, *Offerings and Enrollments in Science and Mathematics in Public High Schools* (Washington, D.C., U.S. Department of Education Pamphlet No. 120, U.S. Government Printing Office, 1957). Contains the latest available information of nationwide enrollment in science in the public high schools. The data is presented in tabular form and comparisons with offerings and enrollments in earlier years are included.

James B. Conant, *The American High School Today* (New York, McGraw-Hill Book Co., 1959). Dr. Conant's penetrating appraisal of and recommendations on the organization and curriculum of the American high school have direct bearing on the science program. His recommendations on science courses for the academically talented deserve especial attention.

Critical Years Ahead in Science Teaching. Report of Conference on Nation-Wide Problems of Science Teaching in the Secondary Schools (Cambridge, Mass., Harvard University Printing Office, 1953). Though some of the data in this pamphlet have now been superseded, the projections for the future and the dicussions of the teacher supply and demand problems which it contains remain quite valid. Also discussed are the preparation of science teachers and science teaching as work and career. (Copies of this pamphlet may be obtained without charge by writing to Mr. Elbert C. Weaver, Phillips Academy, Andover, Mass.)

K. Laybourn and C. H. Bailey, *Teaching Science to the Ordinary Pupil* (London, University of London Press,1957, and New York, Philosophical Library, 1958). Written in England, this book discusses in a forthright manner the special techniques suitable for teaching those students who are neither the most able nor the least able, and who form the vast majority of boys and girls found in the secondary schools. More than five hundred experiments from a broad spectrum of science fields are described and illustrated in detail, with meaningful and helpful commentary freely interspersed among them.

On the Target! High School Science Teaching and Today's Science-Related Manpower Shortage. Pamphlet published in *The Science Teacher* (April 1957). (Reprints available without charge from the National Science Teachers Association, 1201 Sixteenth Street, N.W., Washington 6, D.C.) A concise statement of facts on high school science enrollments and the supply of qualified science teachers. Also comments on the interest of American youth in science and the availability of American resources to improve science teaching, and proposals for action made by the Board of Directors of the National Science Teachers Association.

John S. Richardson, Editor, *School Facilities for Science Instruction* (Washington, D.C., National Science Teachers Association, 1954). Fully illustrated with lay-outs, drawings, and photographs, this is an indispensable guide for the planning of science facilities for all types of high school science courses. The problems discussed range from the selection of a school site with regard to its potentialities for science instruction to the selection and ordering of supplies and equipment.

Science Education in American Schools. Forty-Sixth Yearbook of the National Society for the Study of Education, Part I. (Chicago, Ill.: University of Chicago Press, 1947). The group of chapters in the book on "Science in the Secondary School" provide a useful picture of the status of high school science courses in the last decade. Topics discussed include trends and objectives, content and methods, and the education of high school science teachers. (A new study on the status of science education in the United States is slated for publication by the National Society for the Study of Education in 1960.)

United States Senate, Committee on Labor and Public Welfare. *The National Defense Education Act of 1958* (Washington, D.C., U.S. Government Printing Office, Committee Print, 85th Congress, 2nd Session, September 5, 1958). A summary and analysis of the important National Defense Education Act, which includes provisions for financial assistance for strengthening science, mathematics, and modern foreign language instruction; financial support for the identification and encouragement of able students; support for research and experimentation in the more effective use of television, radio, motion pictures, and related media for educational purposes; and the establishment of a national Science Information Service.

Periodicals which frequently carry articles on the problems of high school science teaching:

American Biology Teacher. (Journal of the National Association of Biology Teachers.) Interstate Press, 19 North Jackson Street, Danville, Illinois.

American Journal of Physics. (Journal of the American Association of Physics Teachers.) American Institute of Physics, 335 East 45 Street, New York 17, N.Y.

Journal of Chemical Education. (Published by the Division of Chemical Education of the American Chemical Society.) Business and Publication Office, 20th and Northampton Streets, Easton, Penna.

School Science and Mathematics. (Published by the Central Association of Science and Mathematics Teachers.) Box 408, Oak Park, Illinois.

School Science Review. (Published by the Science Masters' Association and the Association of Women Science Teachers of Great Britain.) S. W. Read, 31 Grosvenor Road, Chichester, Sussex, England.

Science Counselor. Duquesne University, 901 Vickroy Street, Pittsburgh 19, Penna.

Science Education. (Published by National Association for Research in Science Teaching.) Publication Office, 374 Broadway, Albany, New York.

Science Teacher. (Journal of the National Science Teachers Association, a Department of the National Education Association.) National Science Teachers Association, 1201 16th Street, N.W., Washington 6, D.C.

Scientific Monthly. (Published by the American Association for the Advancement of Science.) 1515 Massachusetts Avenue, N.W., Washington 5, D.C.

MELVIN RONALD KARPAS, Visiting Professor of Social Science, Willimantic State Teachers College (Conn.) received his B.S. in B.A. (Market Research & Economic Analysis, 1954). Ed. M. (Social Science, 1955), and Ed. D. (Sociology and Education, 1958) from Boston University. He has studied at Harvard, and taught at University of Georgia (U.S. Army), Princeton (U.S. Army), Newton Public Schools, and Boston University. He is the recipient of the Franklin and Washington Award, Franklin Medal, C. Hayden Memorial Scholarship, Gutman Foundation Fellowship in Human Relations, NCCJ Scholarship, and a NCSS Fellowship. He is the author of several publications in the area of sociology, secondary education, comparative economic systems, and human relations, and has served as an economic and sociological consultant to two educational foundations. He is a member of Phi Delta Kappa, Pi Gamma Mu, Phi Alpha, American Sociological Society, Eastern Sociological Society, National Council for the Social Studies, National Academy of Economic and Social Science, National Association of Security Dealers, American Historical Association, New England Association of Social Studies Teachers, and the American Civil Liberties Union. Listed in Who's Who in American Education.

LEO E. KLOPFER, Chairman of the Science Department at RHAM High school in Hebron, Connecticut. He earned his A.B. degree in Chemistry at Cornell University (1950) and his A.M.T. degree in Science Teaching at Harvard University (1955). In addition to his high school teaching experience, he has taught science and mathematics in the Army Education Program in Japan (1952-54). His principal research interest is the utilization of historical materials in secondary school science teaching, and he is the author of a series of *History of Science Cases for High Schools*. He is a candidate for the Ed. D. degree in Science Education at the Graduate School of Education, Harvard University.

NOTES FOR GENERAL PROBLEMS OF SCIENCE EDUCATION IN AMERICAN HIGH SCHOOLS

1. David Lilienthal, "Science and Man's Fate," *The Nation*, CLXIII, 2 (July 13, 1946), 39-41.

2. *Christian Science Monitor* (February 2, 1959).

3. Anne Roe, "What Makes the Scientific Mind Scientific," *The New York Times Magazine* (February 1, 1953).

4. *Critical Years Ahead in Science Teaching* (Cambridge, Mass.: Harvard University Printing Office, 1953), 35-36.

SELECTED AREAS:

MATHEMATICS*

E. P. ROSENBAUM
Scientific American

If you have a youngster in high school and are occasionally called upon for help on his math, you may not remember the answers but you recognize the problems. It is the same old math that was taught a generation ago and indeed 200 years ago. The methods and the subject matter have not changed. But a revolution is impending. Deeply dissatisfied with the way mathematics has been taught, several influential groups of educators have begun to experiment with radically new methods of presenting the subject. There are high schools where sophomores are now taking

home problems such as this: Prove that [A, B/l/] if and only if (A $\notin$ l and B $\notin$ l) and $\overline{AB} \cap l = \phi$. (Translated this reads: "Prove that a point B lies between another point A and a line l if and only if A is not a member of l and B is not a member of l and the segment AB does not intersect l.")

There are sharp differences of opinion about whether this sort of thing is helpful or will get very far. But on one thing almost everyone agrees: The old mathematics course must go. It does not tell the students what mathematics is all about. It does not give them any real understanding of the principles of the subject. It is so far behind the times that it leaves out practically all the new ideas and discoveries of the past 100 years. And above all, it has managed to make mathematics about the most unpopular of all branches of learning. Even cultivated men declare their ignorance of mathematics with a defiance akin to pride.

Something has to be done to make mathematics more meaningful and more exciting. At least three major attempts are under way. I shall try to report briefly the rationale and tactics of their approach.

Their underlying theme is to modernize the subject matter taught—more or less (some want to go modern farther than others). Going mod-

*Originally published as "The Teaching of Elementary Mathematics," *Scientific American*, CLXXXXVIII, 5 (May, 1958), 64-73.

ern means mainly two things: pruning out dead wood and introducing some of the new fundamental ideas which within the last century have given more meaning and unity to all the traditional branches of mathematics.

As an example of dead wood one can cite the considerable attention devoted in trigonometry textbooks to solving triangles with logarithms. This was the only method available to surveyors and navigators a century ago; today technicians punch out the answer in calculating machines. Also in the category of dead wood are some of the classic Euclidean "proofs" of geometry; they are not really proofs at all. Today a mathematician attacks these problems with the calculus and the idea of limits, and arrives at truly rigorous proofs.

Mathematics as now taught in the lower schools and even in college seems to be a collection of separate subjects. Each has its own apparently arbitrary rules, taught by rote. But work on the foundations of mathematics in the last century has shown that all the branches of mathematics can be reduced to purely abstract terms, with common properties. As numbers are the elements of algebra, so points and lines are "primitive" elements of geometry, and we can deal with sets of either in the same way—indeed with the same operations. The rules of mathematical logic are universal. These basic ideas and logical processes can be taught even to school children, the modernists believe. In learning them, students will find mathematics more understandable and more meaningful. They will also get at least some acquaintance with modern thinking in mathematics.

The College Board Program

Foremost among the projects for modernizing the teaching of *high-school* mathematics is that of the Commission on Mathematics of the College Entrance Examination Board. This group, set up in 1955, is preparing a considerable revision of the high-school courses. Because the Commission represents a wide range of views, and because it hopes to exert an almost immediate influence on every school in the country, its program is comparatively conservative. It proposes to change the approach and spirit of the algebra course without altering the content substantially. It is revising the geometry course considerably. It wants to make certain changes in trigonometry. And it would add some entirely new courses to the high-school curriculum.

Algebra, like geometry, can be regarded as an abstract deductive system. It is built up from a set of undefined primitive notions and a number of assumed axioms. From these all the other rules and facts can be deduced by logical reasoning. The Commission does not propose to make high-school algebra purely an exercise in abstract deduction, but it believes that students should be led to appreciate the deductive nature of algebra, should learn what the axioms are and how they are used to prove some of the principles. The necessary skills in manipulating alge-

THE CHALLENGE OF SCIENCE EDUCATION

braic expressions will be easier to learn when the reasons behind the manipulations are understood.

What are the "primitive" notions of algebra? They are the notions of number and of operations such as addition and multiplication. The axioms likewise are simple concepts—so simple that they hardly seem to need stating. If a and b are numbers, then a + b is a number. If a = b and c = d, then a + c = b + d (if equals are added to equals, the sums are equal). Then there are the "commutative" laws, a+ b = b + a, and a × b = b × a; the "associative" laws, *e.g.*, (a + b) + c = a + (b + c), and the "distributive" law, a (b + c) = ab + ac.

Although they may appear mere platitudes, these rules form a mathematical system which underlies all the familiar manipulations of elementary algebra and provides a foundation for deriving further principles. Consider the theorem that the sum of two even numbers is also even. To begin the proof, evenness is defined as follows: A number n is even if and only if there is another number p such that n = 2p. The problem is: Given that a and b are even numbers, prove that a + b is even. By the definition of evenness we can say that a = 2x and b = 2y. By the axiom about adding equals, a + b = 2x + 2y. The distributive axiom says that 2x + 2y = 2 (x + y). Since x and y are numbers, x + y is a number. Hence 2 (x + y) is an even number and so a + b is even.

It may strike a layman that high-school freshmen will be neither attracted nor edified by such a laborious proof of a seemingly obvious idea. There are mathematicians who hold the same view. But the proponents argue that in a well-taught course exercises of this kind may become an intriguing journey into the realm of mathematical rigor.

The Theory of Sets

A central feature that distinguishes the "new" algebra is its use of the theory of sets, one of the most powerful tools of modern mathematics. A set is simply a group or collection. The books on a shelf, the people in a room, the letters of the alphabet, the numbers 1, 2, 3, 4 and 5—each of these groups is an example of a set. The set may have only one member or none at all (in which case it is the "empty" set). Or it may be infinite: *e.g.*, all the points inside a circle, or all the positive integers. Mathematicians commonly denote a set by listing its members within braces, *e.g.*, {1, 2, 3, 4, 5}. There is also a conventional shorthand for statements about sets and their members. For example, the statement that 4 is a member of the set A is written 4 $\in$ A; that 6 is not a member of the set is written 6 $\notin$ A. To say that A is a subset of B they write A $\subseteq$ B. The empty set is " ". The letter U denotes a "universal" set—embracing all the members pertinent to a particular discussion; for example, the universal set for plane geometry consists of all the points in a plane.

This unfamiliar language may seem an undue burden to place on beginning students in mathematics, but actually with a little practice it soon becomes easy to read—considerably easier than mastering stenographer's shorthand.

To illustrate the operations on sets I shall mention only three. The union of two sets A and B (written A $\cup$ B) is the set consisting of all the members that are in A, in B, or in both. Thus the union of $\{1, 2, 3, 4\}$ and $\{2, 3, 4, 5\}$ is $\{1, 2, 3, 4, 5\}$. The intersection of two sets (A $\cap$ B) is the set of all the elements that are common to the two sets. Thus the intersection of $\{1, 2, 3, 4\}$ and $\{2, 3, 4, 5\}$ is $\{2, 3, 4\}$. Finally, the complement of a subset A (written $\bar{A}$) is all the members of the universal set that are not in the subset; *e.g.*, if the universal set is $\{1, 2, 3, 4, 5\}$ and a subset is $\{2, 3\}$, its complement is $\{1, 4, 5\}$. These ideas can be presented in a graphic way by simple figures known as Venn diagrams.

Set theory helps to lay bare the unity of mathematics. Algebra and geometry are both concerned with sets—algebra with sets of numbers, geometry with sets of points. The specific operations in both subjects can be considered examples of the general set operation of union, intersection, and so on.

Let us see how the concept of sets can help to clarify the notions of a variable and of an equation in algebra. The College Board group recommends that students be taught first of all that algebra deals with sets of numbers and relations between them. They would then learn that a variable is simply a general name for members of a set. As such it can be represented by a letter. For example, if the set is the series of all the whole numbers, then x can be any whole number, and x + 1 can be 1 + 1 or 2 + 1 or 3 + 1, etc. Further, an equation is a statement about a relationship between members of a set. It may be true or false: 3 + 2 = 5 is true, 3 + 2 = 6 is false, but both are statements. If a statement contains letters, it is noncommittal: x + 2 = 5 is neither true nor false until the place held by x is filled with some member of the appropriate set. Early in the course the students would also learn to work with inequalities, using the symbols > (greater than) and < (less than).

Now any relation can be regarded as a specification for selecting a certain subset from the universal set of numbers under consideration; in set terminology it can be called a set "builder." Thus if the universal set is all the real numbers, the relation x + 2 = 5 selects the set $\{3\}$, or x + 2 > 5 selects the set of all numbers greater than 3. The set selected by a relation is known as its solution set.

The same concept applies to pairs of numbers and can deal with an equation with two unknowns, conventionally represented by x and y. Such an equation of course has more than one solution. For example, the set selected by the equation 5x + 3y = 15 includes such pairs as (0, 5), (3, 0), (1, 3⅓). Now anyone familiar with coordinate systems recognizes this at once as the wedding of algebra with geometry: a pair of number variables (x, y) can define either a point or a line. Thinking in terms of sets, a student can see that the solution of two simultaneous equations of algebra is a matter of finding the intersection of the two solution sets: the solution is the pair of numbers that is common to both sets.

The notion of sets is particularly helpful in dealing with inequalities. The meaning of an expression such as $5x + 3y > 15$ becomes clearer when it is considered as the selector of those pairs of numbers (or points) that lie above the line $5x + 3y = 15$ on a graph. The expression $x^2 + y^2 < 16$ selects the points inside the circle $x^2 + y^2 = 16$. To "solve" this pair of inequalities simultaneously is again a question of finding the intersection of their solution sets.

All these are merely examples to illustrate the College Board group's approach in its proposed revision of the algebra course. Except for the emphasis on inequalities, it does not appreciably change the subject matter of the course, but it offers that subject matter in a new context.

The New Geometry

In geometry the group proposes to change the course radically. In the first place, it wants to eliminate most of the propositions and theorems that students are now required to prove, on the grounds that the students will already have had some training in deductive reasoning in the algebra course and that many Euclidean proofs can be demonstrated more easily by other methods.

The College Board Commission would cut down the theorems to be proved to about 12, in place of the 100-odd in today's geometry books. These 12 would provide a sample of how the facts of geometry can be deduced from a set of axioms. From them the students would develop theorems and facts about triangles, parallel lines, similar triangles and finally the Pythagorean theorem (a triangle is a right triangle if and only if the square of the hypotenuse is equal to the sum of the squares of the two sides). The reason for stopping with the Pythagorean theorem is that it makes possible a shift to analytic geometry—that is, algebraic solution of geometric problems with the help of graphs. The analytic methods show the use of solution sets in geometry.

The Commission does not propose to abandon classical Euclidean proofs entirely. Some problems are in fact easier to solve by the Euclidean procedure than by the analytic method. Students would be encouraged to find and use the most effective approach to each problem.

The recommended short cuts would reduce the time needed to cover plane geometry to less than a year, and the Commission proposes to use the time saved to include some solid geometry. Students would be encouraged to think in three dimensions as well as in two, and they would study the important theorems of solid geometry, mostly from an intuitive point of view.

In the third year, after elementary algebra and geometry, the mathematics course would go on to further work in algebra and some trigonometry, with emphasis on the mathematical behavior of the trigonometric functions (sine, cosine, etc.) and their application to such problems as the study of vectors rather than the solution of triangles. For the fourth year the Commission would offer students who continued in mathematics two half-year courses. The first, called elementary analysis, would deal with the idea of relations and functions from a more advanced point of view. It would investigate the properties of polynomials (algebraic expressions containing many terms) and of logarithmic, exponential and trigonometric functions. The idea of limits would be introduced informally, and students would learn a little calculus, *i.e.,* how to differentiate and integrate polynomials. In the second half of the year there would be a course in probability and statistical inference. Here a student would learn to deal with sets of scattered measurements or observations by means of statistics. He would learn how mathematics is applied to processes governed by chance. The course would demonstrate some of the methods of computing the reliability of a sampling program and of measuring the statistical significance of results. The Commission points out that this material is probably more closely related to the daily lives of people than any other part of mathematics. Also, statistics and probability are becoming increasingly important in science and industry.

Because most of the topics in the probability course have never been taught in high school, the Commission has prepared a new textbook for this course; it is already being used in a few schools. The book presents the subject first intuitively and then from a more formal mathematical standpoint. Set theory is used extensively in the course. I found the text readable and interesting and the material no more difficult than many of the traditional topics in advanced high-school mathematics.

The Commission also would like high schools to offer, for students who take part in the College Board's Advanced Placement Program, a course in calculus and analytic geometry—the usual college freshman course.

Such, in brief, is the College Board Commission's program for reforming the teaching of mathematics in the nation's high schools. It is urging schools and teachers to try all or any part of its ideas, in the hope that the ideas can be tested to see which will work and which will not. It is confident that once students have been introduced to the beauty of modern mathematics, the subject will acquire a new vitality in the schools.

Although the Commission's program is merely a proposal, obviously the College Board is in a position to exercise a powerful effect on the high schools, through its examinations and its close relationship with leading colleges. The Commission has already begun to reach teachers throughout the country by means of pamphlets describing phases of its work, and it will publish its comprehensive report this fall (1958). Very probably its suggestions will generate the writing of radically new textbooks within the next couple of years.

The Illinois Program

Now let us turn to another approach—the program stemming from the University of Illinois, which has had some public attention in newspapers. The leading spirits of this movement are Max Beberman, a teacher in the University High School at Illinois, and Herbert E. Vaughan, a mathematician on the University faculty. They have worked out a program which goes so far in the direction of modernism that it makes the College Board Commission's program look almost antique.

The Illinois experiment began at the University High School, an adjunct of the University's College of Education, in 1952. It has grown into a four-year program which is being tried this year in a dozen high schools in Illinois, Missouri and Massachusetts, and is also being taught to interested employees of the Polaroid Corporation in Cambridge, Mass. With financing from the Carnegie Corporation, the Illinois group has produced a complete series of textbooks and teaching manuals and brings teachers to the University for up to a year of observation and indoctrination in its courses.

The approach to mathematics via abstract generalizations is the very cornerstone of the Illinois program—not merely an exercise in reasoning or a sampling of some of the foundations of mathematics. Pupils in the ninth grade begin their study of algebra with a set of axioms from which they proceed to prove all the rules they must master. The axioms, called "principles of arithmetic," include such things as the definition of zero (any number times zero equals zero), the definition of the number 1 (any number times 1 equals itself) and of course the commutative, associative and distributive laws. The children are led into their rigorous program by easy stages, sometimes in story form. For example, they learn that a number must be distinguished from the symbol used to represent it by means of a correspondence between Ed Brown, a student at "Zabranchburg High," and Paul Moore, a pen pal in Alaska. The two chums have fallen into the unlikely habit of writing each other about arithmetic. Young Paul advances the reasonable proposition that if you take 2 away from 21, you should be left with 1. Ed is startled: is this a joke, or does Paul have something? The text straightens things out by explaining that "21" is a name for the number, not to be taken too literally. It goes on to show that the number 21 can be described by other names: *e.g.,* "20 + 1" or "7 × 3." The pupils proceed to learn that a letter can stand for a number and that it can be given various meanings (*i.e.,* it is a "variable" or "placeholder" for a number). Indeed, the teacher uses blank boxes for the unknown numbers in equations. A substantial portion of the first-year course is devoted to driving this point home. The text observes that a letter plays the same role in a mathematical statement that a pronoun does in ordinary language. The statement "x + 2 = 6" is like the sentence "He was a president of the U. S." Neither is true nor false until a name is put into the proper place. The teacher calls the letters of algebra "pronumerals," and uses this term throughout the course.

A good part of the first year is also given to introducing and exploring the idea of sets. Because of the long pronumeral introduction and the careful development of the set concept, the course covers less ground than the traditional course in elementary algebra.

In the second year the program goes modern for fair. This course is nothing less than "a development of Euclidean geometry which is as rigorous as, for example, that due to Hilbert, and yet which is, we believe, accessible to students who have mastered the first course." It attempts to make clear the nature of geometry as a pure deductive theory which, in itself, has only logical structure and is empty of content until specific interpretations are introduced. Consider, for instance, three postulates which are necessary for deducing the rules of Euclidean geometry. These are: (1) every line is a set of points and contains at least two points; (2) there are three points which do not belong to the same line; (3) every two points anywhere are contained in a line. The teacher brings out that the words "point" and "line" can be interpreted in various ways. The main model used in the course is the "number plane," where points and lines are defined in terms of pairs of numbers (x, y); the course develops all the necessary postulates of plane geometry from this model. But to emphasize that the deductive system can have various concrete interpretations, the teacher examines several models, including one in which "point" stands for businessman and "line" for a partnership. The postulates apply here just as they do in geometry. Suppose there are three businessmen, A, B and $C,$ each of whom forms a two-man partnership with each of the others. Each partnership $(AB, AC$ and $BC)$ is a set of businessmen and contains at least two members (postulate 1). None of the partnerships contains all three businessmen (postulate 2). Every two businessmen are contained in a partnership (postulate 3).

I can make a little clearer how the teacher and children deal with these matters by quoting a recording of a review discussion of postulate systems between Beberman and a bright class:

Teacher: "Where do these postulates come from?"

George: "From the number plane."

Teacher: "What do you mean by that?"

George: "They are properties of the number plane."

Teacher: "If we're talking about the number plane when we say, for example, that each line is a set of points and contains at least two points, is that statement true or false?"

George: "True."

Teacher: "But suppose we are *not* talking about the number plane. Then what about these postulates?"

George: "False."

Chorus: "No!!"

Jim: "You have to give a specific meaning to 'line' and 'point' before you can tell whether they are true or false."

Teacher: "What is a 'model' of a postulate system?"

Jane: "Something that has the properties expressed by the postulates or that satisfies the postulates."

Teacher: "Suppose there are several interpretations for the postulates. You can be talking about the number plane, or about businessmen and corporations, or about class presidents and committees. So what?"

George: "Well, when you try to deduce a theorem from the postulates, you can use a model to find out in some cases that you can't deduce it because it's false for that model."

In the hands of a teacher like Beberman the discussion in a better-than-average class is alert and spirited. A leaflet put out by the Illinois group asserts: "High-school students have a profound interest in *ideas*. They enjoy working with abstractions. . . . Despite the current fashion to point out the usefulness of mathematics in various occupations, most high-school students are not genuinely stirred by such a 'sales campaign.' The goal of vocational utility is too remote to make much difference to a ninth grader. He wants to know how mathematics fits into his own world. And, happily, that world is full of fancy and abstractions. Thus students become interested in mathematics because it gives them quick access to a kind of adventure, which is enticing and satisfying."

The course is frankly experimental. Some of the material has proved too time-consuming or too hard to teach. At the moment the Illinois group is considering whether it should postpone some of the rigorous development to the fourth year. But Beberman starts with the attitude that you don't know what you can teach children until you try.

It is too early to tell how much the pupils eventually get from the course; the first group is just completing the four-year program this term. But I have visited the University High School classroom and can testify that the students seem to carry their burden cheerfully and even enthusiastically. The big question is: How would such a course go generally? Some critics concede that it may work well with a gifted teacher and bright students but strongly doubt that the average teacher or the average class could handle it successfully.

The Critics

There are those who believe that the whole approach of the modernists is fundamentally wrong. One of the most articulate of these critics is Morris Kline, a professor at New York University's Institute of Mathematical Sciences and a popular writer on modern developments in mathematics. He doubts that the abstract approach will get youngsters interested in mathematics.

Kline argues that you do not abstract until you know what you are abstracting from. Professional mathematicians did not arrive at an understanding of the abstract general features of mathematics until after they had explored many of its specific branches. Why should a schoolboy be expected to? Moreover, Kline believes that set theory has no place in an elementary curriculum: its applications there can only be trivial, and even its importance in advanced mathematics is overemphasized.

Kline and others hold that the way to stimulate more interest in mathematics is to make the teaching more concrete rather than more abstract—to relate it more vividly to problems of the real world. Kline would introduce mathematical ideas by means of simple physical experiments and examples drawn from fields such as music and other arts. An active advocate of this sort of approach is the English mathematician and teacher W. W. Sawyer, now at the University of Ilinois. [See previous chapter on page 125—Ed.] Sawyer has built a number of simple devices which he uses to illustrate mathematical ideas.

But the anti-modernists seem to be in a losing cause. They are not organized, nor have they formulated a specific program. Meanwhile the modernist movement is spreading rapidly. At the college level, a number of colleges have adopted a new freshman course designed by a committee of the Mathematical Association of America. The course consists of a half year of calculus and a half year on modern topics, collectively called "discrete" mathematics. It deals with noncontinuous sets (*i.e.*, of discrete numbers or objects) . This branch of set theory is finding increasing application in science, particularly social science, which deals typically with groups of persons, units of product, etc. Even physics is discrete at the atomic level, and matrix algebra, one of the topics in discrete mathematics, lies at the foundation of quantum theory. The new college course is intended to give liberal arts students a taste of some modern mathematical ideas and also to acquaint social science majors with mathematical techniques. The text now in most common use includes symbolic logic, probability, the algebra of vectors and matrices, the theory of games and linear programming.

The Teachers

Mathematics teaching in the nation's high schools and elementary schools obviously will not be changed overnight. Most teachers are not prepared to teach either the modern material or any other new scheme. However, the current great expansion of teachers' summer institutes and in-service courses, financed by the National Science Foundation, private corporations and others, is a golden opportunity. This summer there will be 10 university institutes in mathematics, all teaching mainly the modern topics. Some 40 other universities are offering courses in both mathematics and science, and most of them are stressing the modern approach. It appears likely that most U. S. high schools will have shifted more or less toward the modern approach within five years or so.

Finally I must mention a new project which in the long run may have a stronger impact on mathematics teaching in the U. S. than any of the programs discussed in this article. It is the School Mathematics Study Committee, set up by the mathematics department of Yale University and patterned after the Physical Science Study Committee at the Massachusetts Institute of Technology [see the following chapter, "The Teaching of Elementary Physics," by Walter C. Michels]. The new group will enlist a large number of mathematicians and teachers to consider mathematics teaching in high schools and junior high schools, and very likely it will prepare an elaborate program of textbooks and other teaching aids, as the physical science group is doing. This summer the Committee will hold a conference to review the present experiments and the whole problem. What it will come up with is anyone's guess, but in manpower and financial support it may well carry more weight than any other group seeking to revitalize the teaching of mathematics.

E. P. ROSENBAUM, a member of the Board of Editors of *Scientific American,* was born in New Haven, Connecticut (1916), and attended both Harvard and Yale as an undergraduate and graduated from Yale in 1937. He then taught mathematics and physics at the Milford School, a preparatory school in Connecticut. During World War II, as a Captain in the Air Force, he served at Wright Field as project engineer on the first Air Force radar-directed guided missile. After the war he returned to teaching and administrative work at the Milford School, but then turned to writing, principally for radio and television; he joined the staff of *Scientific American* in 1952.

PHYSICS*

WALTER C. MICHELS
Bryn Mawr College

In the public discussion that has followed the Russians' launching of their satellites, the word "physics" has probably occurred almost as often as "satellite," "rocket" or "missile." Many speakers and writers have implied that the technological progress of the U.S.S.R. can be attributed entirely to the fact that even high-school graduates there have had some 12 years of study in mathematics and six in physics. It seems to follow that all the U. S. needs to do to regain undisputed scientific and technological leadership in the world is to introduce more physics courses and mathematical instruction into its schools. Like all panaceas, this cure for our ills has the attraction of simplicity. It has the further attraction of potent medicine, because physics and mathematics are considered "hard" subjects and we still retain some of the atavistic belief that unpleasant medicines are especially effective—particularly when our children, rather than we, take them.

But before we rush into a program designed only to persuade more students to take more physics, it may pay us to examine physics instruction in the context of American education as a whole, to ask what we want to accomplish and to estimate whether improvement in quantity or in quality is the more important.

At present, between one fifth and one fourth of the students graduating from high schools in the U. S. have taken a course in physics. This is almost the same as the proportion of high-school graduates who go on to college. The agreement is more than coincidental—most of the leading colleges either require or strongly encourage some preparation in physics. The engineering schools, which are the first choices of about one quarter of all the boys in the college preparatory group, almost invariably require physics for admission. Thus a large percentage of the students who take physics in high school are a "captive audience," in the sense that they are required to pass the course to get into college.

Much of the argument for bringing more students into secondary-

* Published originally as "The Teaching of Elementary Physics," *Scientific American*, CLXXXCVIII, 4 (April, 1958), 58-64.

school physics courses seems to be based on the idea that this will ultimately produce more scientists and engineers. It is difficult to predict the extent to which this recruiting effort, even coupled with offers of college scolarships, will succeed. Motivational factors and temporary fluctuations in the demand for professional workers may well be more powerful than educational offerings in shaping the careers of young men and women. Some increase in the number of available engineers and scientists will undoubtedly result from present pressures and from Federal support of education; yet we may be disappointed in the results if we set this increase as our only, or even our primary, goal.

Many scientists and some of our best educators believe that there is a more important reason to increase emphasis on physics in the schools. They believe that our educational system is failing to give most of our population an adequate understanding of the nature of physical science and of its role in the economic, political and cultural life of the 20th century. George Stewart, in his book *Man—An Autobiography,* points out that mankind's way of life has changed more since 1700 than it had from prehistoric times to that date. Our economy of abundance, our increased leisure with its cultural benefits, our rapid communication systems with their effect on international relations, our increasing ability to employ nature for our own ends—all of these are direct consequences of the ideas and the methods introduced by Galileo, Newton and the physical scientists who have stood on the shoulders of these giants. An understanding of science is as important to the businessman, to the lawyer, to the statesman or to any other citizen as an understanding of language or literature or of political and military history. Physics is the most basic of the natural sciences. For this reason, the study of physics is essential to the education of all of the population. The future rests not so much on the number of engineers and scientists produced each year as it does on the intellectual and cultural climate in which they work, and this, in turn, will be determined by our success or failure in making physical science a significant part of general education.

Today's Physics Courses

Physics, like any other subject, can contribute whatever the student is willing and able to take—anything from true understanding down to mastery of a few facts to be regurgitated on an examination. The quality of the course itself will depend on the textbooks, on the laboratory aids available and, most of all, on the ability of the teacher. A truly able teacher possesses both a knowledge that goes well beyond the level at which he is teaching and an understanding of the people he is teaching. From the first of these he draws illustrative materials and constantly changing the methods of presentation, and so makes his subject interesting and exciting enough to fire the imagination of his students. The second quality allows him to tailor his approach to the mental abilities of the students and to choose the phrase or the illustration that will make his meaning clear or will untangle a difficult passage in the textbook. It is because both of these abilities are needed that we are faced with so serious a shortage of science teachers. We cannot solve the problem by bringing into the classroom scientists with an advanced knowledge of their subject but with no feeling for teaching, or by assigning physics to good teachers who know little physics.

In the best of our schools—those that attract good teachers and give them the equipment and time required for a good job—the physics courses contribute to the student's education many things that he cannot obtain from other studies. Here he learns the basic principles that underlie events from the throwing of a baseball to the launching of a satellite; he learns to observe phenomena with care, to extract from them the parts that are most important and to use his observations to test a theory. Because the phenomena dealt with are simple and subject to measurement, he obtains practice in the analysis of their causes—practice which will stand him in good stead when he later faces the problem of analyzing more complicated happenings without allowing his prejudices to influence his interpretation of them. His interest in and understanding of algebra and geometry are likely to be advanced as he sees that these subjects supply valuable tools for the description both of nature and of the products of modern technology. He learns to read carefully, with the idea that every word in a well-written piece of exposition is there for a purpose; at the same time, he learns to communicate ideas in a clear, succinct and unambiguous manner, using graphic methods and mathematical descriptions as well as words. Whether the student is destined to become an engineer, a businessman, a lawyer or a housewife, he or she can be expected to retain these advantages of the physics course long after the detailed facts and the techniques of physics are forgotten.

Few physics courses accomplish all of these things, and many fall far short of the goals that we have set. One reason for their shortcomings is that tradition, economic pressure and mistaken ideas about the nature of science have given us textbooks which sometimes hinder rather than help the teacher. The content of the courses is heavily influenced by the fact that a large proportion of the students taking them are preparing for engineering. These students are interested in the technical applica-
(3) every two points anywhere are contained in a line. The teacher brings
tions of physics. The authors of textbooks and the teachers are tempted to catch their interest by loading the course with technology. Some of the textbooks abound in pictures and diagrams of Diesel shovels, locomotives, gasoline engines and automobile transmissions. No one can object to the use of practical machines to illustrate the principles of physics, but we must protest when the text is so overloaded with these illustrations that the principles themselves have to be treated in a summary fashion, if not omitted entirely. Further, many modern machines are too complex to be described adequately or discussed correctly on the basis of the

physical knowledge available to the students. Under these circumstances, too many physics courses demand only rote memorizations of isolated facts and of formulas into which numbers can be substituted to achieve answers which often mean little to the student who has obtained them.

A second fault of many textbooks is the breakdown of physics into topics or "units"—a misuse of the basically sound pedagogical idea that students should not be confused with too many facts or concepts at any one time. The difficulty is that the texts rob the units of meaning. Sound, heat, light, electricity, magnetism, mechanics and atomic physics are discussed separately as if they had little or no connection with one another. This method of presentation belies the very nature of the subject, for the chief thing that distinguishes the physical sciences from other intellectual activities is the existence of a central theory, based on a very few simple laws and principles, which applies to all phenomena. All of the particular subjects listed above can be closely related on the basis of fundamental principles such as the conservation of energy, of matter and of momentum.

Principles v. Applications

As an example of the failure of the piecemeal approach let us take one of the bulwarks of physics courses—Ohm's law. In the first half of the 19th century Georg Simon Ohm of Germany showed that the voltage required to produce a steady current in any part of a given electric circuit was proportional to the current. That is to say, if the current maintained in a wire is to be doubled, the battery must supply twice as strong an "electrical push." Ohm thus arrived at the concept of electrical resistance and at the idea that the resistance of a conductor depended only on the material and dimensions of the wire, not on the amount of current flowing in it. We now know that Ohm's law holds only for special classes of electrical conductors (e.g. metals). In fact, it is the failure of Ohm's law to apply to the flow of electricity in vacuum tubes and in the materials known as semiconductors that has made all of modern electronics possible. Yet the typical introductory physics course leaves a student with the firm conviction that Ohm's law is one of the most basic principles of physics, or at least that he will not pass his final examination if he cannot substitute numbers in the formula: $E = IR$. If he remembers that E is a voltage, I is a current and R is a resistance, he will be able to arrive at the correct answers to conventional examination questions. But if he tries to apply Ohm's law to any circuits other than the simple ones in which he has been drilled, he will probably go far astray, for he has learned only the letter of the law, not its meaning.

Only a few years before Ohm's law was discovered, the great principle of the conservation of energy (the first law of thermodynamics) had been established. James Prescott Joule of England went on to show that heat, such as that present in a kettle of hot water, and mechanical energy, such as that carried by a rolling stone, were one and the same thing. Further, he found that electrical energy too was connected with these forms of energy. He arrived at the conclusion that a current in a wire produced heat at a rate which was proportional to the square of the current. In the same symbols that we have used to state Ohm's law, we can write Joule's law as $H = I^2R$, where H stands for the rate at which heat is produced.

Joule's law is mentioned in many introductory texts but is seldom given as much prominence as Ohm's law. Moreover, it is almost always presented as an entirely independent principle, connected with Ohm's law only through the fact that both involve current and resistance. Actually Joule's law and Ohm's law both express exactly the same thing. Either can be derived from the other and from the principle of conservation of energy. The electromotive "force" involved in Ohm's law is the energy necessary to move a unit of electrical charge from one end of a conductor to the other. Combining this concept with the fact that the current is merely the rate at which electrical charge is carried through the conductor, we can state Ohm's law as follows: "The energy required to move a unit of electrical charge along the length of any conductor is proportional to the rate at which charge is being moved." For conductors that obey Ohm's law, Joule's law also must hold, because the rate at which energy must be supplied is just the energy per unit charge times the rate at which charge is moved, that is, it is equal to IR times I, or I^2R. According to the principle of conservation of energy, the energy required to move the current against the resistance of the conductor must either be converted to heat or be stored up in some form. When the current in a conductor is steady, there is generally no way that energy can be stored, so the heat given off must be equal to I^2R, in accordance with Joule's law.

It is perfectly true that a discussion of the relation between Ohm's law and Joule's law requires close attention on the part of the student. When he has grasped the reasoning, however, he will have recognized why Ohm's law holds only for steady currents. He will also be able to apply that law to circuits he has never seen before. Most important of all, he will realize that the conservation of energy is a fundamental principle which applies alike to mechanics, to heat and to electricity.

To illustrate further the difference between teaching the principles of physics and merely discussing its application, as courses commonly do, let us take an example in optics. The problem is to explain a virtual image—for instance, the image of an object that you see when you look at it through a magnifying glass. The distinction between a virtual image and a real image (such as is formed by a camera lens) is that the rays of light do not actually pass through the virtual image; rather, the image is a displaced projection formed by extension of straight lines from the eye through the points in the lens where the light from the object itself is bent by refraction. Now many high-school texts deal with this subject mainly in terms of a "lens equation" which provides a formula for determining the distances of the object and the image and the focal length of the lens. The equation says that the reciprocal of the object's distance from the lens plus the reciprocal of the image's distance is equal to the reciprocal of the focal length, thus: $1/u + 1/v = 1/f$. The student must learn this formula and also a set of rules which say that the distance of the object behind the lens is reckoned as *positive*, but that the distance of the image, if it is behind the lens, is expressed by a *negative* number. He may be given a problem asking what the distance of the image from the lens will be when he looks at an object through a magnifying lens with a focal length of one inch if the lens is held three fourths of an inch from the object. If he remembers the formula and the rules correctly, with a little algebra he can arrive at the answer that v is equal to minus 3—the image is three inches behind the lens. By applying another formula he can learn that the image is four times larger than the object being examined.

The formulas are approximately correct for simple lenses, and they are indeed useful for designers of optical instruments. But they are unlikely to have much meaning for a student or to convey anything about the process by which the image is formed. It is very doubtful that his learning to use the formulas contributes appreciably either to his intellectual development or to his understanding of light or of lenses.

Suppose now that the same subject is taught in a course that emphasizes physical principles rather than their applications. The student may first be asked how we know where an object is. After learning that from each point on the surface of an object rays of light diverge in a cone to the eye (a fact discovered by Franciscus Maurolycus of Messina in the 16th century but seldom emphasized in physics texts), and after studying the law of refraction discovered by Willebrord Snell in 1621, the student may apply that law to a simple problem such as why a coin lying on the bottom of a glass seems to rise when the glass is filled with water. By a simple drawing of the refraction of the light where it passes from the water to the air he can predict the displacement of the virtual image. Having done this, he can extend the same process to any lens and locate an image by a scale drawing.

This procedure will probably take longer than memorizing and applying the lens equation, but it should give a student an insight into the behavior of light as it passes through a lens, reinforce his knowledge of the law of refraction and equip him with a method that is applicable to all lenses, not just to the thin spherical lenses to which the simple lens equation applies.

New Approaches

These examples are sufficient to illustrate how greatly physics teaching could benefit from a new approach. Even a good teacher is somewhat at the mercy of the current textbooks and of tradition. A poorly prepared one who has been pressed into physics from some other subject has few weapons with which to fight the trend. He cannot be expected, without more time than is allowed by a busy teaching schedule and an overabundance of record-keeping, to learn enough about physics to challenge the text or to present the subject in an enjoyable and exciting way. If physics in the schools is to be improved to the point that the times demand, we must either find a way of attracting more good teachers or we must see that the poorly prepared teacher is given help.

To draw a wider group of students in the secondary schools into physics we must reduce the number of engineering applications taught in the course and devote the time thus gained to the ideas, methods and history of physics. In the past, attempts have been made to interest more students by giving "descriptive" courses, which present physical phenomena but avoid theory and the use of any mathematics beyond arithmetic or a very little algebra. Some of these courses, particularly when taught with a sense of showmanship, have certainly been popular. No doubt a descriptive course is better than no course at all. But can it give a student the understanding of physical nature and of science that the times require? The distinguishing characteristic of physics is that it involves a central theory, of very broad validity, in terms of which observed phenomena are interpreted and understood. To omit large parts of that theory in order to avoid mathematical difficulties is to deny the able student the intellectual stimulation that is one of the prime rewards of education.

Perhaps good descriptive courses should be developed for the benefit of the lower half of our expanding high-school population, but it is my opinion that we cannot afford such courses for the able students who promise to become the intellectual, political and cultural leaders of the future. For this group there seem to be two possible answers. The first, and best, is to restore solid mathematics courses to the required curriculum for all students in college preparatory courses. The time when we can afford to suppose that intelligent students cannot or will not take a minimum of three years of mathematics in high school is past. The second solution is to include in the physics course whatever mathematics is required for the work, over and above the minimum requirements for graduation from the high school. This would mean that some physics now included in the course would have to be eliminated to make room for the needed mathematics, but we are interested mainly in quality, rather than quantity, in the content of the course. The cost in reduced coverage need not be great if the best secondary-school teachers and the best university physicists can be persuaded to join hands in the discovery of new ways of presenting physics with a high degree of rigor, yet with a minimum of formal mathematics. The history of physics and mathematics supports the amalgamation of the two subjects, for their developments have been so entwined that they are parts of the same intellectual activity. After all, Newton did invent the calculus in order to solve a physical problem.

Training Teachers

The new approaches I have been discussing may sound Utopian, but actually some very promising improvements in the teaching of physics are already under way. Among other things, active steps have been taken in recent years to improve the preparation of teachers. During the past decade a number of colleges and universities have started summer institutes for science teachers. They bring together for six to eight weeks groups of as many as 60 or 70 teachers. The programs include refresher courses, lectures and discussions of current research by active physicists, and laboratory work under better conditions than exist in most schools. The first institutes were financed by a few far-seeing industrial concerns and private foundations; the program is now being supported on a much larger scale by the National Science Foundation, supplying stipends to about 5,000 science teachers each summer, including several hundred in physics. A smaller program, providing teachers with fellowships for a full year of study, has been in operation for two or three years. There is growing support for in-service institutes on campuses near high schools, where teachers study for a few hours each week during the school year. All these training programs are already producing results, and they can be expected, within another five years, to increase greatly the number of well-prepared physics teachers.

To bring physics to many more students in our secondary schools may, however, require more teachers than our educational system and economy can supply. One possible solution would be to enable the good teachers to reach more students by the use of films and television. The Fund for the Advancement of Education made possible a large experiment in this direction. It furnished funds to the schools of the Pittsburgh area for a full-year program of television lectures and demonstrations. The school authorities brought in Harvey White, who had taught physics very successfully to nonscience students at the University of California for many years. He had had experience with television. During the school year of 1956-57 he delivered 162 lectures of 30 minutes each which were carried simultaneously by television to students in 44 schools. At the same time they were filmed. The films are now available through Encyclopaedia Britannica Films and are being used in a number of schools throughout the country. It is perhaps unfortunate that the pace at which this experiment was organized and run did not allow many of the faults of current physics courses to be corrected, but the films should nevertheless be of tremendous help to school systems that cannot find sufficient numbers of teachers who have been adequately trained in physics.

A New Course

A much larger endeavor is being carried out by the Physical Science Study Committee, an organization created by the daring imagination and inspired leadership of Jerrold Zacharias of the Massachusetts Institute of Technology. Supported by the National Science Foundation and a number of private foundations, this committee has brought into its service more than 100 highly qualified individuals, including high-school teachers, research physicists from universities, professional writers, film directors and selected college students. In the past year and a half the committee has considered in detail what might be omitted from the present crowded syllabus to make room for better and more fundamental teaching; it is preparing a radically new textbook; it has made a start on the preparation of new laboratory manuals, teachers' guides and reference materials; it has planned a series of about 70 films; it has designed simple laboratory apparatus, much of which can be built by the students themselves. The course is being developed further by use in a few selected schools, where it has been received with enthusiasm by teachers and students alike. Sufficient materials should be available to allow introduction of the course in many schools by the fall of this year (1958). Five institutes will be conducted this summer to introduce the course to 200 or more teachers.

The work of the Physical Science Study Committee probably represents the most ambitious attack ever made on the problem of presenting any subject to high-school students. It has made use of earlier studies carried out by committees of the American Association of Physics Teachers, the National Science Teachers Association, the American Institute of Physics and others. It has brought together individuals with different backgrounds, including the Nobel laureates Isador I. Rabi and Edward M. Purcell, the Presidential advisers James R. Killian and Vannevar Bush, the writers Mitchell Wilson and Stephen White, the movie producers Frank Capra and Warren Everote, and outstanding teachers such as Judson Cross of Exeter and John Marean of the Reno High School. All those engaged in the enterprise are hopeful that it will produce large results, not only in the teaching of physics but also in elevation of the intellectual level of our schools.

Postscript

Between the time that this article was originally printed and the preparation of the present publication, significant progress has been made on several fronts. The National Science Foundation has expanded its programs for both summer and year-long institutes by almost three-fold. Some 15,000 high school teachers will be attending institutes in the summer of 1959. Dr. Harvey White's experience with television has been applied to the production of Continental Classroom, a physics course being offered over NBC network. Dr. White is being joined by a number of distinguished guest lecturers and it is estimated that more than 200,000 viewers see the program each morning. The Physical Science Study Committee course is being taught to more than 12,000 students in 186 schools during the academic year 1958-59.

WALTER C. MICHELS, a graduate of the Rensselaer Polytechnic Institute and the California Institute of Technology, has been at Bryn Mawr College since 1932, and has been Chairman of the Department of Physics since 1936. Although Bryn Mawr has only 650 undergraduate students, it conducts a full program of graduate work with 200 graduate students, both men and women; the department offers an introductory course in physics that serves at once the future physicist, the major in some other science and the student of liberal arts. One result of Michels' unconventional course was the textbook *Elements of Modern Physics*. Michels served as President of the American Asociation of Physics Teachers in 1956-57.

TEACHERS COLLEGES

HANOR A. WEBB
George Peabody College for Teachers

The university professor was very angry. "My best student, a brilliant young woman, can't get a place to teach! She has just received her Masters degree—her Masters, mind you—in my Botany Department, with a splendid thesis on the histology of the embryo of *Hemerocallis fulva*—day-lilies, as you all know." Several in his audience nodded in appreciation of the compliment.

"But when she applied to our high school for a teaching job, she was turned down! 'No certificate,' they said. When she asked the State Department of Education for a certificate she was turned down again! 'No courses in Education,' they said." His right fist pounded his left palm. "She's my best student, and they won't let her teach!"

In these comments the professor was *so right!*

Neither can this talented young lady (I thought, as I listened) be admitted to the bar, begin practice as a dentist or a physician, register as a nurse for hospital or private service, get a license to preach, join the Musicians' Union, set up an engineering or architectural office, solicit patrons as an accountant, or even open a beauty shop. She can take no employment in occupations in which the public is protected by some form of certificate of training. The lady is a specialist, but not a professional.

The angry professor continued, using the words "dismal teachers colleges," "mandarins of education," "stranglehold," "monopoly," "this education racket," and others as both fists were pounded on the table.

In these comments the professor was *so wrong!*

This tirade, which I heard, has been repeated hundreds—perhaps thousands—of times since standards for teachers were set, and professional schools established to train them. The chief critics, the educational institutions of liberal arts, seem willing to accept the professional schools for law, medicine, engineering, the ministry, and some others; they have disparaged the professional schools for teachers from their beginning. This is one of the facts of life in higher education.

Beginning and Growth

Schools for teachers in North America began in 1839 with a Normal[1] School in Lexington, Massachusetts, now the State Teachers College at Framingham, Massachusetts. The first new building erected for teacher training purposes was at Bridgewater, Massachusetts, in 1846. At the dedication the noted educator Horace Mann[2] spoke these words: "Coiled up in this institution, as in a spring, there is a vigor whose uncoiling may wheel the spheres!"[3]

In the Southern States the "mother of Normal Schools" was Peabody Normal School at Nashville, Tennessee, which in 1875 took over an educational charter first granted in 1785. The many schools for teachers in the region were largely staffed by graduates of that institution, now George Peabody College for Teachers.

From their beginnings the schools for training teachers had one purpose—professionalism. Other institutions had no interest in the teacher. Harvard College, founded in 1636 as the first in its class, enrolled only men, training them in classic subjects for the law and the ministry. Yale College, founded in 1701 and named in 1718, was also a classical school for men. Teaching, however, was a sideline for many lawyers and preachers.

Few women held teaching posts, even in the lower grades, until well past the year 1800. There was a feeling at the time that a woman could not keep order in the schoolroom and remain a lady. A man's training for discipline was that of the barnyard, using methods by which he made calves and colts behave. The results seemed excellent.

A man's intellectual training was chiefly in the classics—a study denied women. He taught the 3-R's chiefly by persistence and sternness. History he had learned in college, and he gave as he had received. He did not trouble himself or his students with the sciences.

Later, as women began to teach in the lower grades, their "maternal instincts" were deemed sufficient guides to their methods. Each lady taught in her own way; there was no uniformity. As a rule the sternest were more successful, and sternness developed with spinsterhood.

Normal schools and teachers colleges were established gradually throughout the nation. Teachers, however, were often appointed for reasons other than their college credits. For decades, county and city superintendents of schools gave their own examinations, when required to do so by legislation. There was naturally great diversity in the questions and the answers. Personal—even financial—influence was frequently a deciding factor in a teacher's employment. In certain districts only a Baptist could teach, a majority of the Board so voting; in neighboring towns, only Methodists could serve. New teachers must purchase an insurance policy from the Chairman of the Board, or take room and meals at the home of his relative. These impositions were not serious financially, as the salaries would not permit too much exploitation. If the kick-back was purely financial, it would not be more than five per cent. The ethics of school employment was too often that of a dirty business rather than a respected profession—for no professionalism had been developed in either employer or employee. Naturally these situations were not universal, but they existed widely.

In all places where education was more than just a racket or a job, a yeast of professionalism stirred. No matter how lax or corrupt a society may be, there are always some whose ideals are high. This ferment had power to rise above the low or mediocre; to reject the cynicism of Shaw[4] and reach for the ideals of Pope.[5]

Three Schools for Teachers

Three types of teacher-training institutions developed over the decades from Framingham, Bridgeport, and Peabody:

(1) Normal Schools with two-year (rarely three-year) programs, granting a license to teach such as the L. I. (Licentiate of Instruction) degree. (2) Teachers Colleges (usually state supported) with four-year programs, giving the Bachelors degree. (3) Teacher-training divisions within an institution of higher learning, called a Department of Education in a college, and a College of Education in a University.

Is it true that survival depends on service? We shall see.

Advantages and disadvantages have been apparent to all who have known these institutions first hand throughout the decades. These will be given concisely for each.

Normal Schools

These favorable aspects may be present:

(a) The atmosphere of learning should be that of the classrooms where children are taught. Discussions of subject-matter topics will always include, "How may these ideas be made clear in the grades (or high school)?" The seeds of professionalism as a teacher may be sown, and may "fall upon good ground."

(b) Training for teaching at definite grade levels may be offered specifically, blending subjects and methods.

(c) Although the two-year program offers only minimum training, the expense to the student and to the state is also a minimum.

(d) The administration and the staff will be dedicated to a single purpose, with those who "fit" doing their best to fulfil a single duty—to train teachers.

These unfavorable aspects may be present:

(a) The two-year program is hardly adequate for training a high school teacher in subjects he may teach.

(b) A preponderance of methods over content may exist, depending on the program of studies, and the personality of instructors. If electives are permitted, students may take the easier courses in methods rather than the more demanding courses with content.

(c) Normal schools are likely to be in small towns, with consequent lack of community resources in cultural matters.

Normal schools as a rule became teachers colleges as soon as practicable, adding two years to their program. In certain states this expansion has been denied for political reasons.

Teachers Colleges

These favorable aspects may be present:

(a) The atmosphere of learning will be that of the classrooms, in both the lower and the upper grades, where young students are to be taught. Discussions will reveal a desire to know what should be taught, and how to teach it best.

(b) The four-year program will be of particular value to those training for high school positions; yet elementary teachers may become more broadly informed.

(c) The larger institution is likely to have more prestige, more experienced and better paid instructors, more competent administrators.

(d) Subject-matter courses have more opportunity to achieve real college standards because of time, facilities, and experience of instructors.

(e) Since all have one goal—training teachers—the instructors in education courses and subject matter courses may have mutual respect, and cooperate effectively.

(d) The administration will be dedicated to the teaching profession as the vocation of the college's graduates.

These unfavorable aspects may be present:

(a) The better-paid positions—presidents and deans—have been inviting to political appointments. In certain states all teachers college presidents have been discharged by each new governor, and new ones installed. Even instructors have had insecure tenure.

(b) Instructors with inadequate college credits have been given appointments through political influence, or have been employed by the president through necessity to fill the position. Often an instructor in a subject could not obtain a certificate to teach that subject in a high school.

(c) Conversely, insructors with many academic credits in a subject but with no teaching experience will be training prospective teachers. This has occurred most frequently when the need to have as many Ph.D.'s as possible on the staff arose to meet the pressure of college accrediting agencies. Has there ever been a clearer example of "the blind leading the blind?"

It must be said, however, that often these academically trained persons are so intelligent and adaptable that, after experience, they become excellent teachers. They forsake their university methods—chiefly lecturing as the sole way to teach—and develop resourcefulness that was never challenged in their graduate courses.

(d) Frustrated pre-medical, pre-law, and pre-engineering students have often sought a place in a teachers college. The instructor who failed promotion to an assistant professorship in the university has also applied. Many of these persons will try desperately to launch university careers in spite of previous failure, and spend the time they should devote to their students' interests in technical researches of their own by which, after publication, they hope to find the university's gate again ajar.

Instructors such as these may not be seriously unfit for a teachers college staff, yet it is difficult not to remember Shaw's biting phrase, "He who can not, teaches."

Colleges of Education in a University

These favorable aspects may be present:

(a) Students may associate with university professors of ability—even eminence—in their respective subjects. They may learn how real scholars speak, think, study, and act.

(b) Association with students preparing for various professions may be broadening. The high calibre of students working earnestly toward the professions of law, medicine, engineering, the ministry, is a challenge to those with other major interests. The presence of students from foreign lands may be stimulating.

(c) University libraries are larger, laboratories better equipped, and there is opportunity to serve in each for experience and small pay.

(d) Opportunities to earn part of one's expenses are likely to be greater in a large institution, and in its city.

(e) The city, and the university itself, will present a variety of cultural opportunities of high artistic and educational value.

(f) In the College (or Department) of Education a more experienced staff of specialists may be present, appointed for professional attainments only.

These unfavorable aspects may be present:

(a) The atmosphere of teaching may be wholly absent in the university's subject-matter courses. To study science—biology, chemistry, physics—with the pre-medics, pre-engineers, pre-research specialists, gives no examples of good teaching topics, materials, and methods for the high school sciences.

In the university the botany professor may not be able to identify common trees; the zoology professor may know few birds by sight; the chemistry professor has little to offer the majors in foods and textiles; the physics professor can not help a young man build a radio set. Outdoor nature is a closed book to many of the university's science specialists. Said Hubbard:[6] "Now owls are not really all-wise—they merely look that way. The owl is a sort of college professor."

(b) Guidance into a career, which the university offers especially in its freshman and sophomore years, is likely to be away from teaching. Subject-matter specialists often recruit the more competent students, with scholarships or flattery, into their own major fields. Young women are led by interesting professor-personalities into such majors as psychology, narrow social specialties, narrow science specialties, none of which are taught at any lower educational level. The blind alleys down which these majors have gone may not be realized until after graduation.

(c) The contemptuous regard whch many university professors hold for a Department of Education is not mitigated by proximity. One of these will criticize the education staff on his own campus as freely as if it were in a distant city. The Angry Professor of our opening paragraph may have been berating members of his own university faculty as "a dismal department."

Many years ago, as a graduate student in the Chemistry Department of the University of Chicago, I remember the springtime queries of my fellow students to each other and to me: "What are you doing after graduation? For the sake of our Department, don't teach! We want to feel proud of *all* our men!"

(d) The staff of the University's College of Education may so resent the attitude of the subject-matter specialists that guidance of students into suitable subject fields as majors or minors is influenced by strong personal feelings. Only after majoring in education may the graduate realize that you don't teach education in high school; that credits in subjects are required for certification. Ironically, there have been instances when the Angry Professor on our first page was the professor of education! *His* bright students could not teach!

(e) While the administration of a Department of Education is usually in the hands of a dean, it is sometimes assigned to one who is only a professor. Either man may be equipped by experience and temperament to battle for the needs of his department, or he may be cowed by the stronger personalities of his fellow deans, the obvious indifference of his president, or set-backs to his recommendations by the university's board or regents. The battle is always fought on each university campus, and Education may not have the heaviest artillery.

(f) Frustrated school superintendents and small-college instructors have sometimes found refuge in a department of education, especially in a State College. These men may be good counselors for young prospective teachers, or they may not.

(g) Every instructor in a department of education has had some subject-matter courses in high school or in college, but it is amazing to learn how many of them have never taught a day at any grade level in public or private schools. Often these persons by reading widely, visiting many classrooms, listening to school talk by the experienced, thereby become good instructors. Earnest persons may successfully teach subjects which they have never taken as formal courses, and may give advice in fields where they have only observed with intelligence.

In summary: It would seem logical that a young person would get the best training for future teaching, and develop the most pride in teaching as a profession, in a state or private teachers college of superior rank. There are a number of these where the subjects are taught by experts, the theory of teaching is presented by persons of experience, the practice teaching is in wholly typical situations—often not found in a Demonstration (or Training) School of a university.

Instructors in these teacher-preparation colleges should have—and usually do have—ambitions for promotion based only on superior teaching by themselves as examples, by participation in educational programs, by writing for educational magazines, by proving their dedication to teaching as a profession, in which they are privileged to serve at a high level. On the other hand, the instructors in the universities know that their promotion to, and upward in, the ranks of professorship depend upon their research publications. They give little thought to students, for "first things must come first." No one will deny that some of the least skillful teaching is in the universities. Yet no criticism is universally true! Certain professors—at the risk of their prestige, perhaps—have had the talent and the desire to teach well, and some of them have written superior text books, thereby becoming disparaged and envied by their colleagues.

The Decline of Teachers Colleges

The peak of teachers colleges in numbers, and probably in influence, came in 1938 when 232 educational institutions were devoted primarily to teacher preparation in the United States.[7] According to Bigelow,[8] in 1938 ninety-two per cent of the institutions that were members of the American Association of Teachers Colleges carried the word "Teachers" in their titles. By 1958 only 124 institutions were devoted primarily to teacher preparation,[9] while in 1956 only thirty-eight per cent of the A.A.C.T. members still retained the word "Teachers" on their letterheads.[10]

This decline in the number of labeled teachers colleges in two decades indicated, according to Bigelow, that "the American teachers college is going to turn out to have been a temporary phenomenon in American higher education. The teachers college as we knew it twenty years ago is on the way to oblivion. . . . This is not a happy prophecy."

What may have been the causes of this striking decline? Certain ones are these, each of considerable influence:

1. *Ambition of the teachers college administration and staffs for the prestige of a university.* This caused changes in titles in California as

early as 1920. Typical is the succession of titles of the West Tennessee State Normal of 1912, which became the West Tennessee State Teachers College several years later by adding two years to the program, then in the '50's, after much legislative battle, becoming Memphis State University by adding some graduate courses in education.

2. *Desire for increased enrollment.* Although from the first many students with no intention of teaching were in the classes, yet the name "Teachers" seemed to turn others away.

3. *Local demands for a liberal arts institution.* The State University was many miles away, was crowded, and had a reputation for campus gaiety. Private colleges had less prestige, and were expensive. A State College near home could serve the intellectual and social needs of young people at low cost.

In each of these three reasons a concept of inferiority implied by the name "Teachers" was evident.

4. *The drumfire of critical addresses and articles*—Angry Professors on the platform and in print—had an effect. Writes Stinnett:[11] "Berating the teachers college has become a litany. It is a remarkable phase of a sustained period of criticism of education and educationists. The lowly teachers college has become a whipping boy for every malcontent who has a pain about education, but can't locate it exactly. [The teachers college] has become a symbol of academic inferiority."

Typical of the "malcontents" is Sperry, who wrote in *Life*[12] a hazy article asking, "Who teaches the teachers?" He took isolated cases, selected for inferiority, to create an unfavorable impression of *all* teacher preparation programs. He mentioned most frequently the state teachers colleges which—as Stinnett points out—now train but five per cent of the nation's new teachers.

Not realizing that he is "beating a straw man" by these references, Sperry admits that teachers colleges took on a task indispensible to an educated citizenry which the prestige institutions of higher learning ignored. But how shallow (he asserts) has teacher training been! He writes more in sorrow than in anger. One wonders whether teachers would prefer to be damned as frauds or pitied as fools!

In his article (cited previously) Bigelow relates the plight of an honors graduate in physics at Yale who was denied a post in a New Haven high school. The Angry Professor was still protesting. One wonders whether the young man subsequently sought a license as an engineer, a lawyer, a physician, or a preacher. If he did, these professions also denied him a certificate.

Teacher Training Must Go On

There will always be teachers. It is an ancient profession as in Greece, an honored profession as in modern Russia, a necessary profession in the United States. No matter where, and under what competitive handicaps, teachers obtain their preparation, certain broad principles of teacher training should apply. Because this volume concerns the sciences, my examples will be the training of science teachers. For all other areas of instruction in subjects taught from grades one through twelve in our schools, the plans differ in details but not in principle.

1. *For the elementary levels.*

(a) Many states demand college credits in science for elementary teaching certificates, also courses in materials and methods of teaching science in elementary grades.

(b) College courses in botany, chemistry, physics, zoology, offered primarily to meet pre-medical and pre-engineering requirements, are not helpful to elementary teachers. These usually count, however, for certification.

(c) College courses in generalized science, offered in many institutions to all freshmen, may be quite helpful to elementary teachers. These courses include topics in astronomy, weather and climate, simple geology, and simple principles of biological and physical science. Of greatest value is a course in nature study, with many outdoor trips and identification of common plants and animals, especially birds.

(d) Specific courses in the materials for and methods of instruction in science in the grades are required by many states for certification. These courses should comprise a full year's study, and be liberal in factual explanations, with a great variety of activities that may be performed in the classroom and out of doors. Certain texts for teachers' use are rich in these features (see the previous chapter on "Elementary Education").

Any institution that claims to prepare elementary grade teachers for their duties should support these courses fully.

2. *For the secondary levels.*

(a) College courses in one major and one minor science field are required for certification in most states. These requirements, however, are almost sure to be minimum; any high school teacher who intends to succeed should obtain additional credits in the sciences.

(b) College courses in generalized science, as offered to freshmen, are of value to the prospective high school teacher because of their broad attention to a number of sciences.

(c) An adequate preparation for high school teaching should include courses in biology, chemistry, physics as foundations. The practice recommended in many places for these subjects is a minimum of three years' study of one (a major), two years' study of another (a minor), and one year's study of a third. This prepares a teacher for service as a broad-minded specialist in a large high school, or for a more general program in a smaller high school. Few indeed are the positions for a narrow specialist on our nation's high schools.

Since the newest teacher is likely to be assigned classes in ninth-grade general science, whatever his specialty, certification for general science is usually given if one is entitled to certification in two of the other sciences.

Future Rise in Standards and Respect

Requirements for certification in science teaching are sure to rise in the coming years. To the complaints that higher standards will accentuate the teacher shortage, the reply is this: let teacher shortages be met by better pay and more respect. How may these be deserved?

Teachers must improve their professional spirit. Their pride in their task and influence must grow. They must hold themselves in high esteem to merit the esteem of others. Many teachers indeed do have this pride, and would not exchange their own prestige for that of any lawyer or doctor in the community.

All teachers, rather than most, should be members of their professional teachers organizations—national, state, and local. They should attend meetings when possible, and participate when invited. They should be regular readers of their Associations' journals, and contribute articles when good ideas come to mind.

All teachers, whatever the level of their special teaching activity, should have this spirit of the professional. The spirit of a science teacher differs in details, but not in principles, from that of each other field. Someone once remarked: "The teachers in our school are like the musicians in an orchestra. Their parts are different, but all use the same notes. The result is harmony."

Have I written wisely or foolishly in these paragraphs? I can only say that I have written from experience of over forty years in teacher preparation. Since 1912 I have been an actor in the drama of the teachers college. I have had eyes to see, ears to hear, a mouth to speak, hands to write, a mind to interpret, and a heart to understand.

Selected Bibliography

D. P. Cottrell, Ed., *Teacher Education for a Free People* (Oneonta, N. Y.: American Association of Colleges for Teacher Education, 1956). Eight leaders in teacher education, with the assistance of consultants and a committee, report on a three-year study (1952, 1953, 1954) of member institutions of the Association. Their chief purpose: "to suggest principles, policies, and possible concrete programs deemed worthy . . ."

American Council on Education, Commission on Teacher Education, *The Improvement of Teacher Education* (Washington, D. C.: The Council, 1946).

W. E. Armstrong & E. W. Hollis, & Helen Davis, *The College and Teacher Education* (Washington, D. C.: American Council on Education, 1944).

E. W. Knight, *Fifty Years of American Education* (New York: The Ronald Press, 1952). "A Historical Review and Critical Appraisal"; see especially Chapter 6, "Teachers and Teaching," 226-287, and bibliography, 286-7.

Cyrus Peirce and Mary Swift, *The First State Normal School in America* (Cambridge: Harvard University Press, 1926). Peirce led the battle to establish the school at Lexington (Mass.) as a three-year experiment in the specific training of teachers, devised the courses and supervised even the daily lives of the twenty-five young women who enrolled. His diary gives details from 1839 to 1841. Miss Swift, aged seventeen, was one of the students, and her diary presents—over the same span of years—the happenings and impressions of the student. Both "journals" emphasize the unbridled criticism against training a teacher professionally. Yet "six of the twenty-five who gathered at Lexington lived to see normal schools in every state in the Union." Interesting, intimate, instructive.

Charles E. Prall, *State Programs for the Improvement of Teacher Education* (Washington, D. C.: American Council on Education, 1946); Prall and L. C. Cushman, *Teacher Education in Service* (*Ibid.*, 1944).

Paul Woodring, *New Directions in Teacher Education* (New York: The Fund for the Advancement of Education, 1957). The broad program —four years of liberal arts, and one year internship in a public school, *not* in a college; stipend during the "professional internship"; Degree: Master of Arts in Teaching. An experiment has also been conducted in certain colleges to combine liberal and professional education within the four-year undergraduate period. The twenty-five individual programs in selected institutions are detailed in the Appendix.

Yearbooks of the American Association of Colleges for Teacher Education (Oneonta, N. Y.: The Association, 1953, 1954, 1955, 1956, 1957). Each yearbook contains varied articles on the functions, programs, policies, problems of the schools that train for the teaching profession.

HANOR A. WEBB, Professor of Chemistry and Science Education, Emeritus, George Peabody College for Teachers, Nashville (Tennessee), received his B.A. degree from the University of Nashville (1908), M.S. from the University of Chicago (1911), and Ph.D. from the George Peabody College for Teachers (1920); Professor of Psychology and Education, West Tennessee State Normal School (1912-1914); Professor of Chemistry and Biology in the same institution (1914-1917); Instructor in Chemistry, George Peabody College for Teachers, 1917-1920; Professor of Chemistry and Science Education (1920-1953), and Emeritus (1953-   ). As the senior author, he authored two general science texts (1925, 1936) and has written numerous articles for periodicals and encyclopaedias, in addition to having been editor of *Current Science* (weekly, Columbus, 1927-1947), the science issues of *Education* (Boston, 1936-1953), The

High School Science Library (annual, 1924-1944) ; and Keystone Visual Units in Science (slides, 1930-1933). He has also served as Chairman, Department of Science Instruction, National Education Association, 1926; Secretary, Nashville Section, American Chemical Society, 1934-1950; President, National Association for Research in Science Teaching, 1937; President, Tennessee Academy of Science, 1946; Secretary, National Science Teachers Association, 1946-1952.

NOTES FOR TEACHER'S COLLEGES

1. The word *Normal* comes through the French from the Greek, and originally meant "a carpenter's square"—hence "a pattern."

2. HORACE MANN (1796-1859), noted New England educator, legislator, administrator; the chief founder of our nation's public school system.

3. Quoted by William G. Vinal in *The Rise and Fall of Ye District School in Plimouth Plantation—1800-1900,* Norwell, Mass.

4. "He who can, does. He who can not, teaches."—GEORGE BERNARD SHAW (1856-1950), dramatist, novelist, playwright, satirist of England, in *Maxims for Revolutionists.*

5. "Let such teach others who themselves excel, and censure freely who have written well."—ALEXANDER POPE (1688-1744), poet of England, in *Essay on Criticism.*

6. Elbert Hubbard (1859-1915), editor, essayist, United States, in *Epigrams.*

7. *Educational Directory, 1937-1938, Part 3.* Washington, D. C.: U. S. Office of Education.

8. Karl W. Bigelow, "The Passing of the Teachers College," *Teachers College Record* LVIII (May, 1957), 409-417.

9. *Educational Directory, 1957-1958.*

10. Bigelow, *op. cit.*

11. T. M. Stinnett, "The Teachers College Myth," *Journal of Teacher Education* VII (December, 1956), 290.

12. John William Sperry, "Who Teaches the Teachers?" *Life,* XXIX (October 16, 1950), 147-154.

COLLEGE AND UNIVERSITY
EDUCATION: SELECTED AREAS

ENGINEERING*

MORRIS A. HOROWITZ
Northeastern University

The employment of engineers, scientists, and related workers has grown more rapidly than the total labor force throughout the period for which data are available. Since 1870, for example, there has been a five-fold increase in our labor force. During the same period, however, the number of engineering and scientific workers increased more than 85 times.[1]

This trend is expected to continue, so that we may expect a greater demand for engineers, scientists and technicians during the years ahead and for other types of workers. In the past there have been fluctuations around this long-run trend line. At times there has been a shortage of engineers, and at other times employment opportunities have failed to keep pace with the number of new engineering graduates. It is important to stress, however, that the long-run trend in engineering needs has been steadily upward.

The United States has become exceedingly sensitive to its need for engineers and scientists since the end of World War II. A great many individuals and groups were concerned about our ability to train enough engineers, scientists and supporting technicians to meet the requirements of our expanding economy and the needs of national defense. By late 1956, 14 non-governmental organizations, nine federal agencies and two Presidential committees were concerning themselves with one or more aspects of the engineering and scientific manpower problem.[2]

Beginning about the time of the Korean War there was a big upsurge in scientific research and development, and the demand for engineers rose sharply. Since the supply of engineers cannot adjust quickly to a sudden increase in demand there were not enough engineers available to satisfy the needs of government and industry. There was great concern over the shortage of engineers and scientists, as indicated by the voluminous material published on the subject, but it was not until October 4,

*This chapter is essentially a summary of William H. Miernyk and Morris A. Horowitz, *Engineering Enrollment and Faculty Requirements, 1957-1967,* a report prepared for the Committee on Development of Engineering Faculties of the American Society for Engineering Education, 1958.

1957, when Russian scientists successfully launched the first earth satellite, that the general public became fully aware of the problem.

While the launching of Sputnik dramatized the shortage of engineers and scientists, the problems of staffing our engineering colleges now and in the years ahead received scant attention. Employment opportunities for engineers were so good that there was a net migration of faculty members from colleges of engineering to industry.[3]

The loss of engineering teachers during a period of rising enrollment in colleges of engineering and institutes of technology created serious problems. In some institutions teaching positions remained vacant, while in others vacancies were filled with part-time or temporary teachers. Many engineering deans of the country were concerned about the quality of newly appointed faculty members. Furthermore, engineering enrollments are expected to rise substantially during the coming decade. This means that there will be a growing need for qualified, career engineering teachers, not only to replace those who are currently regarded as unsatisfactory, but also to provide sound instruction for the growing number of engineering students.

Many engineering educators were well aware that they faced a long-run problem. As engineering enrollments grew there would have to be a corresponding increase in engineering faculties. It was clear that a career in engineering education had to be made competitive in its total satisfactions with non-teaching engineering opportunities.

Because of this, Dean William L. Everitt of the University of Illinois, then President of the American Society for Engineering Education, appointed a Committee on Development of Engineering Faculties late in 1956. This Committee was established to investigate the broad problems of the recruitment, development and utilization of engineering faculties, and a number of meetings were held late in 1956 and early 1957. In these deliberations, the Committee concluded that there was a need for a complete survey of engineering faculty needs, both in terms of manpower and finances, over the next decade. This survey would provide the necessary information upon which various action programs could be based. The staff and facilities of the Bureau of Business and Economic Research at Northeastern University were retained to conduct the survey.

Survey Methods and Results

Early in 1957 the research staff of CDEF prepared the first draft of a questionnaire. This draft was circulated to a small but representative sample of engineering deans for their comments and criticisms. Through personal interviews suggestions for improvement were obtained. Representatives of other organizations interested in the broad area of engineering and scientific manpower were also requested to criticize the draft. The original questionnaire was revised several times and on April 26, 1957, mailed to the deans of engineering institutions in the United States.

Table 1
ENGINEERING ENROLLMENT DATA FOR 1956,
U.S. AND CDEF SAMPLE

	U. S. Totals a (1)	CDEF Sample (2)	CDEF Sample as per cent of U.S. (3)	Expan- sion Multi- plier b (4)
Day College Undergraduate Enrollment	220,237	175,226	79.56%	1.2569
Evening College Undergraduate Enrollment	28,496	25,527	89.58	1.1163
Total Undergraduate Enrollment	248,733	200,753	80.71	1.2390
Public	156,889	132,414	84.40	1.1848
Private	91,844	68,339	74.41	1.3439
Total Graduate Enrollment	25,927	20,305	78.32	1.2769
Public	10,159	7,774	76.52	1.3068
Private	15,768	12,531	79.47%	1.2583

a Engineering Enrollments and Degrees, 1956, Circular 494, U. S. Department of Health, Education and Welfare, Office of Education, Washington: 1957. Does not include Alaska, Hawaii, Puerto Rico; also excluded are five small institutions listed in Circular 494 which have discontinued their engineering programs or which have no undergraduate engineering enrollment.

b These multipliers are reciprocals of the percentages listed in Column 3, and were used to expand the sample returns to represent the population.

The questionnaire was sent to 216 institutions listed in the most recent report of the Office of Education. This was subsequently revised to 210, which figure was used as the total population of engineering institutions.[4] The final results were a 76 per cent return to the original mailing; 160 usable returns were received out of a possible maximum of 210. Eighty-three per cent of accredited institutions and 60 per cent of the non-accredited institutions in the country responded to the survey.[5]

The criteria used for comparing sample returns with the population were enrollment data. A comparison of sample returns with total enrollment figures is given in Table 1. The responding institutions accounted for 80 per cent of day college undergraduate engineering enrollment, and 90 per cent of evening college undergraduate engineering enrollment. The responding institutions also accounted for 78 per cent of total graduate engineering enrollment in the nation.

A further breakdown shows the percentage return from public and private institutions. Eighty-four per cent of the undergraduate engineering enrollment in publicly supported institutions was included in the sample, while 74 per cent of undergraduate enrollment in private institutions was included. For graduate engineering enrollment, 76 per cent of the enrollment in publicly supported institutions and 79 per cent of the enrollment in privately supported institutions were likewise covered by the sample institutions.

A regional comparison of the CDEF sample with that of the total United States indicates that with respect to engineering undergraduate enrollment the sample ranged from a low of 76.6 per cent in the Pacific region to 85.5 per cent in the Southern region. With reference to engineering institutions, the sample ranged from a low of 70.0 per cent in the Southcentral region to a high of 81.1 per cent in the Midwestern region.

A breakdown of returns by accredited and non-accredited institutions shows that of a maximum of 147 accredited institutions 122 (representing an 83 per cent return) submitted usable returns. The sample contained a 60 per cent return of non-accredited institutions. A regional breakdown of these data indicates a sample return of accredited schools ranging from 66.7 per cent in the Pacific region to 92 per cent in the South. Among the non-accredited schools, the sample ranged from a low of 37.5 per cent in the East to a high of 85.7 per cent in the Pacific region.

On the basis of the enrollment criteria, the proportion of institutions, the regional distribution, and the distribution between public and private institutions, the CDEF sample appears to be an excellent representation of the surveyed population. Throughout the report all sample data were expanded by enrollment multipliers. Regional data were expanded

Table 2
ENGINEERING ENROLLMENT PROJECTIONS

	CDEF Engineering Enrollment Projections*		Office of Education Engineering Enrollment Projections	
	Number	Per Cent over 1956-57	Number	Per Cent over 1956-57
1956-57	274,660		277,100	
1957-58	296,213	7.8%	280,000	1.0%
1959-60	334,442	21.8%	310,000	11.9%
1961-62	379,176	38.1%	340,000	22.7%
1966-67	464,711	69.2%	450,000	62.4%

*CDEF sample data projections expanded on basis of actual enrollment in 1956-57 to represent total engineering enrollment in U. S. The expanded CDEF figures in 1956-57 differ from actual enrollments data published by Office of Education because of the omission of three schools which indicated they were dropping their engineering programs, and the exclusion of schools in Alaska, Hawaii, and Puerto Rico.

on the basis of regional multipliers and data for public and private institutions were expanded by separate public and private expansion multipliers based on undergraduate enrollment.

Enrollment Projections

The anticipated increases in engineering enrollment obtained from the CDEF survey and the Office of Education projections are compared in Table 2. On the basis of the CDEF survey, engineering enrollment in the United States is expected to increase by 69 per cent over the next ten years. The Office of Education projection, however, showed a gain of only 62 per cent. For the year 1957-58, the Office of Education had projected engineering enrollment of 280,000, although actual enrollment in that year was 297,100. Thus, the 1957-58 Office of Education projection of engineering enrollment was 17,100 or 6.1 per cent less than the actual enrollment for that year. The CDEF projection (made in 1956-57) was for an anticipated enrollment of 296,200 for the year 1957-58. Thus, the CDEF projected figure was only 900 or 0.3 per cent less than the number of engineering students enrolled that year. For that reason it is believed that the Office of Education projections of engineering enrollment are on the conservative side, and that those made by the respondents to the CDEF survey may be more realistic approximations of the growth in engineering enrollment to be expected during the next decade.

Table 3 shows projected undergraduate and graduate engineering enrollments broken down by public and private institutions, and by day and evening students. Total undergraduate engineering enrollments in 1956-57 amounted to 248,733. By 1966-67 this was expected to increase to 413,883, a gain of 66 per cent. Privately supported institutions anticipated an increase of 48 per cent over the next decade. But these institutions were planning on an increase of only 44 per cent in full-time, day college enrollment while they expected a gain of 61 per cent in evening enrollment. The picture is different in publicly supported institutions. There, an increase of 77 per cent was expected in total undergraduate enrollment. But the large gain (78 per cent) was expected in day college enrollment. Evening undergraduate engineering enrollment in publicly supported institutions was expected to increase by only 67 per cent. It is of interest to note the much more rapid increases expected by the deans of publicly supported institutions than the increases anticipated by the deans of privately supported institutions.

As in the past, most engineering college faculty members of the future will be products of the engineering graduate student body. The projections of undergraduate engineering enrollment may provide a measure of future faculty needs, or the demand for engineering teachers. In a similar way, the number of graduate students—in general, the group from which new engineering teachers might be recruited—gives a measure of

Table 3
PROJECTED UNDERGRADUATE AND GRADUATE ENGINEERING ENROLLMENTS, BY PUBLIC AND PRIVATE INSTITUTIONS

	Undergraduate Engineering Enrollment			Graduate Engineering Enrollment		
	1956-57 Engineering Enrollment	1966-67 Projected Enrollment	Percent Increase over 1956-57	1956-57 Engineering Enrollment	1966-67 Projected Enrollment	Percent Increase over 1956-57
Public						
Day	145,118	258,327	78.0%	6,492	19,317	197.6%
Evening	11,771	19,669	67.1	3,667	6,475	76.6
Total	156,889	277,996	77.2%	10,159	25,792	153.9%
Private						
Day	70,886	102,127	44.1	5,695	9,332	63.9
Evening	20,958	33,760	61.1	10,073	15,704	55.9
Total	91,844	135,887	47.8%	15,768	25,036	58.8%
Total						
Day	216,004	360,454	66.8	12,187	28,649	135.1
Evening	32,729	53,429	63.3	13,740	22,179	61.4
Total	248,733	413,883	66.4%	25,927	50,828	96.0%

the source from which the supply of potential engineering teachers will be drawn.[6]

In the academic year 1956-57 there were 25,927 graduate students enrolled in engineering colleges in the United States. By 1966-67 this was expected to increase to 50,828, a gain of 96 per cent. At the present time, most graduate students in engineering are enrolled in privately supported institutions, with a large part in evening programs. The deans of privately supported institutions expected an increase in their graduate enrollment of 59 per cent by 1966-67. Day college graduate programs were expected to increase more than evening college programs (64 per cent as compared with 56 per cent). Much more rapid growth was expected by the deans of publicly supported institutions. They expected total graduate enrollment to increase by 154 per cent over this ten-year period. And most of the gains were anticipated in day programs, where they expected an increase of 198 per cent. Evening programs in publicly supported institutions were expected to increase only 77 per cent. Thus, by 1966-1967 on the basis of present expectations, publicly supported institutions will be training a larger number of graduate students than privately supported institutions.

Of the 25,927 graduate students enrolled in 1956-57, 22,934 or 88 per cent were candidates for the master's degree, and a majority of these were enrolled in privately supported institutions. Candidates for the doctor's degree in engineering subjects were almost evenly divided between publicly and privately supported institutions. Over the next decade enrollment in the master's programs was expected to increase by 93 per cent, with a gain of 152 per cent in publicly supported institutions and a gain of 56 per cent in privately supported institutions. Total enrollment in doctoral programs was expected to increase by 116 per cent over the next ten years, with a gain of 165 per cent anticipated in publicly supported institutions and a gain of 77 per cent anticipated in privately supported institutions.

It is encouraging to note that a larger relative increase was expected in graduate than undergraduate enrollment. This means that the potential supply of engineering college teachers will have increased, relative to the projected number of undergraduate engineering students. It does not follow, of course, that this will simplify the problem of the recruitment, utilization and development of engineering faculties. Over the next decade it is possible that the demand by non-academic sources for engineers with graduate degrees may increase more rapidly than the expected supply. Industry and government will undoubtedly continue to compete aggressively for the increasing supply of engineers with graduate degrees. It will require equally vigorous recruiting methods by engineering institutions, if they are to attract into the teaching profession on a permanent basis an appropriate proportion of those highly qualified engineers with graduate degrees who have the aptitude for and interest in careers as engineering teachers.

Engineering Faculty Shortage, 1956-1957

For purposes of this survey, engineering teachers were defined as those members of engineering college faculties in engineering departments only. According to the replies of engineering deans, about 52 per cent of the total teaching load in engineering curricula in responding institutions was carried by engineering teachers. An additional 28 per cent was handled by other faculty members teaching basic science and mathematics. The remaining 20 per cent of the course work, which consists of the humanities and related subjects, was carried by non-engineers.

Although the major purpose of the survey was to estimate the demand for engineering teachers, data were also obtained on the actual shortage of engineering teachers during the academic year 1956-57.

On the basis of the survey it was estimated that there were 9,078 engineering teachers in the United States during 1956-57. Fifty-eight per cent of these were in publicly supported institutions and 42 per cent in privately supported institutons. During 1956-57 there was a total of 732 budgeted openings for engineering teachers. These openings represented an actual shortage of approximately 8 per cent of the number of engineering teachers in 1956-57. Publicly supported institutions reported a somewhat greater shortage (8.6 per cent) than the privately supported institutions (7.4 per cent). And among the publicly supported institutions those in urban areas reported a relatively larger shortage. The reverse was true of private institutions.

The shortage of engineering teachers is more than a matter of numbers, however. Many engineering deans have expressed concern about the quality of new teachers they have appointed. Therefore, in an attempt to measure the magnitude of this problem, all deans were asked to indicate the number of positions held by part-time, temporary or unsatisfactory personnel, and to express this number in full-time equivalents. A total of 532 positions, or 5.9 per cent of the combined engineering faculties was reported as falling in this category. Again, publicly supported institutions reported a slightly larger number of positions for which they wished replacements than privately supported institutions.

A more realistic measure of the "faculty shortage" during the academic year 1956-57 would be a combination of the budgeted openings and those positions which the deans regarded as unsatisfactorily filled. The "shortage," as thus defined, becomes fairly significant. With a total engineering faculty of 9,078 the "shortage" of 1,264 represents 14 per cent. In public institutions the "shortage" would be 15 per cent and in private institutions 13 per cent.

Engineering Faculty Projections

Each dean responding to the survey was asked to make two separate estimates of his engineering faculty needs between the academic years

1956-57 and 1966-67. One estimate was the level of faculty needed at specified times within the ten-year period and the second was the total number of engineering teachers who will have to be appointed during this period. The first estimate represents the projected net addition to current faculty. The second estimate represents the total number of appointments necessary to increase the engineering faculties to the projected levels. In any one year the difference between the two estimates is the result of normal turnover resulting from deaths, retirement, migration to industry and similar causes, but excluding transfers from one college of engineering to another.

By 1966-67, the deans estimated, combined faculties would grow from the 9,078 engineering teachers in 1956-57 to about 16,000, an increase of 76 per cent. In order to increase engineering faculties by 6,900 new members over the ten-year period, however, colleges of engineering would have to appoint an additional 9,600 teachers, since during this period an estimated 2,700 will leave the teaching profession for the reasons enumerated above. A comparison of total appointments and the increase in the engineering faculty levels over the ten-year period is given in Table 4.

Whereas the deans of the engineering colleges of the United States collectively expected their faculties to increase by 76 per cent over the next decade, the faculties of publicly supported institutions were expected to increase by 93 per cent, while those of privately supported institutions by only 54 per cent. This would mean a substantial shift in the relative size of the faculty of publicly and privately supported institutions. In 1956-57, publicly supported institutions accounted for 58 per cent of all engineering teachers. By 1966-67 this was expected to increase to 63 per cent.

In order to achieve these faculty levels, however, the number of *new* engineering teachers joining faculties would have to increase 106 per cent. This means that the estimated total appointments over the ten-year period represented 106 per cent of the faculty of 1956-57. In publicly supported institutions there is expected to be a gain of 125 per cent, compared with an increase of 79 per cent for privately supported institutions. As referred to above, the difference between the total number of new engineering teachers appointed and the growth of engineering faculties, will be accounted for by normal turnover.

About two-thirds of the nation's engineering students and faculties are located in two regions—the East and the Midwest. Whereas growth is expected in all regions, the most rapid expansion is expected in the South, Southcentral and the Mountain regions. The East was the only region in which the rate of growth is expected to be below the national average. In part, this is due to the fact that more rapid population growth is expected in the Southern and Western regions. In part, however, it is due to the geographical distribution of publicly and privately supported engineering institutions.

Almost half of the nation's enrollment in private institutions is concentrated in the East. The South and the Midwest, however, account for

Table 4

COMPARISON OF ESTIMATED FACULTY APPOINTMENTS AND TOTAL FACULTY PROJECTIONS,
1956-57 to 1966-67

	1956-57 Faculty		Estimated Faculty Appointments from 1956-57 to 1966-67		Projected Faculty Level, 1966-67	
	Number	Per Cent	Number	Per Cent	Number	Per Cent
Public						
Urban	2,121	23.4%	2,710	28.2%	4,100	25.6%
Non-Urban	3,146	34.7	3,860	40.3	6,000	37.5
Total	5,267	58.0	6,570	68.5	10,100	63.1
Private						
Urban	3,548	39.1	2,790	29.1	5,400	33.8
Non-Urban	263	2.9	230	2.4	500	3.1
Total	3,811	42.0	3,020	31.5	5,900	36.9
Total Public and Private	9,078	100.0%	9,590	100.0%	16,000	100.0%

nearly three-fifths of total enrollment in publicly supported institutions. Since more rapid growth is expected in public than private engineering colleges, the South and the Midwest will account for a growing proportion of engineering students and faculties between 1956-57 and 1966-67. Enrollment is expected to grow more rapidly in public than private institutions in all regions except the Southcentral and the Mountain states.

Salaries in Engineering Education

According to a recent study, the deans of engineering feel that the major reason for the loss of engineering teachers to industry is the salary differential between industry and education.[7] Since the questionnaire was distributed solely to enginering educators, no direct information was obtained on the earnings of engineers in non-teaching jobs. However, the deans responding to the survey were asked to furnish a considerable volume of data on teaching salaries.

Many engineering teachers, as is true of other college teachers, actually spend less than 12 months a year at their teaching jobs. In order to achieve comparability, the questionnaire requested the respondents to report average teaching salaries of engineering teachers by rank on a monthly basis, for the number of months actually spent teaching. The data indicated that the average professor in urban institutions received something in excess of $900 a month, while in the non-urban institutions he is paid slightly more than $800 a month. Highest salaries are reported for professors in private urban institutions. These are followed by the salaries of professors in public urban institutions, and then by professors in public non-urban institutions. The lowest salaries for professors are reported in private non-urban institutions. This same pattern prevails for other ranks, with the differentials narrowing again as we move down the scale from the full professor to instructor. The differential for each rank between private urban and private non-urban is very significant. A comparison of average monthly salaries indicates a difference of $257 a month in favor of the professor in private urban institutions over those in private non-urban institutions. Although the difference is less for other ranks it is still significant.

The respondents to the survey were asked to estimate the minimum salaries that would be required if salaries in their institution were to be competitive with non-teaching opportunities for engineers. In phrasing this question it was recognized that there are many non-monetary satisfactions associated with a career in teaching. Hence, the deans were not asked to estimate the amounts by which their salaries would have to be increased in order to be "equal to" those of non-teaching engineers. Rather, the deans were asked to take into account those intangible satisfactions associated with a career in teaching which cannot be given a dollar dimension, and to estimate the amount by which the salaries would have to be increased in order to recruit a sufficient number of high caliber engineers to staff their faculties fully. The results are shown in Table 5.

With few exceptions, engineering deans reported that their salaries would have to be increased if they were to be competitive with non-teaching opportunities. The average *minimum* monthly salary ranged from $432 for instructors to $758 for full professors, with the average for all ranks at $585 a month. The average *actual* monthly earnings range from $487 for instructor to $889 for professor, with the average for all ranks at $677. The average *minimum* salaries which engineering deans feel would have to be paid if salaries of engineering teachers are to be competitive with non-teaching engineering opportunities are significantly high. On the basis of the deans estimates, the average *minimum* monthly salary for all ranks would have to be increased by 49 per cent in order to make them competitive with non-teaching opportunities. The necessary increases would range from 39 per cent for instructor to 55 per cent for full professor.

The respondents to the survey were also asked to indicate the total annual salary bill for all engineering teachers in 1956-57, and to estimate the total salary bill for specific academic years, taking into account anticipated growth in their faculties and assuming continued improvement in the level of engineering teachers' salaries. Approximately 60 million dollars were required to pay the salaries of engineering teachers in 1956-57, and by 1966-67 the engineering deans estimate that the salary bill will increase to approximately 134 million dollars. This is a gain of 123 per cent. Public institutions forecast an increase in their total salary bill of 142 per cent, compared with an expected increase of 98 per cent in private institutions.

The engineering deans have very likely underestimated the increases that will be required if they are to attract and retain an appropriate proportion of engineering graduates as engineering teachers over the next ten years. The deans have indicated that substantial salary increases will have to be made at the present time if engineering teaching positions are to be made competitive with non-teaching opportunities. These estimates ranged from about 40 per cent for instructors to 55 per cent for full professors. If we add to the $60 million paid to engineering teachers in 1956-57 only 40 per cent of this amount, or $24 million, about $84 million would have ben required, on the basis of the deans' most conservative estimates, to raise salaries enough to make engineering teaching positions competitive at the present time. In addition, the deans anticipated an expansion of 76 per cent over present faculties by 1966-67 and even a proportionate increase of the present salary bill would raise the estimate of $84 million to $148 million. Allowance must also be made for the general increases in wages and salaries which are expected over the next decade, as a result of rising productivity and price inflation. If we use a realistic estimate of an average increase of 3 per cent per year, compounded over the ten-year period, as a result of these forces, this

Table 5

SALARY STRUCTURE OF ENGINEERING TEACHERS IN ALL ENGINEERING INSTITUTIONS, BY RANK, 1956-57

Rank	Average Minimum Monthly Salary (1)	Average Actual Monthly Earnings (2)	Starting Salary To be Competitive With Non-Teaching Opportunities (3)	Difference (Col. 3-Col. 1) (4)	Per Cent Difference (Col. 4 as Per Cent of Col. 1) (5)
Professor	$758.02	$889.10	$1178.65	$420.63	55.5%
Associate Professor	630.50	710.04	921.02	290.52	46.1
Assistant Professor	536.01	605.44	756.95	220.94	41.2
Instructor	482.53	487.14	599.98	167.45	38.7
Average all Ranks	$585.58	$677.65	$872.53	$286.95	49.0%

will add $52 million, bringing the total salary bill for 1966-67 to about $200 million. On the basis of these assumptions, the average annual salary of the expected 16,000 engineering teachers in 1966-67 would amount to $12,500.

Since these are estimates based on a specific set of assumptions, they may overstate or understate the actual salary bill of 1966-67. It appears, however, that a substantially larger amount will be required for engineering teachers' salaries than the deans have estimated if engineering colleges are to be able to attract (and retain) an appropriate proportion of the able engineering graduates of the next decade into the profession of engineering teachers.

Summary

The long run trend in the demand for engineers and scientists is upward. Forecasts indicate that there will be a continued shortage of engineers and scientists over the next decade, with a result of increased enrollments in all engineering institutions. The shortage of well qualified, career engineering teachers is a crucial aspect of the overall engineering and scientific manpower problem in the United States.

In 1956 the American Society for Engineering Education established a Committee on Development of Engineering Faculties to consider the problems of recruitment, development and utilization of engineering teachers. This Committee recognized that a program to strengthen engineering education in the United States must be based on firm estimates of engineering enrollment and faculty needs over the next decade. To obtain such estimates a survey was conducted of all engineering institutions in the United States. The results of this survey show that between 1956-57 (when the data for these estimates were collected) and 1966-67:

1. Undergraduate engineering enrollment is expected to increase by 66 per cent to approximately 414,000.
2. Enrollment for the master's degree in engineering is expected to increase by 93 per cent to about 44,000.
3. Enrollment for the doctorate in engineering is expected to increase 115 per cent to about 6,500.
4. The number of engineering teachers is expected to increase by 76 per cent to about 16,000.
5. If engineering teachers' salaries are to be competitive with non-teaching engineering opportunities, the average engineering teacher will earn $12,500 per year from teaching in 1966-67, an increase of 105 per cent.
6. To achieve this average salary, the total salary bill for engineering teachers in 1966-67 will amount to $200 million, an increase of 233 per cent.

Seventy-six per cent of the deans in degree-granting engineering institutions in the United States responded to the questionnaire. Responding institutions account for 80 per cent of total undergraduate engineering enrollment in day colleges, 90 per cent of evening undergraduate enrollment, and 78 per cent of graduate enrollment.

The survey revealed that during the ten year period, 1956-57 to 1966-67, most undergraduate engineers would receive their training in full-time day college programs. Day college enrollment was expected to increase by 67 per cent, and evening college enrollment by 63 per cent. A greater increase was expected in public institutions (79 per cent) than in private institutions (48 per cent). Graduate engineering enrollment in publicly supported institutions was expected to increase 154 per cent, and in private institutions 59 per cent.

During the academic year 1956-57, there were an estimated 9,078 engineering teachers in the United States. Fifty-eight per cent of these were in publicly supported institutions, and 42 per cent in private institutions. An additional 732 engineering teachers would have been required to bring all engineering faculties to full numerical strength, a shortage of about eight per cent. If replacements for part-time, temporary or unsatisfactory personnel were added to these needs, an additional 532 engineering teachers would have been required. Combining these two figures together, indicates that there was an overall shortage for that year of about 14 per cent.

Faculty requirements in publicly supported institutions were expected to increase by 93 per cent; those in private institutions by only 53 per cent. This would mean a substantial shift in the relative size of faculties in publicly and privately supported institutions. In 1956-57, publicly supported institutions accounted for 58 per cent of all engineering teachers. By 1966-67, this was expected to increase to 63 per cent. In order to achieve these new faculty levels, new appointments to engineering faculties including replacements in publicly supported institutions would increase by 125 per cent, and in private institutions by 79 per cent.

The survey also revealed that there would be some variation in the rate of growth of engineering enrollment by region. Engineering institutions in the East were expected to grow at a slower rate than those in the West. This is partly due to the greater concentration of publicly supported institutions in the Western regions, and partly to the more rapid population growth in the West than in the East.

The results of this study have given the Committee on Development of Engineering Faculties the factual background to study the problems of recruitment and utilization of engineering teachers. The CDEF is presently engaged in surveying by personal visit various engineering institutions, in order to determine how various institutions are meeting their shortages of well-qualified engineering teachers.

Selected Bibliography

William H. Miernyk and Morris A. Horowitz, *Engineering Enrollment and Faculty Requirements, 1957-67* (American Society for Engineering Education, University of Illinois). Results of a survey of the deans of 160 engineering institutions on such matters as enrollment, faculty requirements and projections and salaries.

William H. Miernyk and Morris A. Horowitz, *Salaries and Earnings of Engineering Teachers, 1956* (American Society for Engineering Education, University of Illinois). An extensive analysis of the structure of engineering teachers' salaries and outside income, based upon over 4,000 returns in a mailed questionnaire.

John D. Akerman, "Problems of Industry Raids on University Faculties," *Journal of Engineering Education*, XXXXVII, No. 10 (June 1957), 846-51. Basic cause of raids, it is contended, is the overall shortage of engineers. Suggests that universities and industries cooperate in permitting their respective staffs work in other fields.

American Council on Education, *Expanding Resources for College Teaching* (January 1956). Report of a conference on college teaching sponsored by the American Council on Education, in Washington, D. C., January 19-20, 1956. Discussion on the need for more effective ways of recruiting, conserving and utilizing resources in college teaching.

A. R. Hellwarth, "Migration of Engineering College Faculty Between Campus and Industry," *Journal of Engineering Education*, XXXXVII, No. 2 (October 1956), 157-164. Report by Committee of the Relations with Industry Division of ASEE on a survey to determine the losses of engineering teaching staff to industry and the corresponding gains for the past two school years.

National Manpower Council, *A Policy for Scientific and Professional Manpower* (May, 1953). Contains detailed studies of engineers, teachers and physicists and lists objectives for a cooperative effort involving Government, industry, the educational institutions and professional and other groups which Council believes can provide the nation with the scientific and professional manpower it requires.

Albert Watson Davison, "The Technical Manpower Situation," *Journal of Engineering Education*, XXXXVII, No. 3 (November 1956), 263-267. Discussing the shortage of technical manpower, the author suggests the following as necessary for the education of the kind of engineer and scientist America needs: (a) drastic curricular revision, (b) complete overhauling of the budgeting of students' time, (c) more inspiring teaching and, (d) classification of aspiring students of technology.

Engineers Joint Council, Engineering Manpower Commission, *Engineering Educational Facilities*, Report No. 104 (May 1957). Presents results of a survey of February-March 1957 on the utilization status of freshman engineering educational facilities. Survey shows that from point of view of availability of capacity engineering colleges are not overcrowded and that qualified students are not being turned away.

David M. Blank, George J. Stigler, *The Demand and Supply of Scientific Personnel* (National Bureau of Economic Research, Inc., 1957). An analytical study of demand and supply of scientific personnel over the years: and the conclusion drawn, based principally upon the relatively small increases in salaries, is that there is no shortage of engineers.

"Demand for Engineering Graduates, 1956" (report of Special Surveys Committee of EJC), *Journal of Engineering Education*, XXXXVII, No. 9 (May 1957), 789-797. Results of a survey conducted December 1955, and January and February 1956, indicate that demand for Engineers in 1956 was especially high, and that requirements were significantly higher than accessions.

National Science Foundation, *The National Committee for the Development of Scientists and Engineers* (May, 1956). The report includes the discussion of the general and specific problems in the scientific and engineering manpower development and some suggested solutions.

J. T. Rettaliata, "Scientific Manpower Shortage a Peril to America," *School and Society*, LXXXII (July 23, 1955), 17-20. A general discussion of the scientific manpower shortage: high-school motivation, college motivation, higher education's responsibilities and draft deferments.

Shortage of Scientific and Engineering Manpower, Congressional Hearings before the Subcommittee on Research and Development (April-May 1956). Testimony regarding present and future shortages of technically trained manpower in this country.

Engineers' Joint Council, Special Surveys Committee, *Professional Income of Engineers, 1956*, Report No. 102 (January 1957). Statistical tables on income of engineers, broken down by various industries.

H. H. Armsby, "Growth of Graduate Study in Engineering," *Journal of Engineering Education*, XXXXV (November 1954), 220-6. The special function of this paper is to present a few statistics and estimates, gathered from various sources which outline the growth of graduate engineering study up to the present time.

Ross J. Martin, "Future Engineering Enrollments and the Problems They Pose," *Journal of Engineering Education*, XLVII, No. 2, October 1956, 129-138. An analysis, from statistical tables, of future engineering enrollments, and the problems posed for engineering faculties.

U. S. Department of Health, Education and Welfare, Office of Education, *Engineering Enrollments and Degrees, 1956*, Circular No. 494 (1957). Statistical tables and averages of engineering enrollments and degrees in 1956, and comparisons with previous years.

MORRIS A. HOROWITZ, Associate Professor of Economics and Research Associate in the Bureau of Business and Economic Research, Northeastern University (Boston), worked, after graduating from Washington Square College, New York University (1940), for a number of Federal Agencies in Washington, D. C. as an economist. In 1946 he accepted a Littauer Fellowship at Harvard University; in 1954 he received his doctorate from that University. From 1947 to 1951 he was a Research Assistant Professor of Labor and Industrial Relations at the University of Illinois. When the wage stabilization program was instituted in 1951 he returned to Washington, D. C., first as Director of the Office of Case Analysis and then Vice-Chairman, Review and Appeals Committee, Wage Stabilization Board. In 1953, when the stabilization program was ended, he returned to Harvard, first as a Research Associate in Law and then as Research Associate in Labor Relations. In 1956 he moved to Northeastern University as Associate Professor of Economics. In addition to his teaching and research at Northeastern University, Dr. Horowitz is also actively engaged as a labor arbitrator. He is on the labor panel of the American Arbitration Association and of the Federal Mediation and Conciliation Service.

NOTES FOR ENGINEERING

1. National Science Foundation, *Trends in the Employment and Training of Scientists and Engineers* (Washington: May, 1956), 4.

2. Henry H. Armsby, *Engineering and Scientific Manpower: Organized Efforts to Improve its Supply and Utilization,* U. S. Department of Health, Education and Welfare, Office of Education, Circular No. 483 (August, 1956), 1.

3. There was some movement in both directions, of course, but more engineers were leaving teaching for industrial jobs than the number moving from industry to engineering colleges. See A. R. Hellwarth, "Migration of Engineering College Faculty Members between Campus and Industry," *Journal of Engineering Education,* XXXXVII No. 2 (October, 1956), 157-164.

4. This includes all engineering institutions in the United States, whether or not accredited by Engineers Council for Professional Development, listed in the most recent directory of the Office of Education, and adjusted to exclude those smaller schools which have dropped or planned to discontinue their engineering programs, and excluded engineering schools in Alaska, Hawaii and Puerto Rico.

5. By "accredited institution" is meant an institution with one or more curricula accredited by Engineers Council for Professional Development which is the recognized accrediting agency in the United States.

6. The relationship is not quite so simple. As the number of graduate students increases, it follows that more faculty members will be needed to conduct graduate courses, and to supervise the total graduate training of engineers. Also, the typical graduate class is smaller than the typical undergraduate class, and hence the ratio of faculty to students becomes greater, i.e., a larger number of teachers per student will be required.

7. A. R. Hellwarth, "Migration of Engineering College Faculty Between Campus and Industry," *Journal of Engineering Education,* XXXXVII, No. 2 (October, 1956), 162.

INDUSTRIAL EDUCATION

PAUL E. HARRISON, JR.
University of Maryland

The primary intent of this chapter is to set forth trends which are apparent in industrial education and in particular those of a genre related to contemporary developments in science and technology. Consideration will be given to three programs of education. They are, in order of consideration; industrial arts teacher education, vocational industrial teacher education, and curriculums which are intended to prepare students for professional positions in manufacturing industries or service fields. The latter categories are exclusive of engineers, scientists, or technicians, according to the usual definition.

While there are functioning programs of industrial education which are operative in terms of a primary involvement with the immediate student body, the majority of college and university departments have an assignment in teacher education. The program which is slated to prepare students specifically for industrial situations is generally concomitant to the teacher education purpose.

Industrial Arts Teacher Education

Industrial arts teacher education is primarily concerned with the preparation of such teachers for the public junior and senior high schools. Good[1] denotes the area as follows:

"A phase of the educational program concerned with orienting individuals through study and experience to the technical-industrial side of society."

The significance of education in industrial arts is attested to by the growth of the American industrial system in both size and complexity. This growth has brought an increase and at the same time a shortage in technically oriented manpower. The training and discovery of persons able to operate effectively in scientific and technological capacities can be abetted by curricular strength on the part of the industrial arts program.

Since the early 1900's the rate of increase of numbers of professional people engaged in scientific and technical work has increased at about twice the rate of population growth in the United States. Even as early as the latter part of World War II there was doubt in some quarters as to whether our manpower, in terms of numbers, could meet the demand. The post-war expansion and the Korean conflict aggravated the same points of tension.[2]

Program Functions. The impact of such problems have great import for technically based education. A part of the function at the secondary school level, and certainly there is thus a transfer to the college program, is to develop an atmosphere permeated with aspects of technology and suggestive to the student at either level that here is an area of activity where he may develop certain levels of competency, where he has become oriented and interested. The purpose of the program is to educate, of course, but in the sense that orientation is education, that interest development is education, skill development, and so on. A part of the answer to the manpower shortage likely lies in the industrial arts laboratory where a young man can reveal to himself, concepts and notions which say to him that he is able to deal effectively with situations which involve tools, materials, machines and media as they are found in industry. This point is in agreement with contemporary judgment. For example the President's Committee on Education Beyond the High School notes that:

Revolutionary changes are occurring in American education of which even yet we are only dimly aware. This Nation has been propelled into a challenging new educational era since World War II by the convergence of powerful forces—an explosion of knowledge and population, a burst of technological and economic advance.[3]

While all facets of education have need of criticism, and are duly criticized; there have been major pronouncements in most fields pertaining to the need for a new type of education. The greatest difficulty in many instances has not been with the professional, but with the lay person who must ultimately support the more sophisticated professional judgment.

Public Support. It is unfortunate that professional training; technical, academic, etc. tend to preclude an ability to transmit the background of the discipline to the non-professional. Despite the team approach to research, the higher the level of training, the more heuristic it becomes, and consequently more difficult of communication. Lerner suggests this problem in describing the void between scientists and other citizenry:

"The universe of the scientist is still an expanding one, discontinuous and open. To keep it thus the scientist requires the willingness of an open society to let him follow his nose and give him the right to be wrong, both as a scientist and as a citizen."[4]

It is unlikely that educators, whatever their area of effort, will be allowed free decision as to what shall be taught or even how this should be approached in terms of methodology. Industrial arts teacher education programs will have to indicate, by means of what goes on in the laboratory and the classroom, that they are aware of the problems of the world.

Manpower Problems. It becomes increasingly evident each school term that every member of the instructional staff must assume a responsibility for more and more students. This same factor is operative whether the instruction is being given in the lecture hall or at the technique level. Whether or not we enroll proportionately more percentage-wise of the population, the sheer weight of numbers is likely to double the college level enrollment by 1970 at the latest. This coupled with the demands of a technical-scientific world, will provide a new aura for the ivy covered buildings of the college campus. There is considerable tendency to require an increasing level of preparation for teachers of industrial arts at the college and university level. This is in conflict with the availability of staff and the number of men in preparation at the doctoral level. The consequence of this becomes a new approach to manpower utilization. The new approach can be in a number of areas. For example, the concept of teacher-student cannot but change. In connection with this it is mandatory that the notion of teacher responsibility be enlightened to the point that the student be given more and more responsibility in the learning process. There are so few great teachers. This whole idea suggests that the chasm between teachers and students will be narrower and shallower than in the past. This is in contrast to the proposals which indicate that the only solution is to remove the teacher from the classroom atmosphere with a television screen as the replacement rather than as a facility for furthering the other notion. The suggestion is not for fewer students, this is no longer very possible, but rather for a more wholesome consideration of the constitution of teaching.

Quantity and Quality. It is rather obvious that quantity and quality cannot grow together in this field unless the American public is willing to sponsor education at a higher level. There is evidence in scholarships, alumni support, the support by industry, a reluctant recognition by various legislative bodies, that a higher and higher level of support and new concepts of financing are necessary. Without this many of those who can serve in technical education will be unable to participate.

To this point the problem of quantity and quality has not provided a departure from the same educational standards. But as pointed out in the preceding paragraphs, educational progress will founder without public support. The teacher in turn has an extreme obligation to make his efforts as effective as possible. The fact of financial support does not mean that the individual teacher will not have to redouble his efforts, through improved management of facilities and improved educational techniques. This is mandatory.

There seems to be evidence that groups such as labor and management are willing to expand their efforts and resources in behalf of industrial arts education at several levels. This is suggested in the tendency of individuals and organizations to contribute the efforts of people and the use of facilities for educational purposes. Certain desirable educational experiences are not purchasable in terms of locating them in the laboratory or classroom. For example, the extrusion of plastics as a process can be duplicated for school purposes. How much more complete is the picture, however, if in addition a plastics processing plant is visited. This not only involves the actual processing, but all that is conjunctive to it. It would be difficult to simulate communications, warehousing, transportation, personnel, and other phenomena in any laboratory unless the plant is so categorized. This can be a public contribution. However, it is not likely to operate in an environment where lay people do not realize the potential of their contributions to education.

Technological Demands. The facts of technological demands in terms of industrial arts teacher education have been monumentally posed by Bollinger:

"We are rapidly moving into a highly technical society, a society built upon the applications of science and mathematics to human needs and every day living. To help interpret this world to young America, Industrial Arts teachers may well have to be good technicians as well as able craftsman."[5]

Practices and Trends. Likely the most recent and certainly the most comprehensive review of practices and trends in industrial arts teacher education was edited by Hornbake and Maley[6] for the American Council on Industrial Arts Teacher Education. The investigation was concerned with eight facets of the field. This report will review the two items which were concerned with technical factors, although several suggested implications for this area of the challenge of science to education.

The schools involved in the study reported a diversity of experiences and an extensive amount of coursework in shop activities and drafting. In most instances the technical requirements involved an extensive specialization in one or two shop areas. The students were generally involved in such activities as job planning, production planning, development of instructional aids, and vicarious experiences in planning school shops, equipment selection, materials purchasing, and the care of physical facilities and equipment.

In the context of industry most departments expressed a concern with safety, both as regards direct programming and activity in the industrial arts laboratory. Visitations to industry were cited for this specific purpose. The use of work experience for credit was entered under this category. Technical courses are apparently taught under the assumption that industrial methods and processes are learned by participating in a shop which is appropriately named to a large or small classification of industries or an industry. Obviously this is just an assumption and largely contingent upon the nature of teacher effort.

The study suggested that several categories of work are neglected to the detriment of intellectual development in the sciences and mathematics. No superior practices, the basis for this research, were reported in the academic areas by the 165 schools responding to the inventory. This does not mean that the usual college or university requirements in science and mathematics are not met by students in these fields.[7]

As stated before this study contains much of a nature important to an understanding of present practices in the light of the contemporary challenge.

Educational Program. Other studies and reports have suggested a variety of actions that should take place in teacher education. The primary suggestion is that industrial arts teacher education, in particular, needs to keep up with the various changes in industrial processing, media, research and the like. It is suggested that experimentation in miniaturization, hydraulics, and experiences with the utilization of computers for educational purposes are important.

A considerable amount of attention is directed to the utilization of individualized study. This involves, contrary to many expectations of trends, a greater concern with problem solving techniques and less with rote, per se. Consistent with this is the research reported in the field. There have been graduate studies in recent years dealing more effectively and extensively with aspects of technology.[8]

It is likely that industrial arts can play an important role in the advancement of technological and scientific understandings among students. This assumption is based upon the notion that such a program can be aligned with a world of automation, electronics, experimentation, and technological improvements.

Scientists, engineers, and industrial research personnel work with tools, materials, and products of industry. It is accepted that one's effectiveness in most areas is dependent upon a background of experience. The typical industrial arts program offers very little opportunity wherein scientific and technical data can be applied to the variety of contemporary media. There is little or no opportunity provided to challenge a student in a developmental, constructive, or creative way with regard to the phenomena of electricity, electronics, mechanics, internal combustion, aerodynamics, fluid dynamics, etc.

The industrial arts teacher will normally have taken courses at the undergraduate level in mathematics, physics, chemistry and allied sciences. These, when properly applied would be of considerable value to a forward looking concept of technical education in industrial arts teacher education.[9]

The Future. There are a variety of predictions that could be made with regard to practices in this area of education. Each research considered has set forth certain conclusions. Taken as projective material they provide considerable aim for industrial arts teacher education. A consensus, rephrased would include at least the following notions.

1. There will have to be an ever increasing emphasis upon including science and mathematics among the general educational requirements of the industrial arts teacher.

2. Techniques and teaching procedures will have to be improved at all levels.

3. The methodology of industrial arts education will have to incorporate the methods of industrial research; that is the development of creativity by its partisans and the utilization of all varieties of group and individual processes.

4. A continuing and expanding emphasis upon contacts with industry as a prime representative of technology is necessary that the many necessarily vicarious learnings can take place most effectively.

5. In spite of the shortage of teachers it is important that some selective factors continue to operate that the individuals entering teaching are capable of that which they purport to teach.

6. There is continuing evidence of acceptance of responsibility for the inservice education of teachers. There is also evidence that despite the obvious need for further education many teachers do not participate in programs that would tend to upgrade them. There is, for example, a serious shortage of competent graduate students.

7. There is a growing tendency to enlarge the general shop, laboratory of industries concept. The procedure of rotation through a series of media is beginning to be recognized as a weak program and the multiple facilities are used in a more appropriate fashion as they might be in industry for product development, for example.

8. There is some evidence to suggest that the various "technique" approaches to teaching are dying out, to be replaced by problem solving procedures.

Conclusions. The evidence for industrial arts teacher education suggests that the field is likely in about the same state as other curriculum areas. Much needs to be done to foster development which is consistent with technological evolution. Hornbake has set this forth in concise fashion:

"A primary personal and social contribution of industrial arts education is precisely this dual task; namely the development of technical competence among the millions and the discovery of a variety of technical talents. These goals can be accomplished only if children and youth are provided with representative experiences."[10]

As in the other aspects of those fields dealt with in industrial education, industrial arts at the teacher education level is in a state of flux. The greatest hope is involved in that a number of the researchers and leaders in the field are cognizant of the problem and are working to resolve the flux into programs of activity and learning appropriate to the world scene.

Vocational Industrial Teacher Education

The second category of teacher education to be covered in this report is concerned with the preparation of teachers to provide instruction in trade and industrial education. This category of education has been defined by the American Vocational Association as follows:

Instruction which is planned for the purpose of developing basic manipulative skills, safety judgment, technical knowledge, and related occupational information for the purpose of fitting young persons for initial employment in industrial occupations and to upgrade or retrain workers employed in industry.[11]

It should be noted, that for the purposes of this chapter, consideration is given only to the preparation of teachers at the professional-technical level. With this in mind, it can be noted that the majority of the statements made with reference to industrial arts teacher education have applicability in the appropriate situation.

Staffing Problems. The problem of categories of teacher education for vocational education tends to be much the same as mentioned previously. Teacher supply, as in the other fields, has five major aspects. Each of these is a concern of the teacher education institution as well as that of the various administrative and supervisory groups in the technical, vocational and general high schools where the teachers are employed. The five categories have been cited as recruitment, selection, qualifications, education and training, and certification.[12]

The prime problems in teacher preparation in this area have been set forth by a considerable number of groups. The major emphases in recent discussions have been in terms of the need for technicians. The requirements for teachers have apparently changed but little.

Trade and Industrial Experience. Among the common elements required of all teachers in trade and industrial education is a certain amount of time in the practice of the trade or occupation which he is to teach. In technician training there are a few difficulties presented because of the somewhat different nature of the program. Often the training is unique and does not encompass an apprenticeship or period of early study. The question of "how long" is not generally answerable because of the various local controls involved.

Basic Education. The notion of formal education for teaching in this area has also changed considerably in the last few years. A goodly number of teachers now possess degrees at various levels. There is of course bound to be an increasing percentage with the changes in the nature of the training provided. As the instructional program requires more of the type of training which is only obtainable at institutions of higher learning or in special institutes, it is likely that more and more people will obtain degrees corresponding to their work. The general qualities of teachers in this area are appropriately the same as for any other teacher in a particular school district or system.

Professional and In-Service Education. The professional education of vocational teachers in all categories is being upgraded considerably. As concepts and skills become more complicated it is essential that the professional skills of the teacher match the technical background necessary to impart that which is basic to the student. As a consequence it is likely that there will be a general upgrading of men in the field as well as a stiffening of the requirements for those in training. In terms of the baccalaureate degree the primary bolstering has been in science and mathematics. There have not been large numbers of teachers available for some of the new technical fields. As a consequence a good many of the teachers have had training in the military services. It is likely that the next steps will be in the strengthening of the professional program.

Public Attitudes. Quite appropriately, both labor and management groups have concerned themselves with the nature and state of vocational education in this new era. Walter Reuther, for example, sets forth the notions of a labor group:

"With the spread of automation, there will be a growing need for specialized semi-professional technicians, as well as for professional engineers and skilled workers. The education system of the nations should be preparing now to meet these requirements."[13]

The above remarks citing automation are used to illustrate first, an important point-of-view with regard to our new technology and also to suggest that while a primary activity area is that of the upper echelons of scientists and engineers, the operative utilization of developments as well as the process are contingent upon multiple levels of skill and concept development on the part of people in general.

Education and Technological Development. The problems of vocational education increase considerably as new techniques in industry and science come upon the scene. The design and production of automatic equipment is, for example, but one aspect of its productive use. Personnel are needed to design, as suggested, but also to construct, supervise, and maintain that which is put into operation. This increases our already desperate need for engineers, but even more it demands an engineer of a new type, the systems engineer. Consistent with this pattern of training is that which is most appropriately the function of various technical-vocational schools and educational programs. The technician can be narrowly trained to maintain or operate gear of one type or another, but in order to keep up with the pace set by the professional programs for chemists, physicists, and engineers, the technicians must possess broader skills and more extensive knowledge.

The educational program for such persons is likely to be undertaken by a variety of institutions, both private and public. The unions, private institutes, manufacturing firms, equipment firms, and a variety of other groups will have to stand by the public schools in such training. The assistance may be in the form of specific training programs or it may be the aid of particularly trained individuals, or the materials of instruction. Certainly there is much needed before such programs can be classified as contemporary.

Conclusions. A considerable problem facing all phases of education is that, rather than providing a wholesome and cleancut revamping of education in those areas where such action is necessary, there will be activity at a much lesser level. The mistake may be in the application of remedial measures, as is being done in some instances, or in a dismissal as the time element softens the blow of scientific and technological deficiencies. The area of vocational education has problems, as does the rest of education. Much of it lies at the feet of shadow men—men who are needed, but not available for the important task. With more support, federal and local, it is possible that this first step can be taken and those that logically follow thus enabled.

Education for Industry

The last program to be considered in terms of industrial education in general is titled, Education for Industry, because of the descriptive qualities rather than for its general acceptance in terms of semantic specificity. Such programs are in existence at a number of institutions under varying nomenclature. It does not coincide in purpose or in content with curriculums such as industrial management, industrial administration, or industrial engineering.

The graduates of such curriculums are expected to obtain employment as production or inspection workers in manufacturing industries in particular and then to move into positions of greater responsibility, including supervision. The basis for such curriculums is technical experience leading to competency in this area as well as in leadership and human relations. This is of course in common with other parallel programs. Consistent with other college and university curriculums, the graduates are expected to be competent in communications skills as well as developing capabilities in socio-civic areas. In any instance, the specific courses are not intended to be mutually exclusive: but in general the technical qualifications are obtained through course work in drafting, diverse shops, chemistry, physics, and mathematics as well as from an industrial internship. Additional work in such areas as psychology, sociology, English, speech, and business administration are intended to develop the competencies in human relationships and leadership. Graduates of such curriculums experience the common academic requirements of all university curriculums, such as history, government and politics, and other areas or disciplines.[14]

Such curriculums are largely intended to prepare for responsible participation in the production phases of manufacturing and service type occupations. A significant part of the curriculum involves the completion of two periods of internship in industry. Such periods of actual work experience are visualized as having three major values. First, they enable the student to observe the application of principles and theories that he has been exposed to during the college experience. Second, they provide enrichment for those courses that may follow the various experiences in industry at the occupational level. Thirdly, they provide definite evidence of the ability of the student to obtain employment consistent with his educational background.

The internship activity is of a key nature in any program of this type. It provides an optimal value to the program of learning. The coordinator of work experience, the employer, and the student provide the total frame for the latter, as one who is striving to become a worthy and efficient member of the American industrial scheme.[15]

Conclusions

In terms of characterization, the society in which we operate is often labelled industrial and democratic. Such terms are reasonably accurate, yet the context in which such words tend to be descriptive has altered considerably in the last fifteen years. In order to clarify the connotation, one must use words such as atomic, automated, jet-powered, air-conditioned, and electronic. Any assumption that the lay individual in each instance has much beyond a hazy notion with regard to either the implications or operational aspects of the involved elements would be quite optimistic.

Somehow the vast educational enterprise of the United States must develop more adequate understandings and certainly a contemporary type of know-how which will suggest that American ingenuity has not gone out of style with the Model-T. This is of course much more complicated than in the days of Model-T technology. However, the educational activities of the various school systems should be that much more effective that a significant dent could be made in this vast body of knowledge and technique.

The various educational programs which operate in the broad area of education as derived from an industrial society obviously have unique functions in such a situation. The consensus suggests that much of American technology, as well as that of other nations, whether it exists in the realm of physics, chemistry, or other sciences has need of exploration in terms of application. Certainly the technical laboratories as represented by industrial arts and vocational industrial education can do much to foster higher level activities than are in current practice. There is much agreement that as such programs mature and attempt to better represent our culture there will have to be even further departure from individual activities which emphasize muscular dexterity to the greater challenge of the solution of technological problems. This means that activity will have to be characterized by the words analyzing, appraising, investigating, planning, evaluating, testing, and others which suggest industry and technology rather than the job shop.

Such changes are occurring in teacher education programs and it is here that the change must first be made. Such changes are occurring in secondary education. However, the rate of change is slow in either instance. If there is a challenge that can be clearly stated, it is this: that until technical programs of education become technologically oriented and cease to be manipulatively based, there can only be a level of progress in these areas that is inadequate to the needs of American industry in particular and the culture in general.

Selected Bibliography

An Investigation of the Role of Industrial Arts in a World of Technological and Scientific Achievement, Unpublished Manuscript (College Park, Maryland: University of Maryland, Industrial Education Department, 1958). A report of an ongoing research project devoted to the improvement of industrial arts programs at both the secondary school and teacher education levels.

Henry H. Armsby, *Scientific and Professional Manpower* (Washington, D. C.: U. S. Office of Education, 1954). Discusses the many problems growing out of the shortage of scientific and technical manpower and reports the activities of several agencies in attempting to meet the shortage.

Definitions of Terms in Vocational and Practical Arts Education, (Washington, D. C.: American Vocational Association, 1954). Prepared by a committee on research and publications to assist in the resolution of some of the confusion which exists in terminology in vocational and practical arts education.

Education for Industry, Unpublished Manuscript (College Park, Maryland: University of Maryland, Industrial Education Department, 1957). This manual describes an ongoing program of "Education for Industry" as it has been established at one university.

Carter V. Good, *The Dictionary of Education* (New York: McGraw-Hill, 1945). An authoritative source of professional definitions for the various areas of education.

R. Lee Hornbake, "Time for Progress," *School Shop,* XV, (June, 1956), 7-8. As the title suggests, this article notes the tendency toward doldrums in industrial arts and provides a challenge towards change.

Max Lerner, *America As A Civilization* (New York: Simon and Schuster, 1958). An extensive work providing comment on many facets of American culture.

Manual for Organized and Supervised Work Experience, Unpublished Manuscript (College Park, Maryland: University of Maryland, Industrial Education Department, 1957). Reviews the values of professional education for industry in terms of employer problems.

Meeting Manpower Needs for Technicians (Washington, D. C.: U. S. Office of Education, 1957). Explores various means of increasing the supply of technicians.

William J. Micheels, "Industrial Arts Teacher Education in 1970," *Industrial Arts and Vocational Education* (February, 1958), 29-32. A projection of current data towards determining the nature of program and the needs for such in 1970.

President's Committee on Education Beyond the High School, *Second Report to the President* (Washington, D. C.: U. S. Government Printing Office, 1957). A brochure reporting the status of higher education in America to the President.

Problems and Issues in Industrial Arts Teacher Education, C. Robert Hutchcroft, Editor (Bloomington, Illinois: McKnight and McKnight, 1956). An important volume giving responsible views with regard to issues in the field.

Readings in Education, Gerbracht and Wilber, Editors (Bloomington, Illinois: McKnight and McKnight, 1957), 329 pp. A review of the literature of professional education and related fields with a particular application to industrial arts teacher education. Philosophy, psychology, methodology and curriculum are among those included.

Walter Reuther, *The Impact of Automation* (Detroit: UAW-CIO), 1955. A summary of testimony offered before the Sub-Committee on Economic Stabilization of the Joint Committee on the Economic Report of the United States Congress.

Superior Practices in Industrial Arts Teacher Education, Hornbake and Maley, Editors (Bloomington, Illinois: McKnight and McKnight, 1955). Report of a research reviewing practices and recommendations of desirable practice of 165 industrial arts teacher education departments.

PAUL E. HARRISON, Jr., Associate Professor of Industrial Education, University of Maryland, was born in Illinois (1921), and received his B.Ed. degree from Northern Illinois State University (1942), his M.A. from Colorado State College (1947), and his Ph.D. from the University of Maryland (1955). Taught in Evanston (Illinois) Public School, Chicago Teachers College, Iowa State Teachers College, and for the U.S. Navy Technical Training Program. Has contributed to the *Readings in Education.*

NOTES FOR INDUSTRIAL EDUCATION

1. Carter V. Good, *Dictionary of Education* (New York: McGraw-Hill), 216.

2. Henry H. Armsby, *Scientific and Professional Manpower* (Washington, D. C.: U. S. Office of Education), 3.

3. President's Committee on Education Beyond the High School, *Second Report to the President—Summary Report* (Washington, D. C.: U. S. Government Printing Office), 1.

4. Max Lerner, *America as a Civilization* (New York: Simon and Schuster), 226-227.

5. R. Lee Hornbake and Donald Maley, Editors, *Superior Practices in Industrial Arts Teacher Education,* American Council on Industrial Arts Teacher Education (Bloomington: McKnight and McKnight), 118-119.

6. *Ibid.,* 1-9.

7. *Ibid.,* 120-121.

8. W. J. Micheels, "Industrial Arts Teacher Education in 1970," *Industrial Arts and Vocational Education Magazine* (February 1958), 29-32.

9. *An Investigation of the Role of Industrial Arts in a World of Technological and Scientific Achievement,* Unpublished Manuscript, Industrial Education Department, University of Maryland, 2.

10. R. Lee Hornbake, "Time for Progress," *School Shop,* XV, (June, 1956), 8.

11. *Definitions of Terms in Vocational and Practical Arts Education,* (Washington, D. C.: American Vocational Association), 26.

12. *Meeting Manpower Needs for Technicians,* (Washington, D.C.: U. S. Office of Education), 26.

13. Walter Reuther, *The Impact of Automation* (Detroit: UAW-CIO), 17.

14. *Education for Industry,* Unpublished Manuscript (College Park, Maryland: University of Maryland, Industrial Education Dept. 1957), 2.

15. *Manual for Organized and Supervised Work Experience,* Unpublished Manuscript (College Park, Maryland: University of Maryland, Industrial Education Department), 2.

MATHEMATICS

GEORGE GREISEN MALLINSON
Western Michigan University

The past four or five decades have been midwife to many changes and new ideas in science and mathematics, particularly in areas once thought to be fixed and immutable. Much of Newtonian physics has given way to Einsteinian physics. The theories of relativity and the more recent refutation of the law of parity have given the scientist as well as the layman entirely new benchmarks from which to view the nature of matter and energy. It would, of course, be idle to suggest that such changes and ideas have been viewed by all scientists with the objectivity that these scientists profess to be their trademark. In many cases the innovations came so rapidly that these persons were passed by. With others, the appearance of newer concepts of science threatened to abolish many of the foundations on which cherished theories and hypotheses were based. Hence, they have been often greeted with resistance.

Perhaps among the most startling developments, however, are the changes that have taken place in mathematics. To many persons the field of mathematics has long represented a refuge of stability where objects and phenomena could be counted, classified, measured, and otherwise described. Mathematics was a tool utilized to establish order. However, in many cases, the newer developments in science emerged only because mathematics became the new language of science without which the concepts themselves were inexplicable. Since many of these concepts of science dealt with uncertainty, relativity and probability, the language itself assumed these characteristics. In fact, carried to its ultimate the language and the concepts it explained became inseparable and indistinguishable. Thus, the device which had been used to fix and stabilize, adopted many of the characteristics that it had once been thought to eliminate. A new "speaking" vocabulary developed in the language, including such terms as sets, groups, and limits.

Such developments in mathematics seemed repugnant and heretical to many mathematicians. They represented a degeneracy of the classical standards of mathematics education. Many of the avowed supporters of the new doctrine were looked upon as might be surrealists by the landscape painter. Yet, it was obvious that the new mathematics, as expression of thought, would complement the older concept of manipulation and counting, despite the askance that it was accorded.

The challenge of mathematics to college and university education has come somewhat indirectly in that much of the revolutionary thinking has been done in terms of modifying the high-school mathematics program. Such evidences with respect to high-school mathematics have been appearing for some time. Yet, for such new doctrines to make inroads at this earlier level, the colleges and universities must modify their training programs to prepare a new generation of teachers to handle mathematics in terms of an expression of thought. The challenge can be met, however, only by first analyzing the changes that are proposed for the mathematics programs in high schools. With the information from these analyses, ways can be suggested in which mathematics education in colleges and universities may be modified in order to (1) integrate its structure with the earlier training in mathematics, and (2) provide trained teachers for the high schools.

The New Structure for High-School Mathematics

One of the earliest published efforts to re-examine the traditional program of high-school mathematics came from the state of New York. In 1942 a Mathematics Syllabus Committee was appointed by the Board of Regents to study the high-school program and to prepare new syllabi for the mathematics courses, if needed.

In examining the traditional programs for introductory algebra, geometry, advanced algebra and trigonometry, it was pointed out that in the later courses particularly plane geometry, "little or no attention . . . [was] given to the use of extension of the basic principles of arithmetic and algebra."[1] The Committee indicated also that "as a result, many of these concepts and skills are lost . . . from lack of use."[2]

In an effort to remedy the problem, a new mathematics syllabus was developed in which an attempt was made to integrate plane geometry with arithmetic, algebra and numerical trigonometry in so far as such integration was possible and desirable.

According to the Committee such integration could be brought about by the following:[3]

"Greater use of common and decimal fractions and per cents in mensuration problems

An introduction to the meaning and use of approximate number

Increased emphasis on numerical trigonometry

The use of algebraic symbolism and algebraic proof wherever this is desirable

The use of algebraic equations in the solution of geometric problems

An introduction to coordinate geometry."

Another change suggested was the reduction of the number of geometrical theorems required in the first course in geometry with emphasis of a few related theorems to increase the "understanding of the nature of proof and the meaning of sequential thinking."

The later courses for the eleventh and twelfth grades emphasized further the theme of integration with the combination of intermediate algebra, plane trigonometry and coordinate geometry. At all levels in the syllabus the number of activities involving manipulation were reduced with an increase of problems emphasizing mathematical analysis.

An examination of the postulates underlying these proposals indicates that the extent to which the New York program represents the "new thinking in mathematics" might be questioned. In general, the theme behind the program involves emphasis and de-emphasis, re-ordering and integration of traditional areas and courses. The "modernist" in mathematics might point out that the New York program implies no major change in college and university offerings and hence no real challenge. This viewpoint, of course, can be questioned.

However, there have been a number of other efforts made recently to re-examine the high-school program of mathematics. The Commission on Mathematics of the College Entrance Examination Board developed as a result of the doubts raised in the minds of the Mathematics Examiners. The examiners began to question whether the tests prepared by the Board measured adequately the outcomes of the newer programs of mathematics instruction that had developed in certain of the better schools preparing students for college. The examiners were of the opinion also that the mathematics program in the typical secondary school was outdated. By means of a grant from the Carnegie Corporation, the Commission was established and met for a number of plenary sessions to debate certain implications of the issues just cited.

As a result of its early deliberations these points of view were established:[4]

"(1) In the past thirty years the nature of mathematics as a subject has been substantially altered by the results of mathematical research.

(2) Similarly, the applications of mathematics—oftener than not, the new mathematics—have been greatly extended into such fields as the design of computing machinery; the social sciences, including psychology, economics, and sociology; and industrial quality control.

(3) Modern mathematics embodies both a point of view and new subject matter, much of it simpler than traditional subject matter and provides unifying principles by which the standard core of basic material —algebra, geometry and trigonometry—can be brought into more effective and closer relationship with twentieth century ideas and applications.

(4) Despite these developments, school and college mathematical curricula have been largely unaltered."

In general, the Commission concludes that (1) entirely too much time is spent in high-school mathematics on routine manipulation, and too little time on the emphasis of basic concepts, (2) deductive reasoning, if taught at all is largely confined to geometry and this phase of mathematical experience should be extended into algebra and trigonometry, (3) the student memorizes the steps of deductive geometrical proofs without being expected to reason deductively, and (4) mathematics is often presented as a series of tricks for solving numerical problems, rather than being a creative experience.

The suggestions made for improving the curriculum in mathematics include (1) the elimination of the teaching of mathematical skills for their own sake and the development of the understanding of certain laws of the algebraic system with emphasis on the generality of these laws, (2) change of emphasis on standard topics in trigonometry (i.e., de-emphasis of triangulation with emphasis on resolution of vectors into components), (3) inclusion of solid geometry in the course with plane geometry, and (4) introduction of new topics such as concepts of sets, probability, statistics, abstract algebra, symbolic logic and analytics.

The program of the Commission definitely goes farther than the New York program in that it emphasizes more than integration, namely, the deletion of obsolete materials and the inclusion of modern topics of mathematics. The Commission was, of course, well aware that such proposals would have important implications for college and university programs in mathematics. It would require for programs for teachers, in particular, the deletion of materials taught traditionally in a number of college courses and their replacement with other mathematical learnings. A challenge is, therefore, thrust by the Commission before the mathematics departments of colleges and universities.

In all probability the project for the modification of high-school mathematics that has received the most publicity is the one sponsored by the University of Illinois Committee on School Mathematics. The project is supported by a grant from the Carnegie Corporation, with Dr. Max Beberman of the University of Illinois as a moving force.

As with the two efforts already described, the Illinois program is based on the development of mathematical principles without regard for traditional boundaries of the usual mathematics courses. The course for high-school freshmen ordinarily referred to as "elementary algebra" is called "First Course" in the Illinois system. It consists mainly of algebra, but also includes arithmetic and geometry. The theme of integration of the first course is evident in the later courses also. Students study principles and ideas of arithmetic, algebra and geometry at all levels, the avowed purpose being to develop the concept of unity in mathematics.[5]

There are two basic ideas that underly all instruction in mathematics in the Illinois system. The first is the treatment of mathematics as a form of language. According to Beberman,[6] "numerals are only the names you use for numbers [and] . . . equations are merely statements about numbers that are true if you put in the right names." The conclusion is made, therefore, that mathematics should consist of seeking names for unknowns and variables, rather than dealing on paper only with the barren numbers. Dealing with abstractions therefore, becomes "name seeking" rather than manipulation of numbers.

The second basic idea that pervades the Illinois experiment is the concept of sets. A set is, of course, a collection of items, objects, phenomena, or even philosophical ideas. All the members of the set have at least one characteristic in common. If the student is taught to think analytically in terms of sets, the experimenters claim that the experience teaches him to discover relationships, identify obscure patterns, and perceive new arrangements among items. Such a habit of thinking is believed to cause the student to consider manipulation and detail in the proper mathematical perspective, and to use mathematics as a means for explaining and describing, rather than as a device solely for counting and measuring.

In this program, emphasis is placed on analytical geometry and circular functions of modern trigonometry. Other topics have been dropped, such as the logarithmic solutions of triangles, derivations of mensurational formulae, as well as the substitution of Newton's for Horner's method of finding irrational roots of polynomial equations.

The success of the idea at least in so far as its attractiveness is concerned may be judged by the fact that in the experimental schools involved in the project, enrollments of *non-college-preparatory* students have increased. Further, the students profess enthusiasm with the method for presenting mathematics.

The Illinois program is similar to the program of the Commission on Mathematics in that at the high-school level, new areas of mathematical study involving analysis, are replacing manipulative activity. Both of these programs consist of more than the integration of traditional topics of mathematics into a new pattern. The success in college mathematics of the students from these programs will, of course, be one of the major criteria of the success of the experiments.

A new project underway which may have even more influence than the programs already described should be mentioned here. This project, begun at Yale University and referred to as The School Mathematics Study Committee, is organized in a manner somewhat similar to the project of the Physical Science Study Committee at Massachusetts Institute of Technology. The Committee is planning on recruiting the aid of a number of recognized mathematicians as well as teachers of mathematics in the junior- and senior-high schools for studying the program of mathematics at the levels indicated. The group that is finally assembled will develop textbooks and other teaching materials as has the P.S.S.C. In terms of financial backing, the efforts of this group may "carry more weight" than the efforts of the groups already mentioned.

However, in the back of one's mind there appears to be an obstacle that needs to be overcome. The development of programs like those described demands a program for training mathematics teachers to handle such courses in the high school. Unfortunately, the academic world is not full of teachers of mathematics who are experimentalists and who are able to rise above their past training and develop a "psychological owner-ship" of the new mathematical ideas. Thus, in addition to merely train-ing vast numbers of high-school teachers of mathematics (who are, of course, becoming needed in increasing numbers) the traditional mathe-maticians in colleges and universities must somehow encompass these new mathematical ideas into their thinking and translate them for these future teachers. Obviously the new ideas cannot be "swallowed" by col-leges and universities without some evidence that they are tenable. Yet there is sufficient evidence of their validity to warrant a try. Such will, therefore, demand on the part of mathematics departments in colleges and universities, a philosophy of experimentalism. This philosophy will involve a willingness to accept students who have been through high-school programs oriented around the new mathematical ideas and to train *at least some* teachers to test out the new ideas.

Such is the challenge. What are some suggestions for the ways in which the challenge can be met?

Mathematics Training in Colleges and Universities: A Challenge

The Philosophy of Experimentalism. The writer recently had the op-portunity to discuss with a number of veteran college teachers of mathe-matics certain aspects of the experimental program described above. The comment was made by one, "Oh, we have been doing that for years!" Yet when the essences of the previous programs were discussed, the general opinion was that they were a waste of time since it represented a "lower-ing of standards." The two comments are obviously anomalous. Frankly, there are many mathematicians who are steeped so firmly in the fixed and determinate mathematics that a system involving relationships of the kind discussed earlier is almost impossible to accept. The traditional program of college and university mathematics is thus all that can be digested.

It is, of course, most unlikely that the vast numbers of teachers of high-school mathematics, even those with majors in mathematics, can without help resynthesize their mathematical learning into the modern systems. Even those with keen mathematical perspective do not have available materials and time for such efforts. It is an inexorable fact, therefore, that leadership for the changes must come from the top. This will demand at the college and university level, an eclecticism as well as a philosophy of experimentalism not always evident in the administra-tive complexes of mathematics departments. Yet, by some technique, it will be necessary for these departments, even if only on an experimental basis with a limited number of students, to present undergraduate pro-grams consistent with the new mathematical ideas that are accepted by the various leaders of the new schools. There may be many honest mathematicians who will protest that efforts such as these will produce a group of mathematics teachers who are devoid of understandings of the basic concepts of mathematics. Such persons may need to be reassured that many new ideas have emerged in the fields of science and mathematics as a result of great effort and thought of a few pioneers. When put to test, many of the ideas have proved to have no merit beyond that of the old ideas. A few have been productive of major advances. However, seldom if ever has a new idea proved to be so catastrophic, that those involved in its implementation were instrumental in setting back progress. However, the proposed change of emphasis from "determinism to indeterminism" as exemplified in the new mathematics is perhaps one of the greatest philosophical shifts that has ever been faced.

For those who need comfort from past experience, it is well to note that mathematics has not been completely dormant prior to the rise of the new schools. However, the experiences of many of those seeking new approaches for mathematics have not been fruitful. New ideas have not emerged readily and changes have involved shifts of topics rather than modifications of over-all structure. The changes now proposed are really emergent from past as well as contemporary efforts. The publicity appeared all at once.

Obviously, there is a definite need for a modification of undergraduate training in mathematics in order to provide the contemporary subject matter for teachers. Without major shifts and new emphasis there is little likelihood of the development of modern curriculum offerings in mathematics in the high school. Such changes in the undergraduate programs need to encompass all years of college mathematics. The Commission on Mathematics however has been most emphatic in pointing out that the first two years of college mathematics are especially important in such modifications since the experiences of these years are closest to the high-school program. The tenor of these years in particular is likely to influence greatly both the content and instructional methods of mathematics in the high school. If these two years deal only with deterministic mathematics, with little emphasis on structure and relationship, high-school mathematics is likely to change little, if at all.

The program of mathematics for the freshman and sophomore years of college has been probed by the Committee on the Undergraduate Mathematics Program of the Mathematical Association of America.[7] The efforts of the Committee were coordinated with the efforts of the Committee on Engineering Mathematics[8] and the Committee on Mathematical Training of Social Scientists.[9] These efforts were further supported by those of the Summer Writing Groups at the University of Kansas[10] who sought to produce materials consistent with the recommendations of these committees.

In general, the recommendations and proposals are much too extensive to describe in detail here. However, an examination of the bibliographical citations indicates the efforts made to develop new integrated courses referred to as Universal Mathematics I and II. The emphasis on modern mathematical structure and relationships is evidence in these citations in the sub-titles in apposition. The courses presuppose one year of high-school algebra and geometry and have greater emphasis on analytics and calculus than on the traditional algebra, trigonometry and analytic geometry. The policy of dropping old course titles is further evident in the program now under development at Western Michigan University. The proposed freshman course that replaces algebra, trigonometry and analytic geometry is referred to as Analysis I. The sophomore course that covers elements of calculus as well as the other traditional areas ordinarily following the introductory course is referred to as Analysis II.

The advanced undergraduate program that is proposed at Western Michigan University is based rather extensively on the modern view of the analysis aspect of mathematics rather than on algebra emphasizing mathematical structure. Geometry will change similarly in terms of the study of advanced Euclidean geometry. As yet the program is under development and has not been published.

The recommendations of Syer[11] for the advanced undergraduate program listed below are similar to those of the Commission on Mathematics:

". . . B. *Statistics:* Emphasizing probability and statistical inference.

C. *Applications of Mathematics:* Mechanics (Statics and dynamics), theory of games, linear programming, operations research.

D. *Modern Algebra:* Matrices, Theory of Numbers, Theory of Equations.

E. *Advanced Geometry:* Projective geometry, non-Euclidean geometry, Differential geometry, Topology.

F. *Foundations of Mathematics:* Theory of sets, Mathematical or symbolic logic, Postulates for geometry, algebra and arithmetic, The real and complex number systems . . ."

A comparison of the emphasis of this curriculum pattern with that of the traditional arrangement shows clearly the greater emphasis on mathematical structure with a reduction of emphasis on manipulation and analysis.

The proposal clearly takes cognizance of the needs for teachers of modern secondary mathematics.

The small number of new programs discussed in the literature is evidence of the need for much additional work.

The In-Service Program. The term "in-service" will refer in this report to all work taken in mathematics beyond the bachelor's degree. It is unlikely, obviously, that many of the teachers presently holding bachelor's degrees in mathematics will have experienced a college program with the new emphasis. In effect, therefore, all graduate work in mathematics may be somewhat remedial in nature for some time to come. Within a few years it is possible, however, that the graduate program can be modified so as to be a sequential extension of the new undergraduate program. The remedial aspect of in-service may then be designed chiefly for those whose training in mathematics dates back to the present or earlier.

The responsibility of college and university mathematics departments for in-service training however will not deal with course work ordinarily taken by undergraduates and which these graduates failed to experience as undergraduates. Rather, it will be course work of a graduate nature involving a great deal of ingenuity on the part of the instructors. It will demand efforts from the most able instructors whose aim will be to re-synthesize and fill in the older, somewhat outdated and perhaps incomplete mathematics training of these persons using modern materials like the ones described earlier. Much of such training will be "played by ear" as have been the Summer Institutes sponsored by the National Science Foundation.

The writer is not enough of a modern mathematician to describe the type of pedagogy that will be needed. However, it may be that another Beberman will appear with the key to the renovation process.

An Epilogue: The challenge to mathematics could be solved if, of course, the ultimate goal were more fully validated and accepted or the road had been trod at least once. However, experience is sparse and there are few guideposts to offer direction. Yet, the challenge is present and it cannot be ignored. It must be met by the forward thinking of departments of mathematics in colleges and universities where some of the principles of the new system violate what has been held in cherished regard.

Selected Bibliography

Joseph Breuer (Translated by Howard F. Fehr) , *Introduction to The Theory of Sets* (Englewood Cliffs, New Jersey: Prentice-Hall, Inc., 1958) . The first book on the subject of the theory of sets in the English language, designed specifically for the beginner. Starting with essential ideas developing cardinal numbers, the book progresses through a discussion of infinite sets giving proof of the equivalence theorem and developing transfinite cardinal numbers. Later, the theory is extended to ordered sets and ordinal types, discussing in detail the controversial theorem of well ordering. The theory is next applied to sets of points, types of sets of points, and their use in the study of function. The book closes with a survey of paradoxes and the controversy of formalism and intuitionalism.

Commission on Mathematics of the College Entrance Examination Board, *Concepts of Equation and Inequality—Sample Classroom Unit for High School Algebra Students* (New York: College Entrance Examination Board, 1958) . A sample classroom unit for high-school algebra students, designed to show how the subject matter may be treated in a modern mathematics classroom.

Nathan Altshiller Court, *Mathematics in Fun and In Earnest* (New York: The Dial Press, 1958). Introduces the reader to the implication of reasoning and the relationship between mathematics and genius. The book covers the sociological aspects of mathematics, the meaning of infinite, and the mathematics of aesthetics.

Felix Hausdorff (Translated by John R. Aumann, et al.), *Set Theory* (New York: Chelsea Publishing Company, 1957). An exposition of the most important theorems of the Theory of Sets, along with complete proofs, so that the reader should not find it necessary to go outside the book for supplementary details. The book does not presuppose any mathematical knowledge beyond the differential and integral calculus, but it does require a certain maturity in abstract reasoning.

Kenneth S. Miller, *Elements of Modern Abstract Algebra* (New York: Harper, 1958). A simple, concise presentation of the elementary facts of modern algebra, considering the nucleus of ideas clustered around the concepts of groups, rings, and fields. High school algebra is the only formal prerequisite, although a knowledge of the real and complex number system and some degree of "mathematical maturity" will aid in the complete appreciation of the subject matter.

Paul C. Rosenbloom, *The Elements of Mathematical Logic* (New York: Dover Publications, Inc., 1950). An introductory treatise to mathematical logic. The book covers the structure and representation of Boolean algebras which in turn are applied to deductive systems.

Donald Smeltzer, *Man and Number* (New York: Emerson Books, 1958). An account of the development of man's use of number through the ages, covering the acquisition by early man of a sense of number and of systems of number words, and the transition from early methods of recording numerical information and carrying out calculations to those used at the present time. The aim has been to give a picture of the ways in which man's appreciation and use of number have grown, of the social and practical influences on that growth, and of the place of number in the development of man's way of life.

Hans J. Zassenhaus, *The Theory of Groups* (New York: Chelsea Publishing Company, 1958). Deals with the fundamental concepts of group theory together with a detailed study of the concept of homomorphic grouping. Group-theoretic concepts are developed from the beginning.

GEORGE G. MALLINSON, Dean, School of Graduate Studies and Professor of Psychology and Science Education, Western Michigan University (Kalamazoo, Michigan), is a native of Troy, New York. He received his B.A. and M.A. degrees from New York State College for Teachers at Albany (1937, 1941), and his Ph.D. from the University of Michigan (1947). He was elected Burke Aaron Hinsdale Scholar at the University of Michigan for the year 1947-48, a designation granted to the top-ranked Ph.D. of the year. After teaching in Whitesboro and Buffalo (New York), he served in the U.S. Army (1942-45), and was appointed

Director of Science Education, The Iowa State Teachers College (1947-48) ; in 1948 he joined the Western Michigan University, and has taught in the summer sessions of the University of Michigan, University of Kansas, University of Virginia, Western State College of Colorado, Oklahoma State University, and other institutions. In 1953 he was elected President of the National Association for Research in Science Teaching, and in 1954 President of the Michigan Science Teachers Association; is now Editor of the *Newsletter of the MSTA* and of *School Science and Mathematics;* Chairman, Science Study Committee of the Michigan Council of State College Presidents; a member of the Governor's Science Advisory Board. Since graduating from the University of Michigan, he has published over 125 articles and books, chiefly in the field of science education.

NOTES FOR MATHEMATICS

1. *Mathematics 10-11-12: An Integrated Sequence for the Senior High School Grades.* The Bureau of Secondary Curriculum Development, The State Education Department, University of the State of New York (Albany, New York: The University of the State of New York Press, 1954) , 5.

2. *Ibid.,* 11.

3. *Ibid.,* 11 .

4. *Commission on Mathematics, College Entrance Examination Board.* Unpaged pamphlet (Commission on Mathematics, 425 West 117th Street, New York 27, New York, June 1957) .

5. Garner, John W., "Imagination + X = Learning." *Carnegie Corporation of New York Quarterly,* V (October 1957), 2.

6. ——, "Teaching Mathematics As A Language." *Better Schools,* IV (September 1958) , 6.

7. ——, "Report of the Committee on the Undergraduate Mathematical Program." *American Mathematical Monthly,* LXII (September 1955) , 511-20.

8. Thomas, George B., *et al,* "Report of the Joint Committee of the American Society for Engineering Education and the Mathematical Association of America on Engineering Mathematics." *American Mathematical Monthly,* LXII (May 1955) , 385-92.

9. Bush, Robert R., *et al,* "Mathematics for Social Scientists." *American Mathematical Monthly,* LXI (October 1954) , 550-61; W. G. Madow, *et al,* "Recommended Policies for the Mathematical Training of Social Scientists: Statements by the Committee of the Council." *Items* (a publication of the Social Science Research Council) , IX (June 1955) , 13-6.

10. Summer (1954) Writing Group of the Department of Mathematics, University of Kansas, *Universal Mathematics, Part I, Functions and Limits* (Lawrence, Kansas: University of Kansas Student Union Bookstore, 1954) ; Summer (1954) Writing Group of the Department of

Mathematics, University of Kansas, *Universal Mathematics, Part II, Structures in Sets* (New Orleans, Tulane University Bookstore, 1955) (pagination not consecutive).

11. Syer, Henry W., "Suggested Mathematics Courses for Certification Requirements for Teachers of Mathematics and Science." A mimeographed report (Boston University, October 1, 1957).

BIOLOGY

Carl S. Johnson
Ohio State University

Some biological science is made a part of the general education of nearly every college student. Sometimes it is biology, more often it is either botany or zoology. Separate departments of botany and zoology have become the norm in the larger institutions and the ambition of smaller ones. Once separated they from thenceforth seldom "revert" to teaching biology; it must be either botany or zoology to be "respectable," and each department quite naturally defends its own offering as the more meaningful to the general education of the college student.

The fact that green plants are the penultimate source of all organismic energy supports the claim of botany. Man is an animal is the supporting fact for zoology. Plants and animals are inextricably interrelated; all are life forms; the two "kingdoms" result from man's arbitrary separation of physiologically similar organisms primarily because of structural differences; these are arguments for biology.

Biological Science Ought to Be Included in General Education

There is very little argument that biological science ought not be included in the general education of all college students. This is true even for professors in fields far removed from biological science or any other science for that matter.

Biology, literally translated from *bios logus,* is a study of life. The study of life has tremendous interest and meaning for man; it is both stimulating and functional. It is also broadening. Man is himself an animal utterly dependent on other life forms for his food. In common with all animals he is dependent upon green plants for metabolic energy. Indeed, he is less able than many animals at synthesizing the amino acids essential to the construction of his protoplasm and is thus dependent upon other organisms. The greater part of the mechanically produced energy he now uses so lavishly is also derived from organic products.

The behaviors of matters that characterize the living have long fascinated man. The mysteries of life still challenge man's quest for understanding and will no doubt continue to interest man long after he finally understands the natural processes which make organisms out of inorganic matter. Some opportunity to explore the mystery of life ought to be provided in the general education of college students.

Because man is himself an organism developed from the same antecedents, from the same matter, and by the same processes by which other organisms have been developed, some understanding of other organisms is helpful to man in his quest to understand himself. Man must also live with other organisms, including his own species. In his relationships to his environment, man is governed by the same natural laws as govern all other organisms. Some appreciation of the nature and significance of natural law ought characterize the educated person. For all these reasons some biological science needs be included in general education.

What Course Should It Be For General Education?

That biology should be among the sciences included in general education is seldom disputed. What shall be selected from the broad field of biology? is the bone of contention. What is the best content for providing the basic understandings? Indeed, which are the basic understandings? How are the basic understandings best taught? What courses best provide biological science for general education? Specifically, is it botany, zoology, human physiology, or biology? Or, must it be some of each? These are the difficult questions for those who plan programs for general education.

It is certain that proponents for each course can readily be found. After all, some colleges offer only botany or zoology, some offer biology. Most schools insert human physiology into general education, teaching it as a part of physical education irrespective of the extent to which the same material is covered in required zoology or biology. There are also proponents of more restricted courses, those who argue that courses dealing with a single phylam, or class, or species, or principle, or specialization field are more effective biological science courses for general education than are the general courses. For example, Professor J. G. Edwards of San Jose State College proposed that entomology is superior to general zoology for general education.[1] Similar claims can be found in the literatures for many other branches of biology.

What then do the opinions of biological experts indicate with reference to the nature of biological science for general education? Nothing conclusive. It is possible to marshall supporting statements for that content and for that method of teaching it to the general student that each instructor or department believes is best. The preference is dictated not so much by research or the opinion of authorities as it is by the training and experience of each instructor.

Only one conclusion appears to be unanimously supported: that some biological science should be included in the general education of all college students.

The administrative structure for course development in higher education is such that courses to be offered and the content thereof are the exclusive responsibility of experts in each field. Who else would know as much about the subject matter of the field? This has been the thesis of the subject-matter fields from time immemorial with the tide of opinion with reference to the acceptability of that thesis swinging right and left. At present the tide, already swinging to the support of that thesis, has been "Sputnicked" to remarkable heights. This is the period for the academicians and, with an administrative structure already giving course development to the experts, the system which has produced a great number and variety of courses can be expected to produce still more.

Those responsible for curriculum development then choose from among the courses offered. Rarely do the planners of general-education curricula have much influence, directly or indirectly, over the nature and content of departmentally-developed courses. However, the variety of available courses is so great one might readily conclude that curricula for almost any purpose could be constructed by simple choosing wisely from the variety available.

Unfortunately this is not the case. The most serious problems for general education arise from the same causes which make the course-development process quite good for the training of specialists. As frontiers of knowledge have advanced, the effort required for a person to reach that frontier has increased. As the time and effort for reaching the frontier of knowledge have increased, we have concentrated effort on the area of specialization at the expense of other learnings. In a sense, general education and the training of specialists have become antitheses even though it has remained generally agreed that the foundation for the specialist ought be a good general education.

The effects of specialization upon the nature of general education have been accelerated by feed back. As we have specialized, we have in turn employed specialists to educate more specialized persons. In the case of the biological sciences, this has become true with the passing of the biologists. We no longer have biologists; we have anatomists, taxonomists, geneticists, histologists, cytologists, physiologists. We have specialists in classification, in structure, in function, in phyla, in classes. We thus have entomologists, ornithologists, mammalogists, protozoologists, apiarists, endoctrinologists, neurologists, *etcetera ad infinitum*. It is these specialists we then employ to teach in colleges and universities.

Still more feed back for specialization comes from the emphasis we place on graduate instruction and research. We employ specialists with a view to their training more specialists and to make contributions to special knowledge in their fields through research. In addition to these duties, they may teach the courses designated for general education. We do not seek persons trained for general education as the instructors of the general-education offerings of academic departments, including all sciences, even biology. These offerings are considered beneath the dignity of the specialist, consequently, if he has to teach them, he does not give them the kind of attention he gives his own courses. The system commonly relegates the general-education courses to junior staff members or, and this is often the case in big universities, to graduate students the while they are engaged in research and study in their specializations.

These conditions seem to prevail in the universities and the colleges, because their instructors are university trained, tend to follow. But, because the smaller staffs of the colleges have to spread across several courses, the specialist there has to teach several courses. Somewhat against his desires this may make more of a generalist of him with respect to the field of his department, but it rarely makes a biologist out of a zoologist just as "general physical science" remains an anathema to chemists and physicists.

Which of the presently taught biological science courses are best for the general education of college students? Not usually either botany or zoology because of the effects of specialization upon them. Nor is general biology, where offered at all, the best biological science course for general education simply because it has the acceptable title. Its instructors are not prepared for teaching general biology for students not intending to major in a biological science.

The Best-Basic-Course Problem Also Affects Specialists

The question as to which is best for general education, botany, zoology, or biology, also applies to the training of biological specialists. Shall their basic courses be general biology or shall they be specifically in the field of specialization? Shall the introductory courses be "unified or divided?" to use the language of A. C. Kinsey in his still germain *Methods in Biology*. Among the reasons for the unified course, which he favored for specialists as well as for general education, he listed:

"1. The living world is a unit . . .
"2. The biological viewpoint commands student interest . . .
"3. Science now knows principles common to plants and animals . . . Biology teaching can be unified by building around these broad principles . . . The story of living processes can, for the most part, be presented as a single chain of phenomena which are strikingly alike for both plants and animals . . . The modern sciences [of physiological processes] provide a sound basis for unifying the introductory course."[2]

Kinsey maintained that the survey course in general biology would most often of necessity be a course dealing with the principles applicable to large portions of the world's biota. This kind of course, call it survey or call it biological principles, he recommended as introductory for biological specialists, as basic for biology teachers, whether college or high school, and as the kind of course for the general student.

But, evidence that the problem of making biology well suited for these purposes is one of long standing is furnished with another of Kinsey's statements:

"In the last analysis, the success of any teaching program depends upon the quality of the teachers. . . . Whether the unified biology course can justify its position in the high school *and the college* will depend, in the long run, on whether we can find teachers who can effectively teach it."[3]

Kinsey questioned the value of an increasingly-specialized thesis-centered program for the training of biology teachers, even of college teachers.[4] His viewpoints were supported by many articles and books preceding Kinsey's, e.g., F. H. Blodgett, O. W. Caldwell, A. G. Clement, H. S. Colton, F. D. Curtis, W. S. Gray, S. R. Powers, E. N. Transeau. But his predictions that unified biology would soon replace separate courses in botany and zoology as introductory offerings and that separate departments would be replaced by unified or co-ordinated biology divisions[5] have not been borne out.

Is General Biology Taught in College?

Less than one-half of our colleges and universities offer courses in general biology, that is, a unified biology course for general education. William Goldsmith, of the Department of Biology, University of Tampa, reports this on basis of a random sampling of 87 institutions in 1950.[6] Apparently these "survey courses" were not generally intended for biological science majors because only one-half of the institutions offering them accepted that credit on a biological-science major! This would indicate that the other half did not regard the survey course as respectable for biological specialists. The opinions revealed in Goldsmith's sampling would indicate that the climate of opinion in the biological science field was less favorable for general biology in 1950 than it was in 1935. The causes for this are those already given with respect to the effects of specialization.

What Should Be Done?

It is regrettable that our culture at present places a higher value on physical science than on biological science. It is also to be regretted that the sciences dealing with the nature of matter and the forces, energy forms, and energy transformations should have become separated from the sciences dealing with living things. But, in a value system where material goods are so highly regarded, the separation of physical sciences from biological sciences and the higher regard for the former have been inevitable.

However, within the more valued sciences are many men who perceive a need for more progress in and also better general education in the biological sciences. It is not then necessary to spend great energy arguing the case for the biological sciences. Support for their value to general education is not the limiting factor at present.

Starting with the assumption that biological science will be included in general education, the problems then become: "What should be taught? How shall it be taught? Who shall teach it? and, What preparation should these teachers have?"

The Challenge to Biological Sciences

The challenge to biological sciences is not that of justifying the place of biological science in general education. The challenge is to openmindedly develop more effective ways of providing that part of general education.

The team approach to research and development in new ventures has evolved as a resultant of increasing complexity and specialization. The team approach has proven itself effective whether or not also proven efficient. It would seem reasonable to try the team approach for the development of more effective ways for the sciences, including biology, to make their contribution to general education. There are numerous instances in which specialists in botany and zoology have tried to develop biology courses acceptable to their respective fields. There have been few instances in which this team approach has been extended to include physical or earth sciences. It is doubtful whether there have been any instances, except at the conference or symposium levels, where natural scientists, social scientists, and educators have been yoked to the task of developing general-education courses acceptable to these several fields. This volume is, in a sense, a symposium-level instance, the several fields are all represented. But the development of courses, and their experimental trial to test their effectiveness, still remain to be done. It will probably have to be done at the institutional level, that is, worked out by the staff of each institution.

The development of more effective offerings will have to recognize the potential contributions of all the sciences. Developers must also admit the restrictions imposed by limitations of time. The baccalaureate degree is still a four-year program in most instances. Each of the major science fields, botany, chemistry, earth science, physics, and zoology, would happily provide a year's sequence and only reluctantly accept a single semester as the share of each. There is not that much time. Either the sciences must co-ordinate their offerings or the student must select from among their several general-education offerings.

The most common objection to unified or survey courses is that they are "watered down." This objection is the resultant of determining content primarily on the basis of the total knowledge of the field or fields concerned. When an introductory course is outlined by picking out the more important facts or principles of a comprehensive field, a field wherein it takes years of study to prepare a person for gathering new knowledge for the field, the specialists will remonstrate the omissions of the introductory course. There must be a more effective means of selecting content.

Meanwhile, problems facing man become more complex. It seems that the comprehensive field of biological sciences ought lay aside its own massive compilations of systematized knowledge and, together with other scientists, social scientists, and educators, look at man and the environment in which he now lives. It is certain they could together list quite a number of problems facing modern man. It is certain all the sciences will have contributions to make toward the development of understandings, attitudes, and new ideas helpful to the solution of some of those problems. Since man is an organism, since many of his problems arise out of his interrelationships with other organisms, including fellow man, and since man must accept natural law as the rules of the game, biological sciences will have a large share in the general education co-operatively developed to aid man in solving his problems.

Selected Bibliography

O. C. Caldwell, *et. al.*, "On the Place of Science in Education," *School Science and Math,* XVIII (1928), 640-664. A report to the Council of the AAAS urging broad introductory science for general education.

A. G. Clement, "The Biologic Point of View," *School Science and Mathematics,* XV (1915), 339-341. For the unified approach to biological sciences as seen by a pioneer in ecology.

James B. Conant, *Science and Common Sense* (New Haven, Conn.: Yale, 1951). Includes much from his earlier book, *On Understanding Science.* Establishes a framework for construction of offerings and broad criteria for evaluating their effectiveness.

J. Gordon Edwards, "A Valuable Biology Course For Future Teachers in Elementary Grades and Secondary Schools," *Turtox News,* XXXIII, 1 (1955), 16-19, and XXXIII, 2 (1955), 53-55. Maintains that a course in general entomology is more effective in preparation of teachers, and also for general education of college students, than is either general biology or introductory zoology.

William M. Goldsmith, "Survey Course in Biology," *Turtox News,* XXX, 3 (1952), 66-69, and XXX, 4 (1952), 80-81. Condensation of paper presented to Florida Academy of Sciences in 1950 based on sample survey of 87 colleges and universities. Favors survey course, describes nature of its opposition, and suggests remedies.

Garrett Hardin, *Biology, Its Human Implications* (San Francisco:

W. H. Freeman, 1949). The preface, the first two chapters, and Part V of this textbook defend and demonstrate the essential unity of life.

A. C. Kinsey, *Methods in Biology* (New York: J. B. Lippincott, 1937). Deals with objectives, and method in biology courses for high schools and colleges. Favors unified biology over more restricted or specialized courses in training of biology majors as well as for general education. Comprehensive bibliographies to 1935.

Reginald Manwell, "Where Are Our Future Biologists?" *Turtox News*, XXX, 9 (1952), 154-155. Blames archaic methods and content in biological courses, especially the introductory ones, for lack of vocational interest in biology.

Eric Nordenskiold, *The History of Biology* (New York: A. A. Knopf, 1928). The classic work in the history of biological sciences in particular but showing related development of other sciences.

Wolfgang F. Pauli, *The World of Life, A General Biology* (Boston: Houghton Mifflin Co., 1949). The preface and the first three chapters defend and demonstrate a unified approach to biology.

John H. Storer, *The Web of Life* (New York: Devin-Adair Co., 1953). An introduction to ecology illustrating both broad problems and unifying principles of biology.

E. N. Transeau, "Biology a Single Science," *School Science and Mathematics*, VIII (1908), 775-777. A pioneer in plant ecology he then rejected specialized approaches to biological sciences.

CARL S. JOHNSON, Director of the Ohio Conservation Laboratory (a special summer school for teachers cooperatively sponsored by the five state universities of Ohio, is Assistant Professor in the Dept. of Zoology and Entomology at The Ohio State University. Prior to becoming a member of the Ohio State faculty he taught general biology, human physiology, and ecology at Capital University (Columbus, Ohio). He has also taught one-room rural elementary school, junior high school science and math, and included in his public school work before WW II a tour as high school principal and science instructor. His B.Ed. degree was from St. Cloud State Teachers College (Minnesota) majoring in science education; received his M.A. in science education at Ohio State University and completed a broad interdepartmental Ph.D. in conservation at the same institution in 1951. He is by inclination and training a "generalist" having chosen that unique "specialization" to aid in the closing of the gap between research and education. After spending a year as conservation-education supervisor in the Ohio Dept. of Public Instruction, he started the conservation-education workshop at Montana State University, has been a member of the staff of the Ohio Conservation Laboratory since 1952, and served for five years as chairman of the conservation curriculum committee in the College of Agriculture at Ohio State.

NOTES FOR BIOLOGY

1. J. Gordon Edwards, "A Valuable Biology Course for Future Teachers in Elementary Grades and Secondary Schools." *Turtox News,* XXXIII, 1 (1957), 16-19, and XXXIII, 2 (1957), 53-55.

2. A. C. Kinsey. *Methods in Biology* (New York: J. B. Lippincott Co., 1937), 53-57.

3. *Ibid.,* 246. (Italics added.)

4. *Ibid.,* 257.

5. *Ibid.*

6. William M. Goldsmith. "Survey Course in Biology," *Turtox News,* XXX, 3 (1952), 66-69, and XXX, 4 (1952), 80-81.

MEDICINE

Russell H. Pope, M.D. & Arthur W. Samuelson, M.D.
Bridgeport (Conn.) Hospital

The present position of medical education in the United States, which is the subject of this chapter, is the result of nearly 200 years of education experiment and progress. The first medical school in the North American colonies was established in Philadelphia in 1765 and marked the emergence of American medicine as an independent growth. Dr. John Morgan, who received his medical training in Great Britain, was appointed the first medical professor in North America. He was considered a radical in medical education because of his demands for a good preliminary background of study for premedical students and his attacks on the old apprentice system. These were the first attempts in America to develop a system of medical education.

The success of John Morgan's pioneering in this field speaks for itself. During the next 145 years this spirit of progressive education was carried out in the better American medical schools with excellent results. Many important medical men made their impression on medical education during this time. Nevertheless, medical education was without guidance. Only a few of the best schools demanded rigorous courses of study and some of the good schools required only about a year of instruction. Most of the practical training of a young physician was by preceptorship or association with an older practicing physician. Admission requirements to most schools were non existent and schools have been described as "diploma mills."

It is, therefore, not surprising that the better men continued or expanded their training in the excellent European schools in the latter half of the 19th century. Such men as pathologist William H. Welch and the surgeon William H. Halsted, who deeply influenced medical education and set an example with their aid in the establishment of a superior medical school at Johns Hopkins in Baltimore were in part products of European medical education. Medical education in the 1890's and early 1900's was profoundly affected by the example of Johns Hopkins Medical School. Abraham Flexner called this school the first of a genuine university type. Dr. Flexner, in 1909, made a study of medical education and his report, published in 1910, struck fertile ground. This report was a turning point in medical education in the United States and its explosive effect rapidly destroyed the "diploma mills." The report was supported by the A.M.A. Council on Medical Education which adopted severe standards for schools. In the nineteenth century, 457 schools developed in the United States and Canada. In 1958, there are 78 schools all fully approved by the A.M.A.

The Flexner report emphasized the scientific approach to disease and demanded rigorous training in fundamental human physiology, biochemistry, and pathology. Such orientation leads the student to consider the mechanisms of disease and the basic alterations of anatomy that occur thus permitting a more reasonable approach to therapy. This approach was exceptionally successful in producing excellent physicans and in stimulating related medical professions such as nursing and basic research.

Undergraduate Medical Education

The Medical Schools. There are at present 78 fully approved four-year medical schools in the United States, the two most recent of which are the University of Mississippi and the University of Missouri. Both of these schools graduated their first classes in 1957 prior to which they had been approved two-year schools in the basic sciences. There are also four two-year schools approved in the basic sciences. These are Dartmouth, West Virginia University, University of North Dakota, and the University of South Dakota. In addition there are four new medical schools in the active process of development into four-year schools which include the Albert Einstein College of Medicine in New York, the Seton Hall College of Medicine and Dentistry in Jersey City, and the University of Florida College of Medicine in Gainsville, and the University of Kentucky in Lexington. It is to be noted that all of these schools are jointly accredited by the Association of American Medical Colleges and by the Council on Medical Education in Hospitals of the American Medical Association.

Pre-Professional Requirements. At present, seven medical schools now stipulate a degree as an admission requirement for entrance. These are Einstein, Georgetown, Hahnemann, Johns Hopkins, Kansas, New York Medical College, and Vanderbilt. However, only one (Johns Hopkins) had a hundred per cent of entering students with degrees as of the past year while five of the remaining six schools counted more than 95 per cent of their entering students holding degrees. An additional six schools have only students with degrees in their entering classes despite the lack of a definite requirement. There appears to be a decided trend toward applicants holding degrees despite the increasing arguments favoring lesser time preparation for medical school. Serious consideration must be given to these requirements for admission, when at the same time the duration of post-graduate training in the various specialties is continuing to lengthen.

The Johns Hopkins Medical School[1] has taken significant steps to develop a satisfactory program of training that would shorten the number of years required to train a physician and at the same time to train a more complete and competent medical graduate. The various phases of this program will be discussed elsewhere in this chapter.

Medical Students. Approximately 15,000 students apply for medical school each year and slightly over half of these are accepted by the admissions committees. All these students have taken the medical college admission test and nearly all students accepted attained scores in the upper half. Most educators have found it inadvisable for the schools to dip into the lower scoring half except in unusual circumstances since it is agreed that the medical courses are generally too difficult for this group. Previous experience in accepting low scoring students has repeatedly been found to be unsatisfactory. There is today a significant need to interest more of the top-rank students in the universities and colleges in studying medicine. Most educators again agree that if first rank students are not drawn into the medical field, then the standards of medical practice must of necessity decline. Already a drop in the quality of medical students has been noted. A comparison of the undergraduate classes of first year students in the school years of 1950 to 1951 compared to those of 1956 to 1957 showed the following: In the former group, 40 per cent of the medical freshmen had A averages in their pre-medical courses and only 43 per cent B averages; yet in the 1956 to 1957 freshmen class the number of A average students had dropped to 16 per cent while the number of B averages had risen to 70 per cent. Mitchell[2] stated that the whole future of medicine is dependent on the caliber of men and women brought into it, and that admissions committees should be seeking for the best possible students noting especially intellectual ability, integrity, and strong motivation to study medicine for the right reasons. In addition, it has been proposed that the personal qualities of warmth, depth and breadth of interest, and energy of drive be appraised through procedures best adapted to the individual schools. Wood,[3] in presenting his comments on "The Underlying Cause of Unrest in University Medicine" stated that modern medicine is not an isolated empirical entity but borrows to an increasing extent from sciences and the humanities on all sides (biology, chemistry, physics, sociology, philosophy, etc.). He further contends that the state of unrest and ferment now characterizing present-day medicine has its real cause in the rapidly changing dimensions of medical science rather than the changing patterns of disease, medical economics, group practice, etc. He points out that as depth and breadth of knowledge in medical science have advanced, the most important practical problem confronting the profession today is that created by the excessive length of time required to train a modern physician. Thus he states: "As the depth and breadth of knowledge in medical science have increased, the combined undergraduate-post-graduate courses in medical schools and hospitals have become longer and longer. In this country, physicians and surgeons are considered well qualified to practice only after they have completed four years of college, four years of medical school, and an additional three or more years in a hospital, a total of at least eleven years after high school. When two years of compulsory military training are added many doctors of medicine have reached the age of thirty before they have finished their formal education. Those interested in research have spent a number of their most promising years merely preparing for their careers. This demanding period of training with its inevitable economic strain has become so long that it is beginning to discourage candidates from entering the field of medicine. The relative number of students applying for admission to medical schools as compared to those applying to training in other professions has been definitely on the decrease. Furthermore, there is good evidence that the excellence of practicing doctors is related in part at least to the length and quality of their post-graduate training as house officers. Under the present prolonged curriculum of undergraduate medical education many young physicians are forced for economic reasons to enter private practice before they have gained adequate experience in hospital residencies. Such short cuts to practice in this age of expanding sciences, deprives society of its maximum return on its investment in medical manpower."

While the medical schools today appear to give lip service to the desirability of a broad general education and avoidance of excessive scientific pre-medical preparation, there is reason to believe that medical school admissions committees pay a great deal of attention to the specifically scientific preparation of the medical school applicant. As such, there apears to be a monotonous uniformity in the kind of educated individual who now enters medical school. The question, therefore, arises as to how a greater diversity of education of pre-medical students can be effected so that students entering medical school will be endowed with a variety of educational backgrounds. This problem was discussed by the members of the Workshop Conference on the "Influence of Changing Dimensions of Medical Knowledge upon Medical Education."[4]

One of the major problems discussed by the foregoing group was that of the quantitative content of the undergraduate medical curriculum. It is now required that the student learn vast quantities of detail in all the varied disciplines of basic medical sciences as well as in clinical medicine. It is virtually impossible for the student and faculties to attempt to incorporate all the newly discovered knowledge into the already established traditional information now required in the present curricula.

There should be a definite reduction in the quantity of detailed learning in the basic sciences, now a necessity for all medical students. At the same time there should be an attempt made to offer greater flexibility in content and to relate this to the student's varied interests. He might be permitted extra work in a field of his choice, e.g. histology, while at the same time fulfilling the obligations of a more constricted area of essential basic medical information. This would, of course, relate to the student's interests of the moment and not to his future plans. At a later sequence in his development, the student might then be permitted a greater option to pursue studies related to teaching or research or to a future specialty.

It may be, that the concept of creating a relatively uniform medical graduate able to cope with virtually every common medical problem may have to be abandoned.

In legal education, there is no effort to teach the future lawyer the management of all the vast array of problems that will confront him, but rather to teach him broad legal principles and ways and means of seeking a solution. It may be that future doctors in view of the newer concept of the "team approach" to the patient with various consultants usually available for most problems, may at some later date pursue an educational pattern not dissimilar to that of the legal profession. Only in this way can the tremendous mass of detailed information and in many instances, over-learning, be reasonably tailored to meet the future demands that will be required of the medical profession.

Postgraduate Medical Education

Internship. For the intern year beginning July 1, 1957, to July, 1958, there were 12,325 approved internships offered by 857 hospitals in the United States and its territories. Since there are approximately 7,000 graduates of approved American Medical Schools during this same year, approximately 5,000 intern positions were thus available. This disparity was partly filled by graduates of foreign medical schools accounting for about half the unfilled positions. There are three major type internships offered in the United States: the rotating, mixed, and straight. The rotating internship includes training on various services during the year sometimes with an emphasis on one particular type. The mixed internship includes those with assignments to two or three of the four major services (medicine, surgery, obstetrics, and pediatrics). Straight internships include straight service assignments to medicine, surgery, pediatrics, pathology, obstetrics, and gynecology.

The question as to whether the rotating or straight type internship is to be preferred has been of concern to responsible committees on medical education for some years. In 1952, the Advisory Committee on Internships after a prolonged and careful study submitted a report to the Council on Medical Education and Hospitals in which they favored the rotating internship over the straight as the generally most desirable initial training experience beyond medical school. In a 1955 revision of "The Essential of an Approved Internship" the Council on Medical Education and Hospitals stated "It is the opinion of the council that the best basic education is provided by a well-organized and conducted rotating internship." The council, however, continued to approve straight internships in medicine, surgery, pediatrics, and pathology, believing that such experiences may be justified in these fields for medical school graduates who have definitely determined to follow a specialized or academic career. There has continued to be a great deal of concern over the internship situation and in 1956, communications were sent to the administrators of each hospital offering straight internships requesting a statement regarding the attitude toward such programs and whether or not the straight internship programs should be discontinued. A summarization of the opinions was carefully considered by the Council on Medical Education in Hospitals and presented to the House of Delegates of the American Medical Association in June, 1957. It was agreed that straight internships of superior educational content are justified in the fields of medicine, surgery, pediatrics, and pathology because of their comprehensive scope. However, the council was convinced that a sound well-organized rotating internship best meets the present needs of medical education in a superior manner to the straight and recommended the latter only in situations of superior content and for the purpose of studying new methods of medical education.

Because of the changing values in the undergraduate clinical clerkship, the overall length of medical education, the difficulties in organization and supervision of thoroughly excellent internships whether rotating or straight and many other factors it seems probable that this facet of medical education will continue to be the focal point of considerable attention in the years ahead. Experimentation of the types suggested in the Johns Hopkins program will place a well-organized, well-supervised rotating internship experience within the fabric of the undergradute curriculum itself. This approach differs from earlier situations requiring the internship before granting the medical degree by placing this rotating clinical experience within the curriculum and by conducting the program in the hospital which is an integral part of the medical school.

Residency. There are at present in the United States and its territories some 5,299 fully approved residency programs offered in 26 specialties. The total number of residencies in these programs is 30,595 of which the largest number is in the field of general surgery (5,294 positions). This is followed closely by internal medicine with a total of 5,401 positions and thirdly by obstetrics and gynecology with approximately 2,525. These three specialties account for approximately one-third of the total number offered. The 26 specialties include the following fields: (1) allergy, (2) anesthesiology, (3) aviation medicine, (4) cardiovascular disease, (5) dermatology, (6) gastro-enterology, (7) internal medicine, (8) neurological surgery, (9) neurology, (10) obstetrics-gynecology, (11) occupational medicine, (12) ophthalmology, (13) orthopedic surgery, (14) otolaryngology, (15) pathology, (16) pediatrics, (17) physical medicine, (18) plastic surgery, (19) proctology, (20) psychiatry, (21) public health, (22) pulmonary disease, (23) radiology, (24) surgery, (25) thoracic surgery, (26) urology. In addition to these specialties, residencies are also offered in the field of general practice.

It is to be noted that satisfactory completion of approved residency programs within a specialty enables the resident who has completed these requirements to seek certification by the specialty board concerned. There are at present 19 specialties which have examining and certifying boards approved by the Council on Medical Education in Hospitals of the American Medical Association and the Advisory Board for the Medical Specialties. These boards are in no sense educational institutions and the certificate of a board is not to be considered a degree. It does not confer any personal legal qualifications, privileges, or license to practice medicine or a specialty. The boards do not purport in any way to interfere with or limit the professional activities of any licensed physician, nor do they desire to interfere with any practitioners of medicine in any of their regular or legitimate duties. The purposes of the boards are the following: (1) to conduct investigations and examinations to determine the competence of voluntary candidates for certificates issued by the respective boards, (2) to grant and issue certificates of qualification to candidates successful in demonstrating their proficiency, (3) to stimulate development of adequate training facilities for aid in evaluating residencies and fellowships under consideration by the Council on Medical Education in Hospitals of the American Medical Association, and (4) to advise physicians desiring certification as to the course of study and training to be pursued.

The Role of the Hospital in Post-graduate Medical Education. Since internship and residency represent a progressive continuation of the medical students' increasing development from embryo doctor to fully responsible physician, then the hospitals in which this training is pursued must be considered educational institutions with the same responsibility to the student as was the medical school. The purposes of any educational institution are twofold: one is to provide the easiest possible access to learning for any student and the other to evaluate the accomplishment of a learner and to correct deficiencies as they arise. The hospital as an educational institution must therefore offer something more to the house staff than the repetitive performance of routine duties without intellectual challenge or increasing professional responsibility. If this elevation of the standards of hospital education is to be brought about it must be accomplished by acceptance with in the hospital of educational responsibility, which, once recognized, no longer permits the often heard comment that there is plenty of material for learning if only the interns and residents would take advantage of it.

One factor which has had its effect on hospital education programs is the rapid increase in the utilization of the hospital resulting in a constantly increasing number of internship and residency appointments available to each applicant and a lessening in proportion of approved educational programs. The filling of house staff positions in order to meet service needs has resulted in the appointment of many foreign interns and residents to vacant positions. The increasing number of foreign graduates now filling the gap in our internship and residency programs has given rise to the question of the foreign trained doctors status both now and in the future. The influx of doctors from other countries is being controlled by the Council on Foreign Medical Graduates set up with the sole purpose of evaluating the medical training of foreign-trained physicians.

At present, the only usual formal evaluation of persons in graduate medical programs is that of examination by the American boards of the various medical specialties. While the definite value and importance of these examinations is recognized, the institution which desires to improve itself cannot permit evaluation to wait till after completion of its entire course of study.

An effective instrument to meet these educational responsibilities is required. Specific needs for a distinct faculty organization within each hospital as opposed to the independent teaching of several men is of primary importance. Another need is for the development of curricula and an attempt to incorporate within it a delineated body of knowledge that can be preferred to the student.

In summary, it is increasingly obvious that in establishing graduate medical education programs, hospitals are being forced to assume responsibilities secondary to their primary reason for existence. These are the responsibilities of any educational institution namely; to provide opportunities for learning and to evaluate the accomplishment of the learner during his years of training so as to produce the best possible trained physician in whatever the field of study capable of administering to the patient intellectually, scientifically, and in addition able to practice and understand the art of medicine. This is the legacy that the training institutions can offer to the medical profession as a whole as well as to society.

Current Trends in Medical Education

While the Flexner report and the A.M.A. Council on Medical Education were tremendously successful in producing the type of medical education described above there has been an increasing tendency to modify and reconsider educational problems since World War II. The stimulus for such reappraisal has arisen from the stress of many factors some of which are beyond the control of the profession or medical institutions.

One factor is the marked increase in population of the United States. At present there is one physician per 750 people in the United States. In 45 years the number of medical graduates has increased 117 per cent while the population has increased 76 per cent. In 1957, there were 6,800 medical graduates, about twice the number of deaths. With a projected population of 225 to 250 million people by 1970, this ideal ratio of one physician for 750 people cannot be maintained by the present output of physicians. The recent report from the Department of Health, Education and Welfare indicates that, "To maintain this ratio, the output of physicians would have to expand by 1970 to 8,700 a year from domestic schools, plus another 750 from foreign schools.[5] This compares with the production of 6,800 physicians in 1956. The domestic output would have to rise by 1,900 per year in 1970 . . .

"Therefore, about 1200 (1900 minus 700) additional physicians per year must be produced by new schools. The average medical school graduates about 90 students per year. Therefore, a minimum of 14 and as many as 20 new medical schools will have to be built if the existing number of physicians per 100,000 population is not to fall. To meet this need construction would have to begin in the immediate future and be completed in a few years."

It is estimated that a new medical school, fully equipped but without instructors will cost 50 million dollars. To this must be added the increasing costs of maintaining and improving the present schools in order to preserve the highest level of education, research, and community service. The report of the Department of Health, Education and Welfare goes on to state that "even if funds in the order of 500 million dollars to 1 billion dollars were available immediately for construction of new medical schools, it seems certain that the number of physicians per 100,000 population will decline between now and 1970." These financial problems, although large, are under intensive study by schools and government agencies.

Other problems revolve around alterations in the curriculum. It has been claimed that the scientific emphasis in medical education has lost sight of the patient as a sick person with his own personal problems. Many schools are emphasizing psychiatric education on the one hand and direct experience in the patient's home on the other. The purpose of this is to look at the patient as a whole and appreciate his economic, social, and emotional problems as proper parts of good medical practice.

One current curricula trend is the integration of basic sciences with the clinical services and the abolition of special departments. The basic sciences are not studied as isolated subjects but taught in pre-clinical years where patients are used to demonstrate not only the changes in physiology and chemistry, but the diagnostic and therapeutic aspects of disease. Various schemes and schedules of instruction are being evolved around such considerations in various medical institutions.

One of the more recent developments of interest is the proposed revision of the program of medical education at Johns Hopkins University. The faculty of this institution has reviewed the evolution of the American system of medical education and, in light of its experiences, noted the emergence of certain serious defects. The three most important of these are considered to be (1) the excessive number of years required to train a physician, (2) the dichotomy existing between the liberal arts and the medical sciences, and (3) the noticeable decline of strength that has developed in recent years in the basic science departments of medical schools. The projected Revised Program of the Johns Hopkins University is directed toward counteracting, as far as possible, the three basic short-comings.

In brief, the schedules for the first four years of the medical curriculum in this five-year program would be based on an academic year increased from 32 to 40 weeks, and the fifth year, representing a rotating internship in the Johns Hopkins Hospital, would cover an entire calendar (52-week) year. However, in place of the current degree requirement for admission, carefully selected students may be admitted to this five-year program after two years of college. This will make it possible for the talented student to complete the cycle of college-medical school-internship in seven years rather than in the nine years now involved when a college degree is included as an admission requirement. At the same time this Revised Program offers opportunities for other students to enter after either three or the usual four years of college. A liberal education is provided through interdigitation of courses in the humanities and the medical sciences. Basically this Revised Program will (a) shorten the course of training by one or two years, (b) reduce the total cost of medical education through this time reduction, and (c) increase emphasis on creative and independent study. The curriculum is so designed that during the five years the "iron curtain" barrier between liberal arts and medical science will be broken, it is believed, to the mutual advantage of the medical school and of the rest of the university. It should be reflected in elevation of academic standards of premedical courses in the natural sciences and tend to eliminate current unnecessary duplications of effort. The program is designed to offer unusually favorable opportunities for students to engage in research during their formative years in medical school. Social science instruction in the university should improve through enhanced opportunities in the humanistic and sociological problems of medicine.

The efforts to teach the many features of practice and community service has produced a shift of the main axis of medical instruction and increased the importance of post-graduate education in hospital internship and residency programs as noted above. Nearly all specialty training is performed in such programs, many of which are not associated with universities. Thus, much of the responsibility for specialty training and general medical experience is forced upon the local community hospitals. An important factor in this shift of teaching to community hospitals is the impact of prepayment hospital insurance. Patients having this type of insurance are private patients and not readily available for teaching purposes. One illustration of this impact of current economics is the reduction in ward teaching services so vital to the proper conductance of internships and residency teaching programs, as well as in research and the continued education of the younger attending physicians in hospitals.

This type of problem is one reason for the changing curricula programs in medical schools and is under active study by various groups in the A.M.A. The studies on this and related problems are leading American medical leaders to more permanent and valuable methods of medical education. The overwhelming ability, competence, and dedication of the practicing American physician must not be overlooked in any solution to these problems.

Selected Bibliography

The Advancement of Medical Research and Education (Washington, D. C.: U. S. Government Printing Office, 1958).

Association of American Medical Colleges, *Admission Requirements of American Medical Schools 1957-1958* (Evanston, Ill.).

Kevin P. Bunnell, compiler, *Liberal Education and American Medicine: A Bibliography* (New York: The Institute of Higher Education, Teachers College, Columbia University, n.d.).

Council on Medical Education and Hospitals, Organization Section, "Functions and Structure of Modern Medical School," *Journal of the American Medical Association*, CLXIV (May 4, 1957), 58-61; & "Annual Report on Medical Education in the United States and Canada," *Ibid.* CLXV, No. 11 (November 16, 1957), 1393-1457.

Abraham Flexner, *Medical Education in the United States and Canada* (New York: The Carnegie Foundation for the Advancement of Teaching, Bull. No. 4, 1910).

Health Manpower Chart Book (Washington, D. C.: U. S. Public Health Service, Public Health Service Publication 511, 1957).

H. H. Hussey, "Changing Dimensions of Medical Knowledge," *The Journal of the American Medical Association* (May 3, 1958), 40.

"Internship and Residency Number," *Ibid.*, CLXVIII, No. 5 (October 4, 1958), 521-694.

Sinclair Lewis, *Arrowsmith* (New York: Harcourt, Brace, 1925). An outstanding exploration in fiction of medical education and practice; the "First Year Med" is reprinted in A. C. Spectorsky, Ed., *The College Years* (New York: Hawthorne Books, 1958), 174-179.

E. Gartly Jaco, Ed., *Patients, Physicians and Illness* (Glencoe, Ill.: Free Press, 1958). Selections comprising the results of the latest research, systematic accounts of experience, as well as theoretical speculation about the connection of society and illness, and its treatment.

V. W. Lippard, "Financing Medical Education," *Medical Advance,* V, No. 2 (Spring, 1957).

R. I. McClaughry, "Responsibility of Hospitals in Graduate Medical Education," *Journal of the American Medical Association* (May 31, 1958), 531.

J. M. Mitchell, "Significance of 1956 Institute on Evaluation of Student from Dean's Standpoint," *Ibid.*, XXXII (August, 1957), 552-556.

Sir William Osler, *The Student Life: From An Address to Students at McGill University, Montreal* (New York: Oxford University Press, 1905). This farewell address, by one of the great men in medicine, has been pondered by generations of pre-medics and doctors.

J. P. Price, S. C. Florence, R. A. Nelson, & W. C. Wescoe, "Changing Characteristics of Society," *Ibid.*, (May 3, 1958), 49.

J. M. Stalnaker, "Study of Applicants, 1954-1955," *Journal of Medical Education,* XXX (November, 1955), 625-636.

W. B. Wood, Jr., "Underlying Cause of Unrest in University Medicine," *Journal of the American Medical Association,* CLXIV (June 1, 1957), 548-550.

RUSSELL H. POPE, M.D. (Yale University, B.Sc., 1942, Johns Hopkins Medical School, M.D., 1945), received also specialty training in Pathology in Boston University, 1948-1952, was Instructor in Pathology, Yale University Medical School, is Diplomate of the American Board of Pathology, and a member of the American Medical Association, the Connecticut State Society of Pathologists, and Pathologist of the Bridgeport Hospital (Connecticut).

ARTHUR W. SAMUELSON, M.D. (B.Sc., Yale University, 1945, Tufts University Medical School, M.D., 1948), participated in the rotating internship of the Grace New Haven Community Hospital, and received Specialty Training in Otolaryngology in Boston City Hospital, Kingsbridge V.A. Hospital (New York), and the Lahey Clinic (Boston). Is a Diplomate of the American Board of Otolaryngology (1954), the American Medical Association, AAOO, and on the Attending Staff of the Bridgeport Hospital (Connecticut).

NOTES FOR MEDICINE

1. "Revised Program of Medical Education" (Baltimore, Mary.: Johns Hopkins University, 1957).

2. J. M. Mitchell, "Significance of 1956 Institute on Education of Student from Dean's Standpoint," *Journal of Medical Education,* XXXII (August, 1957), 552-6.

3. W. B. Wood, Jr., "Underlying Cause of Unrest in University Medicine, , CLXIV (June 1, 1957), 551-2.

4. "Influence of Changing Dimensions of Medical Knowledge upon Medical Education," Report of Workshop Conference, Donald G. Anderson, M.D., Rochester, N. Y., and Victor Johnson, M.D., Rochester, Minn., *The Journal of the American Medical Association* (May 3, 1958), 54-57.

5. *Health Manpower Chart Book* (Public Health Science Publication 511, Department of Health Education and Welfare, United States Public Health Service, 1957).

PHYSICS

AMERICAN ASSOCIATION OF PHYSICS TEACHERS

Improving the Quality and Effectiveness of Introductory Physics Courses*

Foreword. Improvement of the quality and effectiveness of physics teaching is a continuing responsibility of the American Association of Physics Teachers, through its meetings and the publication of the *American Journal of Physics.* In furtherance of this objective, and following a recommendation made by the conference on the Production of Physicists held at the Greenbrier Hotel in the spring of 1955, a conference devoted specifically to the improvement of the quality and effectiveness of introductory physics courses was held at Carleton College, Northfield, Minnesota, from September 5, 1956, through September 8, 1956. The conference, which was made possible by the financial assistance of the General Electric Company, was attended by twenty-seven participants (listed at the end of the report), including members of the American Association of Physics Teachers, the American Physical Society, the Optical Society of America, and the Society for Engineering Education, as well as representatives of the research laboratories of industrial concerns employing physicists.

Each participant was asked, in advance of the conference, to prepare a short statement outlining his views on introductory physics courses, the way in which they fail to meet present needs, and how they might be improved. These statements, which were duplicated and sent to all participants, served as a basis for preparing a tentative agenda. No one maintained that our introductory physics courses are perfect or cannot be improved. Although the opinion was expressed by some that these courses, in general, do NOT fail to meet present needs, a majority of the participants recommended revisions in content, in aims, and in emphasis, as well as a reduction in the amount of subject matter covered, with a concomitant increase in depth of treatment.

*"Report of a Conference Sponsored by the American Association of Physics Teachers"; reprinted from *The American Journal of Physics,* XXV, 7 (October, 1957), 417-427).

Summary reports of the conference[1] were published in *Physics Today* and the *American Journal of Physics*. Copies of a more extensive Preliminary Report were sent to each member of the Association and to the chairmen of the physics departments in approximately 600 colleges and universities in the United States and Canada, with the request that the Report be considered by the physics staff and that their considered comments be forwarded to the chairman of the conference.

Most of the participants in the conference reconvened in New York in February, 1957. The preliminary report was revised in the light of the responses that had been received and several sections were substantially rewritten.

The report which follows presents the results of the discussion of the two meetings of the conferees.

Scope of the Conference

The conference was concerned with the single problem of improving the quality and effectiveness of introductory physics courses. This goal is important for its own sake; it can also, as was recognized by the Greenbrier conference, be "a most promising means" of increasing the production of physicists.

Introductory physics courses now being given can be grouped into three reasonably well-defined types.

(1) Courses for physics majors and for engineering students. These are usually courses of ten, or more, semester hours.

(2) Courses for majors in other sciences and for premedical students. These are usually courses of eight or more semester hours.

(3) Courses for nonscience majors. These courses vary in length and difficulty.

Not all present courses fall strictly into these categories; for example, some courses enroll both science and nonscience majors. However, the indicated grouping seemed representative enough to justify its use. It was agreed that all three types of courses were the proper concern of the conference and that the participants should consider both the common features of the various courses and their necessary differences.

Importance of Clearly Formulated Aims

Before any group or any individual instructor can determine the desirable content of a course, the time to be devoted to it, or the manner in which it is to be taught, it is essential that the objectives which the course is intended to achieve should be clearly and explicitly formulated.

Good introductory physics courses have always been designed to give students an acquaintance both (a) with physics as a process of inquiry and (b) with some parts of the body of knowledge and of concepts that have been collected through that process. One cannot accomplish either of these aims without directing some effort toward the other. The relative importance of the two must be determined in accordance with the needs, the interests, and the future plans of the students and with the availability of the teaching staff, laboratory facilities, and instruction time. The fraction of the course that is explicitly and consciously directed toward either goal is not necessarily proportional to the extent to which the corresponding effect is produced.

It was the opinion of the conference that each instructor should try to decide on the relative extent to which he should direct his energies, and those of his students, toward the achievement of each of these and of other aims that seem relevant to the course. Physics teachers have not devoted as much conscious thought to the establishment of specific goals as is desirable.

Some Suggested Objectives

During the discussion of possible objectives for introductory physics courses, the conference adopted the following resolution: "We hold that the goals outlined by the American Institute of Physics Committee on the Role of Physics in Engineering Education are applicable not only to pre-engineering courses, but also to all physics courses, whether for physicists and other scientists or for nonscientists, including those taking integrated courses or general education science courses. We urge that instructors of introductory physics courses consider these goals in the planning of their work." The AIP Committee statement is quoted below from page 17 of the December, 1955 issue of *Physics Today*.

"It is our belief that an increasing segment of the population and particularly those trained in technical fields should have an appreciation for the science of physics and the mode of thought which to such a large degree has been responsible for the phenomenal development of physics over the past decades. Specific contributions which the physicist can, by virtue of his training, make to the education of an engineer are:

(1) The physicist is, almost by definition, curious about the physical world in which he finds himself. It is hoped that some of this curiosity, and some of the satisfaction and enthusiasm resulting from increased understanding, will be imparted to his students.

(2) Along with understanding, the physicist also learns the limitations and scope of his descriptions and interpretations. These often are of a different character from those which the student meets in the solution of classroom problems in engineering and yet an appreciation of them is an essential part of an engineer's education.

(3) He has learned that there are underlying, unifying principles that

can be rigorously expressed in mathematical terms and which are very general in their application. He has found these to be powerful tools in the expression and solution of important problems in nature. Examples of such principles are the laws of the conservation of energy and momentum which he teaches to his students with emphasis on where they apply and on their limitations.

(4) He knows that to appreciate fully the physics of today the student must know something of the historical development of underlying ideas, of the struggles of the past and the great strides that were taken by a few individuals in enunciating these broadly applicable unifying principles. We are firm in our conviction that all engineering students should know something of the origin of the knowledge which they will use so continuously.

(5) The physicist has often pioneered in the development of precise methods of measurement. This he has done to understand nature better and to place his concepts on as firm a base as possible. Thus he, more than other men, has been interested in determining the fundamental constants of nature with the highest possible precision.

(6) The physicist believes that through a proper understanding of the laws of nature, people can arrive at more objective judgments and, it is hoped, keep free of many of the popular fears and superstitions involving science itself today. The history of physics has given ample proof that this can be done and certainly should be one of the objectives of a physics course.

(7) Lastly, a precise understanding of the basic concepts of physics is perhaps the best guarantee that the presently trained engineer will be able to contribute to the technology of tomorrow and herein lies one of the most important contributions the physics teacher can make in engineering education."

Although the objectives listed above were formulated specifically with the engineering student in mind, the conference felt that they were so pertinent to all physics courses that the conference preferred them to other statements of objectives.

It was also pointed out that the precise nature of the subject matter of physics puts teaching physicists in a strong position to introduce the students to methods of clear and concise analysis and presentation, and so to contribute to his development of certain necessary skills. In particular, the student may improve his ability:

(1) To approach and solve new problems, using verbal formulation, mathematical analysis, or experimental manipulation.

(2) To read and comprehend scientific writings at various levels of complexity.

(3) To express himself in clear, succinct, and precise statements, written or oral, qualitative or quantitative.

Content

The selection of content determines in large measure the success with which physics teachers and students will achieve the above goals. The conference felt most strongly that *physics teachers must reduce drastically the number of topics discussed in introductory physics courses.* A more critical and parsimonious selection of content would permit a pace that encourages both reflection on the part of the student and a proper regard for depth and intellectual rigor.

Physics, as a body of knowledge, is now far too extensive to receive adequate general coverage in an introductory course. The instructor must not sacrifice depth and understand by attempting to cover too many topics in encyclopedic fashion. As one of our colleagues has well said: "Let us uncover physics, not cover it."

It was the opinion of the conference that a satisfactory introductory physics course could be constructed around the following seven basic principles and concepts and the material leading up to them:

(1) Conservation of momentum.
(2) Conservation of mass and energy.
(3) Conservation of charge.
(4) Waves.
(5) Fields.
(6) The molecular structure of matter.
(7) The structure of the atom.

Furthermore, these seven principles and concepts outline the minimum content which *any* introductory course must encompass in order to provide a satisfactory treatment of present-day physics.

Lest this list appear too brief, we point out that the discussion of such topics as Newton's laws of motion would ordinarily be included in the development of the principles of conservation of energy and of momentum at an introductory level.

To provide illustrations of the ways in which these topics may be incorporated into a course, the conference requested three of its members[2] to prepare syllabi that would indicate the scope of possible courses, the level of discussion, and the latitude allowed the instructor within the proposed framework. The authors were requested to prepare these syllabi on the basis of existing introductory courses designed for different student groups.

Few instructors will wish to limit their material to the bare minimum of topics. Most will wish to enrich the course by the use of additional topics, chosen from classical and modern physics. The instructor's own interest and special competence, as well as the interests and needs of his students, will best guide him in his choice. Some examples of such topics are:

(1) The electromagnetic theory of radiation.

(2) The second principle of thermodynamics.
(3) The special theory of relativity.
(4) Quantum theory.

The instructor should select sparingly from these topics or from other enriching subjects and should introduce into the course only that material which he can discuss adequately in the available time.

In treating the basic principles and concepts listed above, and in determining additional topics to be included in the course, the instructor may follow various procedures. For example, he may use a historical approach, he may give special consideration to the preprofessional requirements of the students, or he may select topics of particular importance in the development of an understanding of contemporary physics.

It is apparent that differences in content and approach can and should exist among courses. However, whatever the content selected, it should:

(1) Consist of sufficiently few topics so that each can be treated with thoroughness and intellectual rigor.
(2) Present both classical and modern physics as growing subjects, having present-day frontiers in all areas.
(3) Contribute to an understanding and appreciation of the unity of physics.

When to Begin

Because physics is basic to the other sciences, to engineering, and to an understanding of our civilization, a student's introduction to it should come as early as possible. An introductory physics course should be available to all students in the freshman year; physics majors and engineering students should start it then, and others should be encouraged to do so to the extent that this is possible. Introductory physics in the freshman year will in some cases mean that the physics teacher must introduce the student to new mathematical concepts, in particular, certain elementary ideas of the calculus, in order to treat the physical fundamentals with the necessary intellectual rigor. Introduction to the *concepts* of calculus does not imply the mastery of its *techniques*.

Length and Level of Courses

Decisions regarding the length and level of introductory physics courses depend largely on circumstances prevailing at a given institution. Service obligations to other departments may play an important role in decisions reached. It was recognized by the conference that the three types of courses mentioned under the "Scope of the Conference" do not necessarily constitute an ideal program for introductory physics courses. They do, however, represent an existing pattern in many institutions, and with regard to this pattern the conference made the following recommendations.

1. *Course for Physics Majors and for Engineering Students.* The typical existing course of ten or more semester hours, including laboratory, offers sufficient time to fulfill minimum objectives for these students. An additional year of physics should be provided for engineers as has already been recommended by the AIP Committee on Engineering Education. Students majoring in mathematics and students planning to teach physics in the secondary schools should be advised to take this course.

2. *Course for Majors in Other Sciences and for Premedical Students.* The typical course of eight or more semester hours, including laboratory, designed for students in science areas other than physics, provides sufficient time for an adequate treatment of the minimum content.

3. *Course for Nonscience Majors.* For non-science students, the depth and intellectual challenge of the course and the development of student interest are more important considerations than the length of the course. At least six semester hours (one year) are probably required to do justice to the minimum content.

In view of the present relevance of physics for the liberal education of men in every walk of life, all students preparing to teach should take some course in introductory physics. Those preparing to teach sciences should be advised to take a course designed for science majors or for engineers, and not one designed for non-science majors.

At many institutions, the subject matter of physics is included in integrated or survey courses in the sciences. Physics departments in these institutions should take steps to ensure the adequacy of the amount, level, and quality of the physics content of these courses.

The proper coordination between high school physics courses and introductory college physics should also be studied at each institution. If full advantage is to be taken of current efforts to improve the content and the teaching of high school physics, it will be increasingly important that the preparation of the student be taken into consideration in planning his introductory college course.

Instruction

A. *Staff.* Because of the importance of the introductory physics courses and the pedagogical difficulties encountered at this level, it is urged that senior staff members participate actively in the teaching of all phases of these courses—not merely in the large lecture hall. It is recommended that research physicists take turns in handling a small section of introductory physics in both recitation (discussion or problem solving) and laboratory.

Senior staff men should give serious attention to the supervision and training of junior staff members. This is a continuing responsibility since the junior personnel changes rapidly. Such training of younger men is as important and as time-consuming as the teaching of college students. This activity should be recognized by a reduced teaching load for the man engaged in such supervision. Creative contributions to the teaching of physics should be as adequately rewarded as creative contributions to physical research.

Valuable suggestions of ways in which the training of assistants can be accomplished may be found in the report of the Northwestern Conference on the Training of Laboratory Instructors.[3] The quality of available instruction might well be improved by requiring that almost all graduate students, including those on research assistantships and scholarships, participate in teaching.

B. *Techniques and Materials.* In meeting the objectives outlined previously there seems to be no preferred allotment of time among demonstration lectures, recitations and discussions, laboratory hours, quizzes and examinations, and independent reading and other special assignments. However, it is emphasized that demonstration lectures, recitations, and laboratory work are all considered to be highly desirable in all introductory courses, whether for engineers, science majors, or nonscience majors. By careful coordination, the lecture, the recitation, and the laboratory will supplement each other in very desirable fashion in treating subject matter and in meeting the objectives of the course.

1. Demonstrations

Well thought-out demonstrations are effective in illustrating physical principles and phenomena and fixing them in the mind of the student. Demonstration apparatus must be sufficiently large so that all students in the lecture hall can see, and sufficiently simple so that its functions may be clearly understood. Wherever a phenomenon itself may be demonstrated, a film or a lantern slide should not be used. When, however, a physical effect or phenomenon or measurement takes place too rapidly to be seen or involves apparatus too complicated to be understood, films can be used to advantage. Visual aids, such as optical projection and shadow projection, are recommended.

Lecturing to large classes presents special problems. If a physics class is to be expanded, the instructor in charge should be given the necessary means to solve these problems, particularly those relating to the visibility and quality of the demonstrations.

2. Laboratory Instruction

The laboratory is of major importance as a tool of instruction in introductory physics; only through a properly conducted laboratory does the student get the direct contact with physical phenomena that will give him an adequate basis for understanding the nature of physics. There is large room for improvement in the quality and effectiveness of the introductory laboratory. In particular, the use of stereotyped reports fails to make use of the excellent tool provided by the laboratory report for developing the skill of the student in precise and succinct written communication. The conference did not consider the techniques of laboratory instruction because these problems were being studied by a special AAPT Committee on Laboratory Instruction. The report of a conference organized by that committee constitutes the second section of this article. It discusses specific ways in which the laboratory can be made an integral part of the introductory physics course and can contribute to the objectives outlined above.

3. Recitations and Discussions

Recitations and discussion sections conducted in small groups serve to give the student an opportunity to clarify his ideas and to give him training in precise oral expression. The subject matter of the course should not be so crowded that limitations of time prevent adequate direct contact of this type between student and instructor.

4. Textbooks

Although a textbook is usually considered to be essential to an introductory course, the responsibility for teaching rests with the instructor, not with the textbook author. In choosing a text, it is important that the insrtuctor should feel that its general approach is suited to his methods of presentation and to the requirements of his students. It is far less important that the book should cover exactly the subject matter that he plans to teach. He should feel free to omit sections that are not essential to his development of the subject and to add topics not included in the book. Without disparaging the text, he can encourage his students to criticize some of its statements and arguments.

Many good textbooks exist; none that are available can be claimed to be ideal. New and revised texts, both conventional and unconventional, are needed to meet changing conditions. Instructors and publishers can perform a valuable service to the teaching of physics by the production of books which offer original approaches that have been tested in an introductory physics course.

Resolutions of the Conference

(1) We hold that the goals outlined by the American Institute of Physics Committee on the Role of Physics in Engineering Education are applicable not only to pre-engineering courses, but also to all physics courses, whether for physicists and other scientists or for nonscientists, including those taking integrated courses or

general education science courses. We urge that instructors of introductory physics courses consider these goals in the planning of their work.

(2) We recommend that the AAPT actively encourage experimentation with nonconventional courses, and that the Association assist institutions in seeking funds for this purpose when necessary. We further recommend that the AAPT encourage the preparation of the necessary instructional materials for such courses.

(3) We recommend that the AAPT establish a standing Committee on Introductory Physics Courses with the specific functions of encouraging experimentation, of disseminating information, and of developing procedures for the evaluation of experimental programs.

This Final Report has been completed by the Steering Committee for the Conference. The Committee desires to thank all those who have assisted in revising the individual sections of the Preliminary Report.

> The Committee for the Conference
> Walter C. Michels
> Francis W. Sears
> Frank Verbrugge, *Secretary*
> R. Ronald Palmer, *Chairman*

Laboratory Instruction in General College Physics**

Report on the Committee on Conclusions of the Conference on Laboratory Instruction in General Physics

The insight into scientific activity that is afforded by laboratory work is valuable for all college students, and it justifies increased academic support of this aspect of instruction. Students should experience the sorrows and joys of experimentation, rather than merely acquire technical skills. Encouraging results have been reported from institutions using various approaches to the "free" laboratory in which students are largely on their own as to procedure for attaining broadly presented goals. Departments should not allow the pressure of too great a range of material to cause the student to lose sight of the pleasure, excitement, and beauty of good laboratory work. Experiments dealing with "modern" physics are

** Conclusions adopted by the Conference on Laboratory Instruction in General Physics held June 17 to 19 at the University of Connecticut, Storrs. Grateful acknowledgement is made to the National Science Foundation for its grant in support of the conference including publication of forthcoming Proceedings of the Conference. S. C. Brown, H. P. Knauss *(Chairman)*, H. Kruglak, W. C. Michels, C. J. Overbeck, and C. N. Wall, Committee. Reprinted from *The American Journal of Physics*, XXV, 7 (October, 1957), 436-440.

especially attractive. There is an unfilled demand for new and improved apparatus designed to meet the educational objectives of laboratory teaching. Adequate participation by senior staff members is essential for vital laboratory teaching. Various methods of keeping records and reporting results may be used to attain specific objectives. Evaluation of students, of teaching assistants, and of laboratory methods in the light of clearly formulated educational objectives is a problem that must be solved in order to insure that only worthwhile changes shall exert a permanent influence.

I. Objectives

In its preliminary report, the Carleton Conference on Improving the Quality and Effectiveness of Introductory Physics Courses emphasized that the administrative units of the introductory physics course—the lecture, recitation, and laboratory—have common objectives and "supplement each other in a very desirable fashion in the covering of subject matter content and in meeting the aims and objectives of the course."[4] However, the Carleton Conference recognized that the laboratory not only contributes strongly to the principal objectives of the course, but has objectives of its own. The relative importance of the various objectives listed below must be determined locally and with proper regard for the integration of the laboratory work with the entire introductory course.

The objectives of the laboratory may be summarized as *knowing* and *feeling*. Knowing includes, first of all, the direct experience with phenomena which is an essential ingredient of the physics course. The study of physics must not be divorced from the phenomena. The laboratory experiences are the raw stuff from which physical principles and the physicist's description of reality are derived. The laboratory should definitely provide the student with an opportunity to know physics as a process of inquiry leading to theory, not as a mere accumulation of inert information. The student should be encouraged to examine the experiemental evidence for our knowledge; to learn, for example, some of the bases upon which our knowledge of the atom rests. Knowing also includes techniques which the student acquires in the laboratory, particularly the techniques of measurement used by the physicist. At the same time, laboratory work should impart the principles and techniques by which the physicist recognizes the limitations which measuring processes put upon knowledge. There are skills to be learned in the laboratory: verbal ones of writing clearly and precisely and reading scientific writings with comprehension; graphical ones of presenting ideas by means of diagrams, drawings, and graphs; computational skills; and skills of manipulation. Finally, we look to the laboratory to reinforce the understanding of physical principles and to translate them from the condition of abstract statements in the textbook to that of deeply experienced and firmly held knowledge.

Learning to know is not the whole of laboratory work; acquiring the ability to *feel* what the scientist feels is equally important. In the laboratory, the student should be a "physicist for a day." He should encounter the joys and sorrows of experimenting, elation and despair. He should come upon the unexpected, run up blind alleys, and work himself out of tight places. He should experience the sights and sounds and smells and emotions of the laboratory. Having had such experience, the student can claim a kinship, however distant, with the physicist and will have an insight into the scientific enterprise which no amount of mere lecturing can give him.

These objectives of knowing and feeling justify the cost of the laboratory in time and money. Indeed, this Conference recommends that each institution consider increased support of the general physics laboratory for the crucial days ahead. We are convinced that the introductory laboratory that meets these objectives contributes in no small way to the motivation of students toward careers in physics. We also believe that good attitudes toward physics and toward science in general on the part of students with other career objectives often stem from satisfying experiences in the physics laboratory.

II. The "Free" Versus the "Conventional" Laboratory

The Conference noted with great interest the experiments that are being conducted with "free" laboratories in many institutions. As we understand the use of the term, the completely free laboratory would be one in which the students were neither given instructions nor assigned experiments, and in which they were expected to explore subjects of their own choosing by methods that they themselves had discovered or invented, using apparatus that they had designed, built, or assembled from available instruments and parts. The only institution that is known by the Conference to be approximating this practice is the California Institute of Technology. Many other colleges and universities, however, are using plans that are intermediate between the completely free laboratory, and the other extreme of the "cook book" laboratory in which students are given detailed instructions regarding the choice of apparatus, the experimental procedure, the taking of data, and the presentation of the results. In some places, all experiments deal with traditional items, but instructions are minimized and the students are expected to develop their own methods. In others, parts of the laboratory periods are devoted to conventional experiments with quite detailed instructions, while the rest of the time is made available for more independent work. Even those institutions that follow traditional methods show a tendency to omit some experiments even though they have been sanctified by long usage. There seems to be a desire among many instructors for the deletion or drastic revision of at least some of the experiments that are intended to "verify" well-known laws and principles, or are designed to redetermine established constants or properties of materials. Not all the conferees agreed that the deletion of every experiment of this type was wise, and there was no unanimous decision as to which experiments could safely be omitted.

It is clear that the more-or-less free laboratory contributes to the intellectual stimulation of the student and, often, to his emotional satisfaction. On the other hand, it frequently allows him to complete the introductory course without introducing him to some of the more important pieces of apparatus or to some of the basic techniques used in physical measurements. The choices made by departments as to their position in the spectrum of laboratory practices seem to be based less on these advantages and disadvantages than on practical considerations concerning staff time, available laboratory space, and the adequacy of stock rooms and shops. The free labortory makes very heavy demands on all of these facilities. There was almost unanimous agreement among the conferees that the laboratory should require more independent thinking and should serve more as an introduction to research than it has in the past. There was doubt on the part of many as to whether any close approximation to the completely free laboratory was either economically feasible or pedagogically desirable in every institution.

We are in hearty agreement with the spirit of self-examination and of experimentation that is being exhibited by many physics departments. We feel that continued experiments with varied modifications of the free laboratory are needed in the immediate future. We therefore urge the American Association of Physics Teachers to encourage such experiments by using its Journal and its meetings for the rapid exchange of experiences among the various departments. We also suggest that the Association give serious consideration to the possibility of assisting colleges and universities in the raising of funds needed to initiate and evaluate new laboratory practices. The results of these trials in a few institutions may be of value to all students of physics, whether they be potential scientists or those whose judgment and influence in business and the professions may be strengthened by their contact with physical phenomena in the laboratory.

III. Laboratory Course Content

Individual conferees reported the use of introductory laboratory courses in which very different experiments are covered. There was essential unanimity, however, that not all of the topics of the physics course need be covered in the laboratory. In this we agree with the implications of the Carleton Conference Report. We go further, in fact, since we firmly believe that no minimum content can be stated for the laboratory, and that the scope of material covered does not determine the extent to which the primary objectives are met. We urge that departments should not allow the pressure of too great a range of material to cause the student to lose sight of the beauty, pleasure, and excitement of good laboratory work. In this connection, it is pointed out that most students accomplish most when they are stimulated by being faced with the need of determining something unknown to them, in the sense that it is a quantity or result that cannot be found in standard tables.

Many instructors are finding it advantageous to leave to the laboratory the full treatment of some topics that are omitted from the lectures. Among the topics treated thus in various courses are geometrical optics, simple circuitry, calorimetry, Archimedes' principle, and elasticity. Judicious use of this technique may make both the laboratory and the lecture more effective.

We recommend that more experiments in modern physics be added to introductory laboratory courses as rapidly as time and facilities allow. Various members of the Conference reported the successful introduction of cloud chamber experiments, determination of e/m and h/e, and of radioactivity studies.

The continued improvement of laboratories demands the production of new apparatus. The problems connected with their maintenance demand that this apparatus be of high quality and durability. Another committee of the American Association of Physics Teachers is completing an extensive study of apparatus required for the teaching of physics and we trust that both physics departments and apparatus manufacturers will be guided by the report of that committee.

In connection with apparatus problems, there was little hesitation among the members of the conference to use complex pieces of modern apparatus in the introductory laboratory. Examples are the cathode-ray oscilloscope, the stroboscope, and the scaler. If attempts are made to explain these fully most students are soon lost; if they are treated solely as "black boxes" the students may get a wrong impression of physics research. On the other hand, they open new possibilities and give the student a feeling that he is working in an open, rather than closed, science. It is generally believed that the careful and judicious use of "black boxes" is justified, in which the student need not master all their technical complexities at this stage of his training. He should calibrate the device so that he understands its use in the experiment in the light of his past experience. A "black box" so calibrated becomes a "gray box," as one conference member expressed it.

IV. Instructional Staff

The effectiveness of the physics laboratory instructor depends both upon his knowledge of the science and his skill and ingenuity in presenting it. This in itself demands maximum participation on the part of senior staff members. The severe penalty imposed on laboratory instructors by the 3-to-1 academic load credit ratio runs contrary to the purpose and improvement of the laboratory program. Since the continued modification and development of laboratory practices depends largely upon the efforts of senior staff members, we recommend that equality be approached between contact hours and the academic load credit for laboratory instruction. Added benefits from the adoption of this system of teaching load determination would be the provision of more time (a) for effective evaluation of reports and laboratory performance and (b) the counseling of students.

With regard to the use of laboratory assistants for instructional duty, the recommendations of the Overbeck report of the Northwestern Conference on The Training of College Physics Assistants are endorsed. In addition, we strongly recommend that the *most able* and *most experienced* graduate students are none-too-good for laboratory instruction duties, and that they should be selected for this type of duty wherever possible. We agree with the Carleton Conference that the use of research fellows and other fellows for limited teaching duty is recommended as a potential source of valuable assistance in this all-important program.

The Conference considered the values of the various administrative organizations of laboratory staff. The "vertical organization" in which one senior staff member is responsible for all activities of a given section, has the obvious advantage of providing improved student-teacher relationships, but places such great demands for service on the members of the senior staff that it is probably impracticable in many institutions.

V. Laboratory Reports and Records

Institutions use many different types of laboratory reports through which the student gives an account of his laboratory work. They range from a complete formal written report required for each experiment to a laboratory notebook kept permanently in the laboratory. It is clear that no one method of reporting laboratory work is best under all circumstances. The method chosen must depend on the specific objectives set up for the laboratory course.

The requirement of formal written reports has merit in that it trains the student in an important medium of communication. But, on the other hand, this requirement leads to time-consuming operations on the part of student and instructor alike. Compromises may be effected by requiring only a few formal reports during the year, or by replacing the written report by an oral report delivered before both the instructor and fellow students. Notebooks kept in the laboratory do not present all the problems of the laboratory report, but they do require storage space.

Whatever method is used, it is essential that the student's write-up be carefully graded according to a scale of values *known* to both student and instructor. The laboratory report grade may be an important component of the student's laboratory grade, but it should not be the sole component.

VI. Evaluation of Laboratory Work

(1) Evaluation of Students

Proper evaluation of the student's work in the laboratory is highly essential. As was indicated above, only a part of such evaluation should be based on laboratory report grades. Additional methods of evaluation should be employed, such as laboratory performance tests, observation of the conduct of the student in the laboratory, and other tests, including questions concerned with the laboratory on final examinations.

(2) Evaluation of Teaching Assistants

In large institutions where laboratory instruction is carried on primarily by teaching assistants, it is of importance that the senior staff members in charge of the program assume responsibility for an evaluation of the performance of the teaching assistants in the laboratory. Good laboratory teaching by graduate students should be given recognition and poor teaching should be corrected. Methods for the evaluation of teaching effectiveness are discussed extensively in the Report of the Conference on the Training of Laboratory Assistants (1954).[5]

(3) Evaluation of Laboratory Methods

There is need for continuing experimentation in the revision and modification of existing methods of laboratory instruction. There is an equal need for the development of devices which will evaluate the successes and failures of new methods. Without these no sustained improvement in the laboratory instruction can be expected.

It must be emphasized that tests to measure laboratory effectiveness can be designed and interpreted only *after* the objectives of the laboratory associated with any course have been clearly formulated and after the relative importance of the various objectives has been established.

Prepared by the Committee on Conclusions and adopted by unanimous vote of the Conference.

W. C. Kelly
Myron S. McCay
Walter C. Michels, *Chairman,* Committee
on Conclusions
C. N. Wall

NOTES FOR PHYSICS

1. *Physics Today,* 9, 60-62 (November, 1956) ; *American Journal of Physics,* 25, 127 (1957).

2. G. Holton, W. C. Michels, and R. M. Whaley.

3. "The Training of College Physics Laboratory Assistants," C. J. Overbeck, editor (1954).

4. Preliminary report of the AAPT Conference on Improving the Quality and Effectiveness of Introductory Physics Courses, held at Carleton College, September 5-8, 1956.

5. This report is now out of print. The Conference recommended that an effort be made to find funds for a new printing of it.

CHEMISTRY

KARL H. GAYER*
Wayne State University

Introduction

Chemistry curriculums in American colleges and universities have until recently been quite uniform. Part of this uniformity stems from the accrediting policies of the American Chemical Society which deserves much credit for upgrading both the teaching and the facilities of undergraduate and graduate chemistry.

Variation in emphasis does occur with type of institution, faculty research interests, available facilities, etc. The writer's description of both undergraduate and graduate chemistry is influenced by his experiences at Wayne State University, yet he feels it accurately reflects the general conditions around the country. The varied, contemplated changes in teaching make of necessity the trends discussion a general one. This chapter begins with a description of curriculums as they are still found in most schools. This is followed by a discussion of the important changes which have taken place recently or are contemplated for the near future.*

In our discussion of chemistry at the college level we shall first outline chemical education curriculums as they are found in American colleges and universities.

*The author's thoughts expressed are based on his experiences both as a student and teacher. Of necessity the statements herein must be generalizations. His membership on the Undergraduate Advising Committee of Waye State University for a number of years has given him excellent opportunity to become familiar with college curriculums. More recently he had the opportunity of serving on a self-study-committee of the Department of Chemistry. Here a study has been made for the Graduate School, having as its objective an evaluation of Graduate Chemistry. The procedures, goals and accomplishments of the Department were evaluated and compared with those of leading universities and institutes in the United States. From this study (a prototype for other Graduate School departments) plans for future development will be made.

Description of Undergraduate Chemistry

First Year Chemistry

General Chemistry has by far and large become quite uniform with respect to topics covered. This is evidenced by (1) an examination of the course descriptions in college catalogs and (2) from the similarity of numerous textbooks which have been published on the subject. This similarity is further reflected in the ease of transference of college credits from one institution to another. Exceptions to this uniformity are found in such schools as Brown University[1] which has completely changed the course content. While texts vary widely in the mode of presentation, the topics usually covered are the following:

a) The laws of matter and chemistry
b) The gas laws
c) Elementary atomic structure
d) Valence, Elementary chemical bonding
e) The periodic system
f) The chemistry of the non-metals
g) The theory of solutions; elementary acid-base theory
h) Colloid Chemistry
i) Carbon chemistry, inorganic and organic
j) Reaction rate and equilibrium
k) The chemistry of metals and metallurgy
l) Atomic energy-nucleonics
m) Miscellaneous topics such as water purification, chemistry of glass and ceramics, etc.

Variation, of course, occurs in topic content and rigor of presentation from college to college. General chemistry, for example, may vary in rigor from a high school-like presentation to a very theoretical and demanding presentation which assumes the student has had a good high school background not only in chemistry but also in mathematics. This rigorous presentation is intended primarily for students in engineering, physics and professional chemistry curriculums. For students with a thorough high school background the pace is often accelerated to cover the material in one semester rather than the usual two. Again, for the very talented student the course may even be accelerated to cover the material in one quarter as is done at Northwestern University.[2] In some other progressive colleges the superior high school student may be given college credit for general chemistry by passing an examination with a sufficiently high score.

Terminal Type Course. For the non-science major the terminal-type course of one semester or possibly of one or two quarters' duration attempts to give an overall picture of Chemistry. It gives the student a layman's knowledge of chemistry and science in general. Emphasis on theory and problems is considerably less than in the general course. It may be referred to with some justification as a picture-book course. Primarily it is designed for students desiring a broad liberal arts background in various fields. This course often includes other sciences such as physics, geology and biology. Obviously, when it takes on the aspects of general science, its chemistry content is considerably curtailed. In keeping with its aims sometimes even its title is changed to Physical Science.

Terminal-type courses are also designed to meet the needs of particular groups such as students with nursing and home economics majors. These courses usually emphasize the application of chemistry to the particular field. Again the topics covered are similar to those of the general course except for the fact that theory and mathematics have reduced emphasis. For nursing students stress is on the application of chemistry and biochemistry to hospital and clinical work. For home-economics majors food and "pots and pans" chemistry applications are the important topics. These courses usually assume the student is without a high school chemistry background and needs chemistry as a tool to understand better his chosen profession.

Second Year Chemistry

Qualitative Analysis. Qualitative Analysis is a study, largely laboratory, of the chemistry of the more familiar metal elements and of nonmetals in the form of radicals such as sulfate, nitrate, carbonate ammonium, etc. The basis for study is the chemical identification and separation of the elements in water solutions. While the procedures have lost a considerable amount of their former practical importance, the course still serves as an important tool in teaching laboratory technique, descriptive inorganic chemistry of the elements, reaction rate and equilibria.[3] Here the student gets his first real opportunity to develop laboratory techniques in handling and separating chemical substances. Qualitative Analysis being classified as an analytical course is somewhat arbitrary and stems from its being offered in the sophomore year, the traditional time to teach analytical chemistry. Actually, it can be considered an inorganic or even a general chemistry course as is evidenced by its increasing appearance in the first year in conjunction with General Chemistry.

Quantitative Analysis. Following Qualitative Analysis in most college curriculums is Quantitative Analysis. As the name suggests, this is the study of the techniques of measuring the quantity of a chemical substance present in matter. It is primarily a course of exacting techniques. In Qualitative Analysis the student had only to make separations. In Quantitative Analysis he must in addition measure the amounts of the separated substances. He becomes familiar with the general classical methods of determining substance quantities. The course is usually divided into two parts: gravimetric and volumetric analysis. The former depends principally on determining quantities of pure substances by measuring weight; the latter on determining quantities of pure substances 'by measuring volume. From the measured quantities the student calculates the amount of a particular constituent. The course may be either one or two semesters in length, depending on the coverage.

Third Year Chemistry

Organic Chemistry. Organic chemistry, the study of the element carbon, owes its title to the mistaken belief that the chemistry of living processes differed from those of the lifeless or inorganic world. Since the time of Wöhler (1828), it has become known that chemical reactions of carbon in life processes are not fundamentally different from those of other carbon reactions, or for that matter, of any of the elements. The difference is only in degree. Organic chemistry usually taken in the third year involves the study of the mechanisms by which one atom of carbon combines with other carbon atoms or with other elements. It also involves a study of the relation of the physical and chemical properties of the compounds to their structural components. The numerous carbon compounds (hundreds of thousands) restrict the descriptive part of the course to learning the general properties of types of compounds. These properties are then related to the electronic configuration and functional groups associated with the carbon atom. Again the course may be one or two semesters in length. Laboratory work accompanies the lectures.

Fourth Year Chemistry

Theoretical Chemistry. Theoretical Chemistry, also known as Physical Chemistry, is a rigorous, thorough, mathematical study of the laws of chemistry and matter.[4] More specifically it involves a study of the laws of gases, liquids, and solids, the various types of solutions, chemical reactions (equilibria), thermodynamics, reaction rate, colloid systems, electrochemistry, etc. The theoretical concepts taken up in lectures are demonstrated to the students by comprehensive laboratory experiments and problem assignments. This course, usually a year in length, is the most difficult of the undergraduate chemistry curriculum and the most important. It probably best measures a student's ability in chemistry and, for that matter, in the other physical sciences. Two versions of the course are often offered in universities. For the premedical or predental student with a major in chemistry, its length is usually one semester. Its mathematical requirements have been reduced so that a student with a minimum knowledge of physics and no calculus can enter. For the chemistry or science major the one-year course has calculus as a prerequisite.

Other Undergraduate Chemistry Courses

Inorganic Chemistry. This course, often given in the junior or senior year, studies fundamental principles of inorganic chemistry. It relates electron structures of the elements to their position in the periodic system and to their physical chemical properties. It is, on one hand, amplification of the descriptive portion of first-year General Chemistry and, on the other, a complimentary study of Physical Chemistry. It is inorganic chemistry at considerably more advanced level that than studied in General Chemistry. Physical Chemistry is usually a desirable pre- or corequisite. It is a one-semester course, one half of which is devoted to principles and the other to using these principles in explaining the descriptive chemistry. The lectures are often supplemented with laboratory work in which preparation techniques for inorganic substances are studied along with the chemical behavior of these substances.

Biochemistry. As the title implies this is a study of the chemistry involved in life processes. It is fundamental to medical students and is therefore more often taught in medical schools rather than in Liberal Arts colleges.

Glass Blowing. While this is not a chemistry course as such, it is important to the student who plans to do chemical research. Glass blowing is essentially a laboratory course in which the student learns the technique of making glass apparatus.

Industrial Chemistry. This is a study of industrial processes and unit operations. It is probably more closely associated with chemical engineering.

An examination of college catalogues reveals other chemistry titles such as food chemistry, natural product chemistry, etc. To describe these is beyond the scope and purpose of this work.

Goals of Undergraduate Chemistry

The Undergraduate Chemistry Department has for its goals one or both of the following:[5]

(a) Preparing the student for graduate work in chemistry. (Openings in chemistry for bachelor's degree candidates are becoming quite limited) .

(b) Acting as a service department in offering the chemistry courses necessary for majors in other fields, such as, Dentistry, Medicine, Pharmacy, Engineering, Nursing, Home Economics, etc.

While the coverage of material in the service courses parallels that of the professional chemistry curriculum, the rigor is often much less. "Watered-down" so to speak. Whether a student follows the professional chemist's or pre-medical curriculum, he studies courses in General, Analytical, Organic, and Physical Chemistry.

Category (a) above probably requires some amplification.[6] It is the more *important* function as far as the chemical profession is concerned. To prepare for graduate work in chemistry, the student is encouraged or required, to study as wide a variety of courses in the various divisions of chemistry as possible. Furthermore, he is encouraged to take as many courses in mathematics, physics and related sciences as will fit into his program. The better students in many colleges and universities are encouraged to participate in seminars and research. This enables those interested in chemistry as a profession to become acquainted with their vocation during their senior year. The more theoretical background that can be obtained in undergraduate work by the student, the higher the plane on which he does work in his graduate study and the greater his potential as a research worker or teacher.

Chemistry at the Graduate Level

Graduate chemistry has as its goals the prepartion of the student for work as a professional chemist in fundamental research, development research, management or teaching. Graduate chemistry is an advanced study based on concepts learned at the undergraduate level. It entails the solution of an original research problem as evidenced by a suitable thesis or dissertation in addition to the regular graduate courses. The courses studied in graduate school round out the student's knowledge both generally and specifically. Most of the course work is taken in one or more of the several areas where the research specialization lies. In addition to the chemistry course work, the student may pursue study in such related fields as mathematics, physics, biology, etc. Graduate study in chemistry requires a knowledge of foreign languages such as German or French and in recent years Russian.

Doctoral study is designed to bring the student's knowledge of chemistry to the frontiers and thereby enables him to make his own contribution. The doctoral degree usually requires about three to five years' study, which is more than twice the time for the Master's degree. Corresponding to the longer time of study involved, the research accomplished for the doctoral is of a more difficult and demanding nature. This research usually leads to one or more publications in national and international journals. Upon completion of graduate study, the student, as the name of the degree implies, is a philosopher of chemistry. He is competent to carry on discourse from a theoretical and practical standpoint in various areas of the subject and knows at least one area intimately.

Teaching chemistry at the college and university level and industrial research, at least of the fundamental type, usually require a Doctoral degree. Management positions especially in chemical and related industries are increasingly being filled by Ph.D.'s.

The Master's degree demands similar scholastic and research achievements but of a less comprehensive sort. It provides enough theoretical background for some of the non-fundamental research positions in industry. Laboratories concerned primarily with product improvement and development may be staffed in part with people with a Master's degree in Chemistry. Small colleges not emphasizing chemistry and many secondary schools have Master's degree holders on their faculties.

The Pattern of Graduate Study. Pursuing chemistry study on the graduate level, the student must successfully meet certain requirements.[7] First, the better graduate schools reluctantly accept undergraduates who have an undergraduate honor point average much below "B." To screen applicants for admission some schools use either the Graduate Record Examination or their own entrance examinations. After a period of time (several or more semesters) after the essential course work has been completed, the student is usually required to pass a comprehensive departmental examination which admits him to the Ph.D. candidacy. This examination covers the several areas of chemistry. Failure to meet certain minimum standards at this point usually concludes his graduate work, the "failures" often being "washed out" with a Master's degree. Concurrently with his course studies, the student must pass examinations in two languages, one German and the other usually French. As mentioned before, in recent years Russian has become important.

Following the successful completion of the departmental written examination the student must pass an oral examination by his doctoral committee. Is the student in all respects of Ph.D. caliber? Specifically it serves to test the candidate's ability to think his way through problems and questions.

Completion of the qualifying exam usually permits the degree candidate to pursue his research problem without interruption except for his attendance at advanced lectures and seminars. Finally he must defend his dissertation before his committee.

Teaching and Research Fellowship Programs. Very important to graduate study in chemistry are the teaching and research fellowship programs. Their importance is multifold; First, it gives the student the financial aid necessary for him to attend school; second, the teaching program gives him highly valuable experience; and three, most universities could not carry out their undergraduate chemistry teaching were it not for the graduate teaching program.

The research fellowship program enables the student not particularly interested in teaching to pursue his studies and research at a maximum pace. Research fellowships are usually awarded to advanced students who have demonstrated their research capabilities. First year graduate students, it is felt, can derive more benefit from serving on the teaching program. Most of these fellowships are at present sponsored by industry or government agencies. Their stipends vary widely. A common value is in the neighborhood of $2,000 per year. In addition to the stipend the grant allows for payment of tuition and purchase of equipment necessary for research. In recent years this type of subsidy has become increasingly popular with students. It has tended to jeopardize the teaching program in that it has siphoned off the not too numerous graduate-student teachers. The problem is at present unresolved.

In the teaching fellowship program the student is a member of the junior faculty. He is given charge of laboratory and drill (quiz) sections.

The experienced fellow is often given lecturing duties in his last years as a graduate student. The program gives the senior faculty member time to devote to other scholarly pursuits. In large universities it enables the undergraduate to have some of the benefits of personal attention which otherwise would not be given. The stipends have been comparable to those of the research fellowships. Recently it has been found necessary to increase their values in order to compete with the research grants. The student on a teaching fellowships requires a considerably longer time to complete his degree requirements. Hence the lesser popularity of this subsidy. Finally, it must be remembered that this program is the training ground for the future chemistry faculties of colleges and universities. A major number of university chemistry teachers have obtained their only teaching experience and training as teaching fellows.

Current Trends in Chemistry Teaching

The college chemistry curriculums described above remained essentially unchanged for some fifty or more years in the United States. Since the close of World War II, however, changes have been proposed both in particular courses and in the college chemistry program as a whole. Changes in individual courses by faculty have of course always occurred within modest limits. Recently, (even before "sputnik") changes and talk of changes have become more prevalent. Whole undergraduate programs have been revised in well-known colleges and universities.[8,9] In chemistry, these changes have been motivated by a number of factors, among which are:

1) Chemistry has become a much more exacting and mathematical science.[10]

2) Information in chemistry has doubled every thirteen years during the last half-century, while the present standard college curriculums have been without real changes for the same period of time.

3) From a chemistry practice sense, old line courses have become somewhat obsolete. These somewhat outmoded courses have not been without value, however, in that they still serve as teaching tools.

4) In recent years, there has been a tendency on the part of colleges to give more recognition to good high school chemistry teaching. The university student thus must increasingly depend upon his high school background for his knowledge of general chemistry fundamentals. This will eventually force all secondary schools to improve their chemistry teaching.

Undergraduate Trends. General Chemistry is becoming more and more a theory course. The descriptive inorganic chemistry is being replaced by theory. It is axiomatic that as amount of information increases beyond the bounds for the mind to assimilate it, generalizations must and will be resorted to. General laws replace the singular descriptive facts.

There is an increasing tendency to combine principles of General Chemistry with Qualitative Analysis in the first college year. Former topics of General Chemistry are now being covered in Qualitative Analysis. The overlap would thus be eliminated, for instance, in the Theory of Equilibria and solution chemistry.

Both Qualitative and Quantitative Analysis are becoming less and less sophomore year courses. Qualitative Analysis, as mentioned above, is being taught in the freshman year while Quantitative Analysis may eventually be postponed to the junior or senior year where it would be modernized and up-graded in rigor to give the student modern knowledge of Analysis from the instrumentation standpoint.

Organic Chemistry, traditionally a junior-year course, is being given increasing attention in the sophomore, or even in the freshman year, as at Brown University. This is quite feasible since organic chemistry is a very systematic study, primarily memorization and has the capacity of arousing beginning students' interest.

By demanding that chemistry students have better high school mathematics background, the instruction of calculus can be started in the freshman year. This enables some of the better schools to give the year of Physical Chemistry as early as the sophomore year. There is also a tendency to tie more closely Physical Chemistry to Analytical, Organic, and Inorganic Chemistry. This may be done by making Physical Chemistry a pre-requisite for senior courses in Inorganic and Analytical Chemistry.

Course Offerings. Pervading all these changes is the goal, especially in the more progressive schools, of forcing the chemistry student to get more of his knowledge through independent study and research assignments. By supplementing instruction courses with practical assignments during the junior and senior years, the student becomes increasingly self-reliant and his university study tends to approach that of the practicing professional chemist. There is again increasing tendency to get away from "spoon feeding." There is considerable thought being given to decreasing course offerings and to making the remaining courses more rigorous. In the future only the fundamental-type courses would be required. The better student thereby emerges from a four-year education in chemistry on a much higher intellectual plane than in the past. Limiting the course offerings enables the better student to pursue more easily limited research in the senior year. At present the demands upon a student's time for formal course obligations permit too little time for pursuing independent study. Carrying the results of this further, it makes possible more effective, original contributions to chemistry at the graduate level.

Similarly, at the graduate level, there is a tendency to decrease course offerings and to make the advanced degrees in chemistry more of a research degree. There is a tendency to require the graduate student to have only those courses which serve as tools for his carrying on research i.e., chemistry theory, physics theory and mathematics. A limited number of other courses would be elective.

It is axiomatic that the scientist can rise only as high as his theoretical and mathematical background will take him. Therefore, we can expect the present emphasis to continue. Undergraduate and graduate education in chemistry will have more theoretical aspects. This, at the expense of the descriptive.

Inorganic Chemistry. With the achievement of nuclear fission in the early nineteen forties inorganic chemistry began a "Renaissance."[11,12] A whole new area of inorganic chemical research became important during World War II. This coupled with the greater application of theoretical principles has created renewed interest in a somewhat dormant field. Another factor in the renewed interest are such substances as inorganic polymers, chelate compounds, organo-metallic-compounds and the metallurgy of the less familiar elements such as titanium and hafnium etc. The study of inorganic chemistry at both the undergraduate and graduate levels has taken on renewed importance. There is at present a marked shortage of inorganic chemists which will not be alleviated in the foreseeable future.

Standards. Beginning with the upsurge in enrollments there will be an increasing number of schools upgrading their standards. Only those students who are scholars in the best sense will be able to get into the better schools. This is especially true in the science areas such as chemistry where laboratory space is already critically short. For example, before a student can register for General Chemistry at Wayne State University he will have to have had high school chemistry. Lacking this he will have to demonstrate proficiency in the classical high school mathematics areas by passing an examination.

Similarly, at the graduate level where high standards have existed all along, the requirements will rise. Only the gifted student will be accepted for admission. Competition will increase.

Post-Doctoral Program. The past decade has witnessed a great increase in post-doctoral studies in chemistry. It is problematical that this increase will continue. At the very least it will continue at the present level. Among the factors which have contributed to the emphasis of this aspect of graduate study are the following: 1) the tendency to pursue research on a "team" basis, 2) The complexity and long range nature of many research problems, 3) the increase in grants for pure research and 4) the need for the "super" specialist.

The trends in chemistry teaching summarized above indicate a varied approach to the problem of change. The reader is referred to the bibliography for a detailed discussion of the subject. Some of the ideas proposed are, of course, radical and will prove impractical. The recent events in international politics, and the stupendous scientific achievements, will undoubtedly bring many more proposals for up-grading American schools. We must make certain that our plans in science and education are efficient. Chemistry at the university level must change to meet the challenge of our times.

Selected Bibliography

L. B. Clapp, "Reducing Duplication in High School and First-Year College Chemistry," *Journal of Chemical Education,* XXXII (March, 1955), 141-3. Ways are suggested for coping with the problem of duplication in high school and first year college chemistry courses. The relative merits of proposals are discussed.

A. J. Currier, "Trends in Chemical Education," *Journal of Chemical Education,* XXXII (May, 1955), 286-9. A number of problems related to teaching chemistry are reviewed. Recent trends in chemical nomenclature, organization of topics, laboratory versus textbook methods, etc. are discussed.

F. Daniels, "Teaching Physical Chemistry—Forty Years of Change," *Journal of Chemical Education,* XXXV (July, 1958), 322-3. A noted authority, F. Daniels, reviews the changes that have occurred in the teaching of physical chemistry.

T. P. Fraser and others, "Review of Recent Research in Teaching Science at the College Level," *Science Education,* XXXIX (December, 1955), 357-71. As the title suggests, this is a report of research underway to improve the teaching of the various sciences.

S. Geffner, "Summary of Remarks to N.A.R.S.T. Meeting, New York City, April 18, 1955," *Science Education,* XXXX (March, 1956), 137. A summary of ideas for improving the chemistry syllabus in New York City schools. College level chemistry courses are suggested for superior high school students.

E. S. Gilreath, "What Topics Belong in a Modern Course in Qualitative Analysis?" *Journal of Chemical Education,* XXXIV (August, 1957), 391-2. Emphasizes the need for interpreting chemical reactions by means of orderly, sound modern concepts. Qualitative analysis must be modernized to survive.

E. L. Haenisch and E. O. Wiig, "American Chemical Society and Training in Chemistry," *Association of American Colleges Bulletin,* XXXXII (May, 1956), 321-36. A review of the accomplishments of the American Chemical Society in setting up standards for training chemists in American colleges and universities.

M. C. Hannon, M. C. Markham, "Research Program for More Effective Teaching of College Chemistry," *Journal of Chemical Education,* XXX (September, 1953), 475-7. Describes a plan for using research problems as a tool in teaching chemistry at the undergraduate level.

W. F. Hart, "Evaluation of an Undergraduate Chemistry Curriculum," *Journal of Chemical Education,* XXXI (July, 1954), 361-4. An intensive self study has been made of the undergraduate chemistry curriculum at Lafayette College. Recommendations for improving the chemistry curriculum and undergraduate study in general are made.

Journal of Chemical Education, "Symposium of New Ideas in the

College Chemistry Curriculum," Part I, XXXV (April, 1958), 164-77; Part II (May, 1958), 246-60. A series of papers by leaders of the chemistry teaching profession which were read at the symposium held at the 132nd meeting of the American Chemical Society in New York City, September 1957. The papers deal with a wide variety of new ideas in teaching undergraduate chemistry.

M. Kilpatrick, "Building a Chemistry Department," Part I, *Journal of Chemical Education,* XXXI (May, 1954), 247-50. The undergraduate chemistry curriculum at Illinois Institute of Technology is described.

M. Kilpatrick, "Building a Chemistry Department," Part II, *Journal of Chemical Education,* XXXI (June, 1954), 313-16. The graduate chemistry program at Illinois Institute of Technology is described. Admissions, course work, requirements, etc., for the doctor's and master's programs are discussed.

C. R. Meloy, "Some Trends in Teaching of General Chemistry," *Journal of Chemical Education,* XXXI (August, 1954), 424-5. A study was made of the topic content of General Chemistry courses and of the time devoted to each topic.

A. P. Mills, "Graduate Chemistry Course Offerings in American Colleges & Universities," *Journal of Chemical Education,* XXXII (September, 1955), 472-3. The graduate chemistry course offerings along with their credit values in American colleges and universities are tabulated.

D. G. Nicholson, "What's Happening to Descriptive Inorganic Chemistry?" *Journal of Chemical Education* XXXIII (August, 1956), 391-2. The changes in freshman chemistry texts are discussed. The writer indicates the trend is toward more physical chemistry.

R. S. Nyholm, "Renaissance of Inorganic Chemistry," *Journal of Chemical Education,* XXXIV, (April, 1957), 166-9. The factors contributing to renewed interest in inorganic chemistry are reviewed.

J. R. Sampey, Jr., "Chemical Research in Liberal Arts Colleges in 1952," *Journal of Chemical Education* XXXI (January, 1954), 14. The prevalence of research in Liberal Arts Colleges has been studied.

E. G. Rochow, "Coordinated Program for Teachers of Chemistry & Physics," *American Journal of Physics,* XXI (October, 1953), 559-60. Suggests a more logical distribution of topics studied in elementary chemistry and physics courses.

KARL H. GAYER was born in Cleveland (1913) and obtained his elementary and secondary education in the Cleveland Public Schools. He was graduated Magna Cum Laude with the B.Sc. Degree in 1943 and the M.S. Degree in Chemistry in 1944 from Western Reserve University. In 1948 he received the Ph.D. Degree from The Ohio State University. From 1932 to 1942 he was employed in the Cleveland Experimental Laboratory of the E. I. du Pont de Nemours Co. While at The Ohio State University he held teaching fellowships and an instructorship. From 1948 to 1954 he held an Assistant Professorship at Wayne State University, and from 1954 to the present, an Associate Professorship. Since 1954 he has been head of the Inorganic Group and Executive Secretary of the Department of Chemistry. He is a member of Phi Beta Kappa, Sigma Xi and Phi Lambda Upsilon, also a member of The American Chemical Society, The Metropolitan Detroit Science Club, The Physical and Inorganic Chemists Club of Detroit and The American Association of University Professors. His research interests are in general, inorganic and physical chemistry in which areas he has published during the past decade. He is co-author of the widely used "Manual for General Chemistry." During the past several years he has been a member of faculty committees studying the role of physical education in the Liberal Arts College and co-author of a department report evaluating graduate school chemistry.

NOTES FOR CHEMISTRY

1. Leallyn B. Clapp, "The Brown Experiment in Chemical Education," *Journal of Chemical Education,* XXXV (April, 1958), 170.
2. L. Carroll King, "A Special Course for Superior Students," *Ibid.,* XXXV (May, 1958), 250.
3. E. S. Gilreath, "What Topics Belong in a Modern Course in Qualitative Analysis," *Journal of Chemical Education,* XXXIV (August, 1957), 391-2.
4. F. Daniels, "Teaching Physical Chemistry—Forty Years of Change," *Journal of Chemical Education,* XXXV (July, 1958).
5. E. L. Haenisch & E. O. Wiig, "American Chemical Society and Training in Chemistry," Association of American Colleges *Bulletin,* XXXXII (May, 1956), 321-36.
6. M. Kilpatrich, "Building a Chemistry Department," *Journal of Chemical Education,* XXXI (May, 1954), 247-50.
7. M. Kilpatrick, "Building a Chemistry Department," *Journal of Chemical Education,* XXXI (June, 1954), 313-36.
8, 9. "Symposium of New Ideas in the College Chemistry Curriculum," *Journal of Chemical Education,* XXXV (April, 1958), 169-177; (May, 1958), 246-260.
10. E. G. Rochow, "Coordinated Program for Teachers of Chemistry and Physics," *American Journal of Physics,* XXI (October, 1953), 559-560.
11. D. G. Nicholson, "Trends: What's Happening to Descriptive Inorganic Chemistry?", *Journal of Chemical Education,* XXXIII (August, 1956), 391-2.
12. R. S. Nyholm, "Renaissance of Inorganic Chemistry," *Ibid.,* XXXIV (April, 1957), 166-169.

ZOOLOGY

W. R. Murchie
University of Michigan
Flint College

A description of the role played by zoology in American college or university programs would require a virtual outline of basic objectives in upper level work. Limitations of space will permit only a cursory examination of the problems, and that in general terms. To approach completeness, the study should encompass small institutions with one or two professors "doubling in brass," as well as densely populated university departments. Beyond pointing out these variations in physical size and composition, our analysis would account for the variety of purposes toward which the curricula are directed; in short, the educational objectives.

Course offerings, their arrangement, sequence, and general content, provide a useful common denominator for our study. Even superficial examination reveals an apparent and rather surprising similarity throughout the field of zoology, seemingly independent of department size. A good share of this homogeneity is real. Interpretation of the why and wherefore of this constancy will show forces acting within or imposed upon departments of zoology. Some of these may be shown to be conservative, tending to shape and hold the science matrix against rapid change while other pressures may induce important changes in policy or direction in a relatively short time.

Although the mechanical bases of zoological curricula are largely the same from school to school, student requirements have caused a diversity in course content and intent obviously related to the subsequent use each individual may make of his training. Space does not permit description of the complete spectrum in these programs. It is possible, however, to arrange the components of the zoological curriculum into a few categories at least for purposes of discussion. With student requirements as our base, we find a major dichotomy between the pre-professional and non-professional (sometimes mis-labeled "cultural") programs. For the former, with fewer students but a far greater number of courses, the objectives can be identified generally as medicine (or quasi-medical fields), teaching, research, and industry or technology.

The Pre-Professional Program

It has been an almost inviolate rule that the training of zoologists must include certain cognates; botany, chemistry, physics, geology, mathematics, and psychology are commonly listed. Applications of zoology so often require the use of facts and principles drawn from these subjects that alliances with other branches of natural science have evolved into a variety of interdisciplinary courses. General biology, for example, has wedded zoology and botany into a single unit. Here, the degree of fusion varies from mere botany-zoology sequences to complex courses based on pure principles. Other combinations include biochemistry, paleontology, physiological-psychology, and so forth. Implications of this development for the budding zoologist are considerable, especially in industry, technology, or wherever application to human need is paramount. Each technological advance compounds the need for further interrelationship. In future years, we are told, legal controversy may rage over fixation of responsibility for damage caused by atomic radiation of various sorts. In this, lawyer and biologist, geologist and physicist, chemist and sociologist, will find need for a common ground of understanding. We may question whether existing patterns of training will suffice to bring about the required meeting of minds. Who will make the decisions and on what bases?

It is safe to say that leadership will fall to the synthesist; the practitioner of his science whose strengths include competency, not only with the skills of his discipline, but in certain bordering fields as well. As a matter of fact, except for an academic career of teaching or research, professional survival demands that a zoologist stand astride one or more areas beyond his own subject. The purist need not fear, however, that all classical courses in zoology will be immediately chopped up and grafted here and there to other disciplines. Fusions arise out of need and survive on demand; artificial assemblages will probably collapse as rapidly as constructed.

Course sequences in zoology *per se* are guided intradepartmentally by the idea of a student's growth in the field. Those responsible for arranging such programs like to think that each student majoring in zoology will be able to examine a living system from several aspects. In very general terms, these viewpoints may be classified as: (1) morphological, (2) physiological, (3) developmental, (4) systematic, and (5) ecological. Comparative anatomy, genetics, embryology, histology, invertebrate zoology, physiology, ecology, parasitology, limnology, and entomology are examples of specialized units through which this analysis is accomplished. A large number of courses exist in addition to those listed. These usually deal with specific animal groups, organ systems, biological principles, or develop some special methodology. Examples would be herpetology, histological technique, cytology, endocrinology, acarology, evolution, and others. The further hope resides with the professor that, beyond developing a usable vocabulary and reasonable technical skill, his students may gain some comprehension of the history and philosophy of science. These objectives are still imperfectly met in the undergraduate program. We must hasten to point out that many of these concepts are too complex for inclusion in an introductory course and are usually treated more extensively in advanced work. The species concept, for example, is sketched in briefest outline for the beginning student. For some time it has been customary to insist on "distribution"; either as a straight catalogue statement or by means of courses included as prerequisites. Unless specific requirements are laid down as part of a degree program, students who normally elect courses steeped in morphology or physiology may avoid contact with a number of important concepts drawn from, let us say, ecology or systematics.

To suggest change and experimentation would at this point seem proper. Indeed much has been done of that nature. Here and there the encyclopedic trend of presentation has been curtailed and emphasis placed on the generic or conceptual approach. A so-called rational method, involving more student participation, has been substituted for the lecture type. Although information must perforce move more slowly when class discussion becomes the means for developing ideas, proponents of this method feel that total comprehension by the student is greater.[1]

Course form and content have also been modified in some schools. Thus, condensations of material from two or more areas may afford more hours for peripheral subjects. Telescoping of this sort might include comparative anatomy with embryology, general anatomy with histology, or histology with embryology, to give but a few examples. Another possibility is the extraction of certain elements from existing courses; these segments may then be formalized as courses to stand alone. Teaching aids in the form of films, models, radio, television, and similar devices, have been utilized quite heavily to extend the range of student experience. Perhaps the most successful and widely used devices for developing self-reliance have been seminars, honors work, special problems, and reading courses. Very often these are well designed to amplify the student's knowledge beyond the limits of the classical courses. Programs of this type require staff members with peculiar talents in systematizing material, thus, lack of success may not lie in the method but rather in the execution. Moreover, innovations demand organization, organization requires time, where staff-time is scarce, the fringe courses suffer.

There are other inherent limitations imposed by the academic system which slow the acceptance rate of curriculum change. For example, courses must be convertible; credit must transfer from school to school. Recognition of changes in traditional courses comes slowly when the medium of information is a grade transcript. If the new course requires additional staff and equipment, reluctance to move rapidly is understandable. No less important are the limitations imposed by the lack of suitable textbooks, manuals, and the shortage of sympathetic instructors. All of these forces are more or less internal controls. They are stronger in the smaller schools with limited numbers of instructors but operate similarly in large universities.

Pressures external to any department of zoology may be equally conservative in their effect, or, at times, may cause precipitous changes in direction. Results of an educational operation, regardless of approach, must be relevant to the student and the society buying his services and experience. The necessity for a consistent level of achievement in entrance or aptitude examinations by graduates in zoology from a specific department cannot be ignored. The kinds of information required by such tests and the method of extraction, will help to fix the original mode of presentation. Prerequisites for entrance to graduate study, medicine, dentistry, medical technology, fisheries biology, physical therapy, and nursing serve to stabilize the undergraduate zoology curriculum. The latter is not isolated in tradition and purpose; it represents response to available texts, equipment, capabilities of staff, and demands largely beyond the control of the department proper. The pressures will not bear equally on all types of departmental organization, thus, changes in professional or graduate school admission policy could require modifications in the zoology program of a small college. Despite these vicissitudes (or perhaps because of them!), professors in smaller institutions have been able to make significant pioneering changes in zoological pedagogy; a fact bearing testimony to the excellence in teaching so often found among such men.

Teacher Training Program

In the furor periodically stirred up in connection with teacher training, charges and countercharges flow freely. Criticism on each side of the issue may become sharp and intemperate. The conflict is too complex, the points of argument too numerous, to be discussed here. Nevertheless, as zoologists have a large stake both in producing teachers and in training products of our high schools, they must consider the needs of the student-teacher in curriculum planning.

A complaint commonly directed toward zoologists (as part of the scientific community) is their lack of awareness of the kind of information which can be put to use by grade or high school teachers. A rebellion has occurred against the encyclopedic approach; that is, presentation as mere recitation of facts. Science, it is hoped,[2,3] can be taught as a method, as a way of life pervading all human activities. A second desired requisite of the science curriculum is that it should integrate scientific thought for the potential teacher. Thus, zoological principles of significance to psychology, should be distinctly labeled as such so that the student will perceive the interrelations which do exist between these subject fields. On the basis of our earlier discussion, it would appear that the zoologist has no serious quarrel with these contentions. Sharp differences in opinion do exist, however, between the academician and professional educator concerning the *method* of training zoology teachers. Extreme positions may be held by both sides. The zoologist has contended that educators demand too much course work in education *per se*. He further believes that the method of presentation is overemphasized at the expense of basic course material. His suspicions are aroused by a syllabus; he sees in this the contention that anyone can teach natural science, given enough teaching methods and a few facts! The educator is likewise unconvinced that the scientist views the problems correctly. He points out that high schools are bulging with students who cannot be turned out into the street. Of this horde, too few have any interest in science, fewer still have the desire to continue into college and professional training. So, the educator argues, high school teachers who merely pass along the facts they have learned in college, in the way they learned them, serve far too few, and those badly.

Between these extremes we find the majority of scientists and educators, making some effort at least, to understand the problems of their opposite number. Success has obviously not been universal. It will require more imagination, more unbiased analysis, more interlocking of pedagogical theory, before a precision instrument of education becomes a reality. Without attempting to present a template for teacher-training in zoology, we may profitably examine some of the peculiarities of life science in contemporary curricula.

Generally the basic course in zoology is descriptive; the scientific method itself is usually "described." By the very nature of this arrangement, experimentation as a method comes through to the student as a blurred image. Observation is emphasized in introductory work, certainly, but how rarely can the subtle process involved in establishing hypotheses be transmitted to minds schooled in conformity! Our scientist will undoubtedly point out that unless his students receive a reasonably substantial diet of facts, the scientific method, as a process, may become sterile and withered. He will endeavor to show that zoology must provide certain tools in its role as a pre-professional subject. However, if he does not demonstrate the relevance of zoological principles to all science, and indeed to all culture, he may expect a hostile reception. As we shall point out presently, this is not a problem for the zoologist to solve in isolation; educators share some of the responsibility.

Life science commonly forms a convenient first course in science, perhaps the only course. Inasmuch as mathematics is not a prerequisite, biology may be "easier" for the general student than physics. Young people do have an enormous appetite for information about themselves and living things. The relative ease with which interest is aroused, may give rise to the belief that life science is not difficult to teach. Moreover, it is given early in the student's career, quite often without experience in chemistry, physics, or mathematics. A course in science geared to this level does not, so it may be reasoned, demand highly-trained personnel.

An unfortunate consequence of this is the "reconstruction" of instructors from other disciplines who may have had two or three courses in science. To expect such an individual to impart the philosophy and method of science is pure delusion. Generalizations are risky, but there are, I feel, no hazards in stating that the basic ingredient required for successful presentation of zoological science, whether fact or method, is the confidence the teacher has in his own abilities; confidence which comes through reasonable competency in subject matter.

New ideas are set forth frequently, by scientists and educators alike, in the attempt at a curriculum design which will serve the needs of the greatest number of schools. To illustrate, we may examine one such science curriculum offered as a recommendation by F. G. Watson.[4] The program requires a one-year basic course in each of the following: biology, chemistry, physics and mathematics (analytic geometry or differential calculus). Highly recommended semester courses are to be chosen from physiology, chemistry (organic and quantitative), integral calculus, algebra, or geometry, atomic physics, geology, astronomy, and meteorology. The following courses are listed as "desirable": genetics, ecology, cytology, histology, biochemistry, physical chemistry, calculus, electronics, history and philosophy of science. It is suggested that, including the required basics, science courses would total 56 semester hours. This proposal would move the student through science on a broad front. Zoology, for example, is submerged in the total program and a "strong major" in one area would not occur. A major under this plan would involve election of courses to provide strength in biology-general science or chemistry-physics-mathematics sequences.

A little study will convince anyone that no great change is required in existing programs, insofar as zoology is concerned, to meet these requirements. Few zoologists, moreover, would deplore the addition of enough physical science to prepare a student to teach general science. It is likely that products of this type would be received with considerable joy on the part of high school administrators.

The motivation on the part of zoologists in embracing change toward a more generalized program should not be construed as completely altruistic. Scientists must maintain the initiative in science education. Abdication of ths responsibility to professional educators can lead to practical difficulties at college and high school levels. It is not unusual for students majoring in humanities or social studies to embrace life science as a first minor. This may be encouraged due to available job openings, locally or regionally. Because such students conform to the letter of the law, and, with limitations of space, equipment and time to be found in many high schools, they may be able to accomplish as much as the "strong major."

There are, in my estimation, several things which must be firmly established as educational policy if science teaching is to be improved. The zoologist-scientist must ask himself whether it might not be better to have teacher-training broader in scope, with the final product more science-oriented. The educationist must in turn be educated. In response to budget limitations, or as a result of public pressure, he may press football coaches, English, or home economics teachers into science courses. Administrators must be made aware of the damage which can be done through such misidentification of science. The zoologist is perhaps justifiably criticized for producing teachers who are able to identify an insect, name the phyla, or demonstrate the circulatory system, but with little real feeling for the scientific method and its implications to mankind. Conversely, to expect that scientific principles can be imparted by someone with the barest grasp of the subject, is misleading. Progress here, as in so many areas, will come through cooperation.

Counseling in Zoology

There are three aspects of college teaching which are difficult to define; these are: counseling, assistantships, and staff research. All in varying degree, are inherent in a successful program. Counseling must begin with the earliest hours of a student's career. To the greatest extent possible, zoological science, its objectives, career opportunities, and limitations must be identified for the neophyte. No mechanical device nor testing procedure can do this job satisfactorily. The brief distillation of achievement tests does not inspire potential leadership in the unfired clay of youth. Complex, formalized procedures are unwieldly artificialities employed to handle situations where large numbers of students must be counseled rapidly. Under the best of conditions, they are but shadowy outlines of the reality they mimic, i.e., the close human contact of maturity and youth.

By far the greatest number of students advised by zoologists has an objective which may be declared professional. Whether it be community service or the market place, zoology at the undergraduate level is a means of preparing for a rather specific niche. It is manifestly impossible for the advisor to be sure-footed among all possible career objectives; it is rarely necessary that he be so informed in the early stages. More important is the challenge posed by the counselor in forcing the student to reexamine his objectives. Attention must be directed to the implications of the program chosen by the student. To permit a potential scholar to enter a profession based on pure technology, without challenge, would suggest to me that counselling would be far better placed in the hands of a battery of secretaries. We must not condone the sowing of confusion nor erection of false barriers; neither should weak students be advised to reach too high. Above all else, no potential leader should be lost because he aimed too low!

A rather discouraging development in counseling has come about in recent years, at least insofar as zoologically-oriented professions are concerned. As greater reliance has been placed on career counselors, recommendations to entrance committees, medical colleges, and graduate departments have declined in importance. This has had a tendency to lower the value of student-faculty relationships and reduces the value of experience. Soft recommendations on the part of careless professors has no doubt contributed to this condition. Such offenders should be aware that they afford no real service to their department, the professional school, nor the student falsely encouraged. Counselors in zoology, and other areas of natural science, must strive constantly to identify and inspire the best of our high school graduates. It has been suggested that university departments might activate advising programs within the high schools. The benefits likely to accrue to the profession through such a program are interesting to speculate upon.

Undergraduate Laboratory Assistantships

Utilization of undergraduates as laboratory assistants in zoology has roots deep in the American system of college teaching. Few practicing zoologists do not bear the happy scars of the experience. Colleges large and small make such opportunities available. Among the more important benefits to be derived from a teaching assistantship are: (1) the chance to analyze concepts and to verbalize them before other students; (2) the sense of accomplishment in "getting across" a point; and (3) apprenticeship in laboratory housekeeping. All these help to build confidence and polish skills as no formal course can. It must be emphasized that the profit to the student is very personal. In large measure, the sense of gain depends upon the relationship he is able to establish with other students, the subject, and the professor in charge. Poorly organized, uninspired programs are mere drudgery; a "putting in" of time. The converse of this may be an exciting adventure in academic responsibility.

Research and the Teaching of Zoology

Research activity is commonly hailed as a segment of scholarly life upon which considerable emphasis should be placed. The expectation follows that productivity will prove to be a vital force in the class room. Here again, generalization is impossible. Preoccupation with research by all members of a three man staff might prove disastrous, but the large university without a preponderance of research-oriented individuals would be anomalous. Research activity carried as an ornament will be negligible in effect. Either the results, or some expression of research as a "way of life," must be transmitted if pedagogical values are to be realized. In many cases, the rituals of zoological research are singularly unimpressive to undergraduates. Even the results may appear blunt to the young person expecting as he does some flashing insight into natural law. The impression of progress is there, nevertheless; a contagious spirit of seeking truth. It is probably the impression which is important. That his instructor has a passionate interest in water requirements of the meadow vole, the Collembola of a forest floor, or the innervation of a gland, may strike the student as odd. He is, however, aware of a more important fact; a challenge has been sought and met.

Obstacles to research activity, particularly in smaller colleges, have been listed many times; the usual reasons given are: lack of money, time, administrative support, space, or equipment. Of these, heavy course loads, committee work, and other extra-curricular demands appear to share the most blame. Occasionally an instructor will admit to a lack of interest or desire to do research regardless of the fact that he may have the physical means for it. There can be no questioning the fact that many zoologists have been forced by circumstances to postpone investigation. They have a sincere conviction that their first responsibility lies toward students placed in their hands for training. It would accomplish little at this point to look back and scold administrators for not at least making an effort to improve these conditions. The decades immediately behind us, up to the time of World War II, were far removed in conditions and outlook from those following the war.

Nevertheless, adjustment of course loads and relief from administrative duties of a routine nature can be of tremendous value. Administrators should be encouraged to explore means of accomplishing these things above all else. To rush about searching for outside means of supporting research is an unmitigated waste of time unless the staff member has the time, energy, and ability to carry out the job. The basic responsibility lies with the instructor and potential researcher. He must be aggressive in preparing himself to utilize opportunity, whether in the form of grant, fellowship, assistantship, or workshop. But regardless of the availability of outside support or no, he must organize time, even in bits and dabs, and channel his curiosity into productive imagination. Conversion of research experience to classroom will more often than not be quite indirect, for which reason its value may be underrated. The best synopsis of this need in college science has been given by Alfred North Whitehead:[5] "Education is discipline for the adventure of life; research is intellectual adventure; and the universities should be homes of adventure shared in common by young and old."

The Non-Professional Program

Inasmuch as there are really few offerings in zoology for a non-major beyond the usual introductory course, the term "program" is perhaps a misnomer. General zoology courses are, however, well populated with students from other disciplines (for some of whom we must admit the subject represents distribution credit in science). The number of these students is great enough to warrant consideration of their needs and the course types which have been put together for them. As a point of reference, the typical two-semester course in zoology may be used. One semester of such a course usually deals with an overview of the animal kingdom, drawing heavily upon evolutionary concepts and a wide variety of biological principles such as parasitism, metamerism, polymorphism and the like. The other semester is commonly devoted to structure, function, and development as exemplified by the vertebrate animal. Courses of this type are still widely taught in the United States. With some deletions, this basic plan may be compressed into one semester and is then often followed by a semester of botany. Such sequences are frequently labeled "general biology." As mentioned previously, all shades and differences may occur in the degree of fusion between these two areas.

Where preprofessional training remains as our objective, the general zoology or general biology course contains no inconsiderable vocabulary. This emphasis on verbiage, so say the critics, devitalizes the course for the non-major and clots the flow of information concerning major concepts. An insistent call arises for integration of zoological principles with other disciplines and the exposition of man's place in nature. Enthusiasm for change in this direction has not come exclusively from outside zoological departments. Cooperative efforts have been attempted; labeled variously as core curricula, basic science, general education, and so forth.

The academician does regret the opinion that his subject "lacks cultural aspect" or is "too professionally oriented." His feelings are certainly justified when a toothless survey course is described as having more cultural value than the regular offering in general zoology. Equally deplorable are the reasons sometimes given for the creation of such surveys: (1) they are cheaper to operate, or (2) they are easier. In general the non-major zoology course should require understanding of principle with less emphasis on detail and mastery of techniques. Total costs should be higher (non-laboratory courses excepted) but per capita outlay is likely to be less than that of the regular course. All phases of laboratory, lecture, and recitation must be highly organized. Presentation of material in an offhand cut-down version of the regular course will not accomplish the desired ends. Well designed, spirited courses in what are referred to as "integrated principles of zoology," may permit elbow room for work with smaller preprofessional classes while retaining a firm grip on the entire program in life science.

No conformity in subject matter need be stated for non-professional courses. We have noted that where prerequisite and sequential needs obtain, consideration must be given to certain facts, concepts, and techniques. These demands are considerably more subdued in the course oriented for non-majors; even though the objectives were universally established and agreed upon, the means of accomplishing those ends would differ tremendously from school to school.

Program modifications may also extend to upper division work. It is possible in such instances for a general purpose introductory course to be given, followed by offerings in such areas as eugenics, human physiology, conservation, history of science, and heredity, among others. These are predominately non-laboratory courses having few or no prerequisites.

There are many, including the writer, who feel that when the integrity of such a program is maintained, *i.e.*, as upper division courses, it may have great value for majors and non-majors alike. Zoological science need not be irrationally sequential. It is rather unfortunate if a student, having taken physical science during his freshman year, must return to an introductory course if he wishes to elect some zoological science in the third or fourth college year. At the present time it is quite fashionable to state that introductory courses in zoology need not attempt comprehension of the entire subject. We have already touched upon search and experimentation currently underway to insure subject competency without literally swamping the student in facts. This, it must be remembered, concerns the potential zoologist; how much more critical is the situation of the non-major!

Curiosity is not the exclusive property of zoologists, nor of scientists in general. The academic life of generations has hopefully sought to challenge students to rework curiosity into scholarly imagination. The unique attribute of zoological science, exclusive of method, is its greatest subject of study—man himself. The descriptions of our zoologist-scientist as they relate to natural law and man, remove doubt, place knowledge at the right hand of belief. But activation of natural law to the uses of mankind is rarely within the province of science. This responsibility continues to lie with merchant, artist, mechanic, and politician. If the scientist would guide the footsteps of these, he must fix carefully on the means of transmitting his information. It will not do if he trains only those like himself. Students of perception will approach science for tools with which to crush bigotry and erase blind dogma. Here should lie at least one objective of the non-professional program in zoology. This is an immediate service of science; a deeper view of scientific responsibility was written by William H. Prescott[6] over one hundred years ago:

"But if the processes of science are necessarily slow, they are sure. There is no retrograde movement in her domain. Arts may fade, the Muse become dumb, a moral lethargy may lock up the faculties of a nation, the nation itself may pass away and leave only the memory of its existence, but the stores of science it has garnered up will endure for ever."

This remains the challenge to science.

Selected Bibliography

Ludwig von Bertalanffy, "Philosophy of Science in Scientific Education," *The Scientific Monthly*, LXXV (November, 1953), 233-239. A thoughtful consideration of integration between the several science areas; suggests that knowledge of the all-pervading symbolism of science may serve as the common denominator.

E. J. Boell, "Science and Liberal Education," *American Institute of*

Biological Science Bulletin, VIII (January, 1958), 18-20. A strong appeal for the return of natural science to the true liberal arts program; a role which the writer contends is an historical and logical place long held by science until it became misidentified with technology.

Sheldon G. Cohen, "The Undergraduate Student and Scientific Research at a Liberal Arts College," *American Institute Biological Science Bulletin,* VIII (April, 1958), 16-18. Outlines the belief that undergraduate students may be encouraged to participate in research, as well as pointing out possibilities for research activity with few facilities and modest budgets if there is desire.

Harry J. Fuller, "College Teaching: Its Present Status and Its Improvement," *American Institute of Biological Science Bulletin,* III (January, 1953), 20-22. An appeal for consideration of excellence in teaching as a criterion in rank and salary advancement; also considers nature and role of motivation in improving college teaching.

R. E. Gibson, "The Arts and Sciences," *American Scientist,* XXXXI (July, 1953), 389-409. An attempt to relate science to the rapidly changing needs of society and to the technology which serves society.

Thomas S. Hall, Chairman, *Improving College Biology Teaching* No. 15 (Washington: National Academy of Sciences—National Research Council, 1957). A generalized review of the current scene with suggestions for improvement. An extensive review of pertinent literature is appended.

Ernest Nagel, "The Methods of Science: What Are They? Can They be Taught?" *The Scientific Monthly,* LXX (January, 1950), 19-23. A suggestion that the methods of science can be taught but that it would be difficult to formalize methodology apart from subject matter.

James A. Peters, "A New Approach to Teaching Freshman Biology," *American Institute of Biological Sciences Bulletin,* VII (June, 1957), 14-17. A somewhat generalized account of the mechanics involved in developing life-science courses involving substantial amounts of student participation.

S. R. Powers, "An Evaluation of Science Teaching in Senior High School," *The Scientific Monthly,* LXIV (May, 1952), 273-279. An outline of the aims of high school science with suggestions as to the approach which might be used in teacher preparation at the college level.

Fletcher G. Watson, "Course Requirements for Future Science Teachers," *The Scientific Monthly,* LXXXV (December, 1957), 320-323. A direct statement of curriculum design for potential teacher candidates; emphasis upon giving broad coverage in science training rather than the development of strong majors within one area.

W. R. MURCHIE received his undergraduate training and baccalaureate degree at Marietta College, Marietta, Ohio (1942). The M.A. and Ph.D. degrees in zoology were granted by the University of Michigan in 1948 and 1954 respectively. From 1946 through 1950, he served as

Instructor in Biology at Marietta College. For two years (1954-1956) he held the position of associate professor of biology at Thiel College, Greenville (Pennsylvania). In 1951, he returned to Michigan where he is now associate professor of zoology at The Flint College of the University of Michigan. His research and major publications relate to the biology of terrestrial earthworms.

NOTES FOR ZOOLOGY

1. R. B. Green, "Education in the Sciences," *The Scientific Monthly*, LXXI (1954), 40-44.

2. E. J. McGrath, "Science in General Education," *The Scientific Monthly*, LXXI (1950), 118-120.

3. Van Cleve Morris, "Training of a Scientist," *The Scientific Monthly*, LXXXV (1957), 126-129.

4. F. G. Watson, "Course Requirements for Future Science Teachers," *The Scientific Monthly*, LXXXV (1957), 320-323.

5. Alfred N. Whitehead, *The Aims of Education* (New York: New American Library, 1954), 102.

6. William H. Prescott, *History of the Conquest of Peru* (New York: The Modern Library), 826.

CONSERVATION

CHARLES W. MATTISON
Conservation Education Association, Inc.

One major obligation of American citizenship is to know and understand the basic facts concerning the wise use—conservation—of the natural resources. Dr. R. H. Eckelberry, The Ohio State University, puts it this way . . . "Conservation enters into the daily lives of all our people at all ages. It is, therefore, an essential factor in citizen training and a matter on which all must be informed."[1] Hugh Bennett, formerly head of the Soil Conservation Service, U. S. Department of Agriculture, has said, "Education is a prerequisite to conservation. It is essential to an intelligent understanding of the value and importance of natural resources in terms of individual and national life."

Colleges Are The Key

Ignorance of natural resource facts leads to a sense of futility and consequent irresponsibility toward the resources. This in turn, is a threat to individual and community welfare and to national survival. Unless ignorance is replaced by enlightenment, this threat to the individual and even to the Nation will be translated into reality.

A prime responsibility for replacing conservation ignorance with enlightenment rests with our colleges and universities. Dr. Paul B. Sears says, "Anything vital that happens in the colleges will eventually be reflected in society at large. When the proportion of college graduates was still small their influence in public affairs, religion, and the schools was nevertheless very great. The present insistence upon college education for the masses is proof of the prestige of the college in our society. For these and other reasons, education at the college level is of strategic importance in the conservation movement."[2]

The Job Is Not Being Done

Are the colleges and universities meeting the challenge? No, according to my interpretation of a report by Drs. Charles E. Lively and Jack J. Preiss who have made the only comprehensive study of conservation teaching in our institutions of higher learning. These authorities found that only one-half of our colleges and universities now teach conservation.[3] The U. S. Office of Education lists 1,886 schools of higher learning in the United States. Therefore, no more than 950 colleges and universities are offering instruction in conservation. In many of them the programs are inadequate, directed primarily at students preparing for professional careers either in or related to conservation. There is little conservation in general education, although some few colleges do include it in their offerings. Teachers colleges in general seem to be above the average in the amount of conservation included in their instruction.

Conservation is significant to any or all of the disciplines being taught in our colleges and universities. In the natural sciences, it is of course more obvious than in the physical sciences. However, of what value is a physical science without the natural resources to back it up? How much instruction can be given in a physical science without considering the natural resources with which that science is involved? Can we send a rocket to the moon without natural resources? Can the atom make electricity without the natural resources that have been refined to release its energy? The answers are obvious.

Understanding Is Essential

The fact remains that most students complete four years of intensive college training without any understanding of natural resources, their interrelationships, their condition or inventory, their social and economic importance, their conservation. And they can't contribute to the best use and conservation of the resources unless they have these understandings. Sound action whether on the land or at the voting polls is impossible without knowledge. Since conservation action is often a joint public-private venture, intelligent voting, a civic duty, is as important as intelligent practices on the land.

Our colleges are the last step in formal education. With college education common to to-day's masses, our schools of higher education can, if they will, mold a conservation consciousness into a great part of our population each year. That they do so is essential to both our conservation and education systems—to our future as a leader among Nations.

Leadership Is the Solution

Admitting that the colleges and universities are not discharging their responsibility in conservation education, where does the fault lie? Certainly we cannot condemn them alone. The American people must bear a big share of the blame in their prodigal use of the natural resources. With the Nation's birth, our resources were looked at as inexhaustible and at no time in our history has there been public demand that conservation be taught at *any* educational level. So right here we get to the age old question, "which came first, the chicken or the egg?" The lack of demand for conservation education results from a lack of knowledge of the natural resources and their socio-economic importance. The lack of knowledge comes from a lack of education in the use and care of those things which are basic to our standard of living. The solution seems to be one of leadership in education, a role that the colleges and universities have many times seized in other fields of learning. Here is another opportunity for those institutions—seize the leadership in conservation education!

Just what should the American people know about the resources if their attitudes and actions toward them are to change for the better? What should the colleges teach them?

Action Is The Goal

There is general agreement that the ultimate goal of the Conservation Education Association—whether in the schools, colleges, universities, or with adults—is action by every person to conserve the natural resources through protection, wise use and by making them fully productive. There is little point in concentrating on improvements of attitudes toward the natural resources unless those attitudes result in *action* to properly conserve those resources. Everyone, young or old, rich or poor, country dweller or city resident, *can do something,* depending upon the needs.

Four Major Objectives

Now, what should we know about the natural resources before we can take intelligent action concerning them? There are four major understandings that are absolutely necessary.

The first of these is self-evident. We must know what the natural resources are and understand that all of the renewable natural resources—soil, water, forests and wildlife—are closely related.

Following our ability to know and fully understand the natural resources we must then learn just how they affect us—how they influence our day-to-day living—what they mean in our standard of living. As we proceed along this path of learning, we become aware that there can be no standard of living—in fact, there can be no life—without the natural resources.

At this point in our education, we should pause to reflect on the third essential part of our conservation knowledge—what is the current condition of the respective resources; what will their productivity be in relation to the needs of expanding populations and economy and a continually rising standard of living. We should be disturbed because we will find that we are not yet giving the resources the care and protection needed to make them meet our expanding needs.

Some Progress Made

Fortunately, we have made some progress—more and more of us are recognizing the needs and doing something about it. Many farmers are using soil conservation practices on their own lands; in the United States, we are fortunate to have our great national forest system where 181 million acres of wildlands are under good management; in Canada are the Crown forests; under a tree farm program a substantial acreage of private forest land in America has been put under management which was almost totally lacking only a few years ago. And there are some teachers—too few, unfortunately—who on their own responsibility are teaching conservation in our elementary and high schools.

Yes, there has been progress! But even so, millions of tons of topsoil are still lost daily through erosion; too much forest land is still without adequate protection and good management; wildlife management is handicapped because of those who believe protective laws and game farms are all that is needed to insure better hunting and fishing; many farmers and other owners of small forests still fail to recognize timber as the profitable crop it is. Many streams and rivers still are polluted and unfit for either human or industrial uses.

What Is Conservation?

The 4th objective. Now just what is this conservation on which we need to put more emphasis? Actually in its true meaning, conservation is the use, care and protection of the land and other renewable natural resources so that their productivity will be guaranteed for human use for all time. For the non-renewable resources—minerals—it is careful exploitation and utilization so that they will last as long as possible or at least until satisfactory substitutes are found. These definitions recognize "stewardship" as essential to conservation. They include three principles —"use," "care," and "protection" which are the major principles involved in conservation.

Those are the objectives of conservation education. The areas or subjects that must be considered in reaching toward those objectives are—
Soil
Water
Forests, grasses and other plants
Wildlife
Minerals
Closely related to all of them are the recreation and scenic resources.

Our Basic Assets

These are the basic assets of all peoples. Natural laws govern them to a large extent. Their long time value to man is dependent upon his observance of those laws. "When man destroys natural resources, he injures himself."[4]

Air and sunlight are the two indestructible natural resources. They will always be abundant unless some unforseen spurt of man's so-called ingenuity—misuse of the atom as a possibility—either contaminates them or actually reduces the amounts reaching the earth. However, even now we do abuse these great resources. Smoke of cities, burning refuse near Newark Airport, New Jersey, and poorly ventilated homes are examples. Our conservation problem with air and sunlight is to use them in the interests of all people rather than any particular group. Continued abuse has serious health and morale implications.

Non-Renewable Resources

Coal, oil and other minerals are destructible and non-renewable natural resources. They will eventually be exhausted or out of man's economical reach. To conserve them, we must use them for their best purposes and make them last as long as possible. There must be discrimination in their use because once gone, they never can be replaced.

Renewable Resources

Soil, water, plants and animals are destructible but renewable natural resources. Soil, the basic resource, can be quickly destroyed through careless land use. It can be renewed only over many years' time. Water is easily polluted but supplies can be maintained and even renewed under careful use of the watersheds. Plants and animals reproduce themselves and supplies can be maintained under a system of management which uses the growth increment while leaving the principle or endowment healthy and intact for continued reproduction.

Two Teaching Methods

Among the methods of teaching conservation in the colleges and universities, two bear some elaboration. The first, through special conservation courses, usually elective in general education, reaches few students. However, it does have the advantage of presenting conservation as a subject, concentrated into an orderly and continuous teaching program for at least one semester or longer. Analysis of one such course shows that it begins by relating the natural resources to human welfare and exploring the meaning of conservation. It follows the exploitation of our natural resources from the Nation's birth to the present time showing how fast increasing populations have affected resource use. The effects of natural resource abuse and waste in other nations are studied. Land use patterns are discussed and the significance of various kinds of land ownership to resource use and conservation are brought out. Debates are built around public vs. private ownership of certain types of lands in the public interest.

Soil as the basic resource is carefully scrutinized while soil groups and practices to maintain soil fertility are explained. Factors affecting soil erosion are examined and the social and economic significance of erosion are emphasized. Sound soil conservation principles and practices are described.

The course presents water, with which most people associate no conservation problems, in its proper perspective as Richard E. McArdle, Chief, U. S. Forest Service, once described it.

"All of our agriculture, all our industry, our great cities, our jobs, our homes, our very lives depend upon having an adequate and continuing supply of this substance. Without water this country of ours would be a barren waste. Without it, not one of us could live more than a few tortured days.

"There's no doubt about it; usable water is becoming harder to get. And it's costing more too. When a city has to reach out farther and farther for water, there must be larger and larger expenditures for reservoirs and pipe lines. When water carries a high silt content there is extra expense for silt removal. When the usable life of a reservoir behind a dam is shortened by accumulation of silt, the cost of water obviously is increased.

"I think we need to get down to earth with this water business. I mean that literally, because it is our treatment or mistreatment of the earth, our management or mismanagement of the land and its resources, that actually determines what kind of water we get and how and when we get it."

Uses of water are listed and problems of supply and pollution are discussed. Irrigation and drainage in relation to land reclamation are considered and their relations to wildlife brought out. Hydroelectric power with all its political and economic implications is analyzed. Navigation with its attendant problems of channel siltation is surveyed. Flood alleviation is studied and the respective values of mechanical and vegetative control are discussed. Watershed protection and management are emphasized throughout all the water discussions. The Tennessee Valley Authority is presented with all its pros and cons.

Forests, with their universal appeal, are studied exhaustively. Values and services of forests are described. The extent of forests is mapped out and their many products, both tangible and intangible, are listed to show the social and economic importance of our woodlands both public and private. Commercial forest land is compared to non-commercial forest land and the amounts of both are noted together with their patterns of ownership. The status of forest conservation is detailed to show its many ramifications. Finally, national, State, local and private programs of forest conservation are presented and thoroughly discussed.

Outdoor recreation as it affects or is influenced by the several re-sources is considered as a specific part of the course.

The western grazing lands are examined as to their significance in the western livestock economy. The need for carefully controlled grazing to protect strategic watersheds and other land values is emphasized.

Field trips are used liberally throughout the course so that students can observe resource conditions on the ground and come to their own conclusions as to conservation needs.

The second method of including conservation in general education at the college level is through integration into established courses. While including less detail than a special conservation course, this method does reach many more students. Essentially it follows somewhat the same out-line as does a conservation course although the conservation content will no doubt be less. In many instances only the resource or resources of local significance are stressed.

Main Goal: Teach Conservation

There's divided opinion among both educators and conservationists as to the ideal method—special courses or by integration in established courses—of teaching conservation at the college level. In the final analysis, method is a responsibility of the college administrators and their teaching staffs. In any event, it is not the real point at isue when there is so little conservation included in the curriculums of our colleges and universities. The real point is "How can we get adequate conservation into those curriculums?" The Conservation Education Association, Incorporated, the only organization in the world devoted exclusively to the promotion of conservation teaching at all education levels—elementary, intermediate, high school, college, and adult—suggests this way of doing it—

"The administration at the college or university level must be sym-pathetic to the needs of conservation education in order that adequate training is made available to teachers and college students generally. The president, or his delegated representatives, should direct the program to provide this training. A duly appointed conservation committee (made up of representatives of the college administrative and teaching staffs) should study the curriculum and develop—

"1. Adequate integration and coordination of the conservation train-ing that is to be offered in each of the various departments.

"2. Courses in conservation of natural and human resources for pre-service teachers and, through extension or summer school, for in-service teachers.

"3. Methods of research in specific problems in conservation of both human and natural resources as they affect the local area.

"It is assumed that in teacher education institutions general education will include some conservation understandings. It is recommended, how-ever that *all* students be required to enroll in a 'general' course in Conservation of Natural Resources for a full school year. This assumes, of course, that the faculty in charge of this program will be fully conversant with the conservation fields and that the administration will be sympathetic and actively support the teaching program. It is further assumed that a continuing evaluation program will be carried on to increase the effectiveness of the conservation education program on the basis of the actual accomplishment of teachers in the field. Further, all interested groups must work together to find ways of increasing the motivation of students to study in this field, and to increase the enthusiasm of their professors.

"Teacher education institutions should solicit the support of, and develop strong interrelationships with local, State, national, and even international agencies—whether public or private—in the initiation, development, and improvement of the conservation education program.

"Responsible, talented, local laymen, skilled in fields related to natural resources, or demonstrating successful resource management, should be used in the teacher education program in conservation.

"Teacher education programs should provide for the following:

"a. Acquaintance with natural resource problems and practices pertinent to local, State, national and world welfare.

"b. Acquaintance with, and training in, the techniques for teaching conservation.

"c. Acquaintance with available public and private programs, services, and sources of information and assistance.

"d. Training in integration of conservation subject matter into varied subject matter fields.

"e. Camp and field experiences.

"Mechanics of instruction in teacher education for conservation teaching should provide for the following:

"a. The ecological approach in the basic biological courses, the inclusion of field and laboratory practices, and attention to the appropriate season of the year.

"b. The ecological approach in basic social sciences to include field experiences and applications.

"c. The integration of conservation into all possible subject-matter fields, with particular emphasis in the general education courses.

"In view of the rapid growth in outdoor education, school camping, and recreation programs, additional conservation content should be provided in those programs concerned with preparing leaders in these fields.

Will Colleges Meet The Challenge?

"It is the consensus of the Conservation Education Association that such conservation education in our colleges and universities will develop a citizenry sufficiently aware of its environment to resolve, democratically, its resource problems."[5]

The challenge is to our colleges and universities. They can advance the day when all Americans will have a real concern for the use and conservation of their natural resources—their soil and water, their forests and ranges, their wildlife, their minerals—yes, and even their sun and air. When that concern is aroused, the future of Americans yet unborn will be more secure. The potential of those unborn Americans is 300 million people or more by the year 2000, only 42 years away. Everyone of them, like those in the world before them, will depend on the natural resources for their standard of living and their security—their very existence. The schools of higher learning can help to arouse that concern. Will they do it?

Selected Bibliography

Durwood L. Allen, *Our Wildlife Legacy*. (New York: Funk & Wagnalls Company, 1954). Here are listed basic principles, which if applied promptly and vigorously, can insure a sound wildlife conservation program in America.

American Association of School Administrators, *Conservation Education in American Schools—29th Yearbook* (The Association, Washington, D. C., 1951). Defines conservation education as a broad area of school responsibility and guides the school administrator in developing an effective conservation teaching program.

Association for Supervision and Curriculum Development, *Large Was Our Bounty: Natural Resources and the Schools* (The Association, Washington, D. C., 1948). Reports on what schools are doing to make wise use and conservation of the natural resources a part of formal education. Includes examples of effective teaching.

Ward P. Beard, *Teaching Conservation: A Guide in Natural Resource Education* (Washington, D. C., American Forestry Association, 1948). Gives the teacher basic facts about the natural resources and lists educational methods and principles for teaching conservation.

Charles H. Callison, *America's Natural Resources* (New York: The Ronald Press Company, 1957). Basic facts about the respective natural resources, presented in a non-technical manner.

Marion Clawson & Burnell Held, *The Federal Lands, Their Use and Management* (Baltimore: The Johns Hopkins Press, 1957). A study of the use and management, from an economic viewpoint, of the Federally owned lands—the national parks, national forests, wildlife refuges, grazing districts, submerged areas off the continental shelf and similar areas.

The Conservation Foundation, *Resource Conservation Education*. A collection of papers (The Foundation, New York City, 1952). A series of 35 articles dealing with conservation education through formal education, public education, private and public conservation activities, information programs and education associations.

The Conservation Foundation, *Resource Training—for Business, Industry, Government* (The Foundation, New York City, 1958). An examination of the need for a sound knowledge of the natural resources and their conservation by professional men and women—lawyers, doctors, engineers, teachers, and others. It includes suggestions for giving this knowledge at the college and university level.

Tom Dale and Vernon G. Carter, *Topsoil and Civilization* (Norman, Oklahoma: University of Oklahoma Press, 1955). A report on the fall of nations and civilizations which neglected to care for their natural resources.

Forest Service, *Timber Resources for America's Future.* (U. S. Department of Agriculture, Washington, D. C., 1958). The most comprehensive survey ever made of America's timber resources and future timber needs.

Bernard Frank and Anthony Netboy, *Water, Land, and People* (New York: Alfred A. Knopf, 1950). An analysis for the layman of the nature, significance and costs of our national water resources and their problems; the efforts which have been made to meet those problems; and recommendations for future action.

Charles E. Lively and Jack Preiss, *Conservation Education in American Colleges* (New York: Ronald Press Company, 1957). Reports on the status of conservation teaching in our colleges and universities.

C. W. Mattison, *The Conservation Education Association—Its Objectives and Its Work.* A paper given before the Northeast Wildlife Conference, Montreal, Canada (The Conservation Education Association, 1958). A report on the activities and objectives of the Conservation Education Association, Inc., from its founding in 1953 through 1957.

CHARLES W. MATTISON, President, Conservation Education Association, was born in Sullivan County in the Catskill Mountains in New York (1904). Before entering college, he taught in a country school in the Catskill Mountains; then studied forestry at Cornell University where he was graduated (1928). Thereafter has worked continuously in his profession of forestry in all parts of the United States. Since 1946 has directed the school and college cooperation program of the Forest Service, U. S. Department of Agriculture. Is a founder of the Conservation Education Association, Inc., and in 1957 was elected to the Presidency for a two-year term; is a member of the Society of American Foresters, the Soil Conservation Society, the American Forestry Association, and the Conservation Education Association.

NOTES FOR CONSERVATION

1. R. H. Eckelberry, *The Case for Conservation in a General Education.* An article from a collection of papers. (The Conservation Foundation, New York City, 1952).

2. Paul B. Sears, *Conservation in General Education at the College Level*. An article from a collection of papers. (The Conservation Foundation, New York City, 1952).

3. Charles E. Lively and Jack J. Preiss, *Conservation Education in American Colleges* (New York: The Ronald Press, 1957).

4. Ward P. Beard, *Teaching Conservation* (The American Forestry Association, 1948).

5. Conservation Education Association, Inc., *State Level Coordination of a Conservation Education Program* (The Association, 1955).

SOCIAL SCIENCE EDUCATION

Roy G. Francis
University of Minnesota

Science in a Democratic Society

Science may be regarded as a particular body of knowledge. It would be erroneous to hold that science alone constituted 'knowledge.' Instead, it must be admitted that science, when it refers to "content," is a special kind of knowledge. There are many ways to characterize this kind of knowledge. For one thing, it is 'general' rather than 'particular.' It comprehends statements of relationship, which is merely a special characteristic of 'laws.' It is "verified" knowledge, where "verification" involves a set of rules connecting theoretical concepts to a world of essentially kickable facts.

Science is, of course, "logical"; at least, we would hesitate greatly to refer to science as being "illogical."[1] However, being logical does not differentiate science. St. Thomas Aquinas was most certainly logical. All of the scholastics were. Theology is not a "science." Moreover, "science" is empirical; today, at least, we require that a science be testable. This, in turn, requires some commitment to a limiting state of affairs. Unfortunately, the meaning of "empirical" is not uniformly shared. In some ways, the highly contrived laboratory is the basis for "empirical" research. In some ways, the highly unique world of the flow of history is also "empirical." And some scientists would argue that they do not care about the world of everyday experience; they are concerned only with making general statements. This issue will haunt us again.

Of immediate concern is the answer to this question: "Who possesses this knowledge?" When we recognize that "knowledge is power," we must become aware of who the bearers of scientific knowledge are, and how they relate to those who do not share this special knowledge. Insofar as all holders of scientific knowledge are, in our society, the products of our educational system, it is clear that we must maintain a democratic access to education, lest particular power be the property of a special elite. Of this, more later.

Now we must distinguish between two forms of the "scientific enterprise." These two forms are commonly called "pure" and "applied" science. The words are not too happily chosen. The idea behind "pure" science is the discovery of new knowledge, the broadening horizons of human control. It is "pure" in the sense that it is not contaminated by any other value than those relating to the acquisition of knowledge. "Applied science" refers to the uses to which scientific knowledge is put. In one way, we may characterize the mood of science as 'doubt'; one doubts one's theories, one's data, one's conclusions. When these are no longer doubted, but held to be the bases for inferences to be also regarded as true, we no longer have the mood of science. We are in the realm of application. Notice here, again, the mood of belief, of faith. The notion of doubt is gone. The formula is correct; the span will not break under the predicted stress.

The public has practically no contact with pure science. It knows only "applied science"; and this it knows inadequately. To the public the "engineer" and the "doctor" are scientists. If we admit "application" as science, they are; but if we limit science to the acquisition of new knowledge, they are not. Of consequence is the likelihood of the public to confuse the true spirit of scientific inquiry with the elaborate technology associated with the application of scientific knowledge.

In one sense, of course, the public is right. It cannot really be held that "all knowledge is power"; a more acceptable statement would be "all usable knowledge is power." Of course, one may talk about the implicit power in much of pure science, but this does not detract from the point we seek to make.

The ways by which we distinguish "pure" and "applied" science are not the only ways to proceed. We may have pointed out the vastly different value premises involved in the two forms of activity. In some ways, we may agree with the idea that the scientist holds that "knowledge is good." Yet this is neither necessary nor sufficient to explain the sacrifices made by a man becoming a scientist. He may well be motivated by a desire for prestige or (however wrongfully) for riches, as well as "to do good." In no circumstance, however, may he violate those values which may be called "intellectual honesty." The entire enterprise rests upon the premise of complete commitment to honesty.

The applied scientist, however, is involved in the world of action, not merely contemplation. One may study a door, its composition, the arc it cuts when being opened, the pressure required to move it, the motives for using it: without committing the student to using it. The application of this knowledge involves a value premise which is not contained in the prerequisites of science. It is true that one may engage in inquiry because he wants to use that knowledge: this may be needed to "get the scientist in action," but is unnecessary for science itself. The application of science, however requires specific value commitments.[2]

It is for this reason that the question of how the applied scientist relates to the rest of society is important. An entire culture may explicitly or implicitly agree with, say, the premises of its medicine men (scientists or shamans). Then society will applaud the work of the expert. But what if the scientist holds values shared by only a small minority of the society: what, then, are the implications of the value commitment?

In a society as materialistically oriented as is that of twentieth century America, those values supporting technical applications of physical science are shared by most of the public, if only implicitly. The problem of the social sciences, however, is quite another story. Here we may view antagonistically arranged parties. In the matter of social organization, for example, that which is beneficial to 'management' may be harmful to 'labor' (from their individual points of view). Knowledge as to how to foment race violence may never be put to use; but there are those to whom such knowledge would bring some rewards.

To put the argument another way: the engineer shares the dominant views of a society, and hence finds universal acceptance. It also denies him the view of 'value' as being problematic. The 'social engineer'—the industrial psychologist, the family counselor, the social worker, the minister, the jurist—however, may or may not share the values of his society. Indeed, the values of the majority may be the cause of the problem of concern to the 'social engineer.' He is, then, not given the universal acceptance his counterpart is given. However, he is made painfully aware of the problem of values. And these problems he "feeds back" to those disciplines upon which his activity rests.

A common argument tossed at the social scientist is that "one can never truly have a science of human behavior." This argument is, of course, entirely specious. If one is unable to discover uniformities of behavior, then behavior ought to be random—and describable by probability theory. Obviously, the discovery of uniformities of behavior is precisely the general task of science. The argument appeals, basically, to a notion of engineering as science. A correct formulation would be: there can never be a completely controlled society; precise human engineering is not possible.

Objections to this statement may be raised. However, we must admit the probability of incomplete learning. The subsequent generation never exactly duplicates the preceding one. While one may execute or exile those unable to learn the rules of the game, it is doubtful whether such a procedure would be generally acceptable. Moreover, complete control over the human behavior patterns would require complete knowledge which, in turn, implies the absence of anything new. As long as there are changes, human responses will not be entirely predicted. The formulation against social engineering is true at least for a dynamic society.

The objection to social science, however, runs a bit deeper. There are those who object to the idea of social engineering. To have a "science of society" implies to some the existence of a super "scientist" who would pre-empt the right to make decisions. People would be 'forced' to behave in specific ways. "Social Planning" is a common epithet; "a new priesthood" is a rarer one. It is quite apparent that such fears are based upon the old idea which equated science with engineering. When, however, science is regarded as a part of knowledge and the decision to use it is seen to require an additional value premise the fear of social science is removed.

Science: Content and Method

Science consists of a set of integrated statements or propositions. Some of these statements define the concepts to be used. Some are statements about the world. Some are conclusions drawn. Some are the rules by which conclusions are reached. Some contain reference to the procedures of the particular science. In short, some scientific statements contain elements of content—the subject matter of the discipline. Others refer to the method of inquiry which justifies the content-statement. Science encompasses both.

The person who does not participate in the "sub-culture" of science, for the most part, does not properly appreciate the importance of the distinction. He either is unaware of the difference or will be enamored of the technology of inquiry, giving prestige according to the complications found in the tools of the trade. For the most part, however, he will have had some contact in high school at least and will remember some vague notions about "osmosis," "capillary action," "convection," "distance to the moon" and other contentual items.

The 'social scientist,' however, is most likely to be aware of the problems of method.[3] To be sure, those in the classical sciences may take method for granted. It is exasperating to the social science professor to hear a major in, say, chemistry make disparaging remarks about sociology without any sophistication in scientific method. It is also exasperating to hear sociology professors insist that they are scientists and immediately expose their ignorance of scientific method. Many undergraduates have wearied of the courses in sociology, for example, in which some reference to "scientific method" is made. In the next section we will examine the reasons for these references. Now, we should be careful to observe what the implications of an insistence upon 'method' are.

The idea of 'method' as central to inquiry is the fruition of Baconian empiricism. Francis Bacon argued against the appeal to authority for 'proof.' He argued against 'logical proof.' He insisted upon developing a science from "facts"—the world of direct observation. Yet he recognized the limitations of direct observation; man learns improper ways of viewing the world. His language limits him. He must, somehow improve his contact with the world, to rid himself of any subjective source of bias.

One obvious way is in the improvement of one's "instruments." In astronomy and biology, for example, this implied the development of the telescope and microscope. Additional problems of light, refraction, and optics in general were soon to emerge. But the idea was simple: get an instrument which, independent of its user, would give precisely the same "facts." 'Facts' could then be organized into general categories; after observation came classification. The process was called "induction." After one had developed a set of classes, these could be regarded as "facts" and "one could proceed to a higher level of abstraction."

In the social sciences the task was greater. Words were more often used than physical instruments. The semanticist held the promise that the perfection of words would lead to correct apprehension of the world, and hence a correct social science. He failed to see that meanings are not intrinsic to the world, but grow out of the interaction of human beings. Nonetheless, a great deal of effort was expended in the direction of improvement of instruments. Much of it, of course, was necessary and worthwhile. This was not the only aspect of science to be included in 'method.' "Methods of analysis"—particularly statistical methods of analysis—were to be developed. These were to force conclusions from the data independent of the individual scientist. The "facts" and "scientific method" alone were to "cause" the conclusions. The individual scientist was to have no part in the activity; he was a "replaceable part." No proposition should depend upon the particular individual making the inference.

So far so good. But one difficulty lay submerged. The difficulty was this. This orientation to science lent itself to sterility—recognizing that that which is sterilized may be pure.

Science is not a slowly evolving process. There have been specific events, events which could not have been predicted from what was known before they occurred. The emergence of Pavlov and his experiments with dogs is such an event. This particular experiment in the conditioned reflex generated a number of similar experiments, and soon degenerated into a vastly sterile preoccupation with the Norway white rat and his skills at outguessing the psychologist. Sigmund Freud is another example. It is easy, and hence commonplace, to test out the logical implications of some of his ideas. It requires more skill to submit them to empirical test. But no aspect of scientific *method* will show why he proposed the *set of premises* which he did. Method may test out the consequences of a particular set.

Accordingly, we must recognize that science, as a search for new knowledge, involves a creative human being. The creative scientist cannot be passively oriented to the world, waiting for facts to speak to him. He must be actively oriented to the world. He must attempt the arrangements and rearrangements which lead most directly to the desired results. The Baconian insistence on method is not the whole story. This insistence generated a great skill on testing hypotheses, on the development of a technology. Unfortunately it begged the question of where hypotheses came from.

It is easier to teach 'method' in the sense of given rules and instruments than to teach the "creative" aspect of inquiry. As a matter of fact, it is probably impossible to teach students how to be creative. There can be no logic of creativity for logic is an argument proceeding from premises to conclusions. If the conclusion is "creative," it existed in the premises, and hence was only apparently creative. What can be done, perhaps, is to keep alive the inquisitive point of view that most children seem to possess.

This is done at no little cost. Insofar as the teacher is a custodian of students, he is required to be concerned with 'order.' He cannot tolerate an absence of discipline. Science requires the unique harnessing of creative imagination with rigorous self discipline. The creative individual is something of a rebel. The task is somehow to orient the rebellion towards socially approved ends. The creative student is likely to be asking questions the teacher cannot honestly answer. The desire for order may suggest a dogmatic reply. This is the burden of the teacher who would do more than turn out mere technicians.

The Importance of Scientific Method in Teaching Social Science

The problem of learning "scientific method," or, for that matter, for developing the scientific mood, is not uniform throughout the various disciplines. The classical sciences, for the most part, do not, today, require the same emotional response as do the contemporary sciences. The student in physics, to illustrate the point, probably does not experience any *emotional* disturbance in learning the differential equation from which the 'law of free-falling bodies' may be inferred. Very few students in astronomy today are distressed to learn that the earth is not the center of the universe. For the most part, students do not enroll in such courses with definite notions contrary to what will be learned. They may have a preconceived notion that the material will be difficult and this may turn out to be a self-fulfilling prophecy. But there is very little which must be unlearned prior to learning a correct idea.

The contrary obtains in the social sciences and, to a degree, in the so-called 'life' or biological sciences. A real problem exists in the teaching of the social sciences.[4] It has been said that "every man is his own sociologist." The statement means, of course, that our culture includes a set of beliefs about human behavior. Normally, the source of those beliefs is an identifiable person.

We get our 'explanations' of behavior from our parents, our teachers, our ministers and our friends. These more or less "common sense" formulations are likely to disagree with those from the formal school sciences. To a large extent, our folklore is committed to a quasi-biological theory of behavior. We have assertions which hold that "blood is thicker than water," "blood will tell" and other arguments which are based on the family as a biological unit. Indeed, there is no accidental connection between a racist doctrine and a social structure which allocate prestige along family lines as we see in the South. Other "explanations" may involve some idea of a "supernatural" source for behavior.

It is abundantly clear that the theories offered by a social scientist differ, or may differ, markedly from that which the student is trained to believe. Stating it specifically, a student may be forced to choose between two competing statements about the world. One is offered by his course in social science; the other he brings from his common-sensical environment. How to choose? An instructor may simply appeal to his authority in the school system; but this is at best a short-lived advantage. It is at this point that the student must understand "scientific method"; this constitutes a major reason for teaching 'method' to students. Like the student, a scientist must make a decision about a statement about the world. 'Scientific Method' may be regarded as the rules in terms of which one may make a judgment about a given proposition. It is a systematic rationale for making choices. It is a protection against any instructor holding only his structural authority as a basis for belief.

There are other reasons, perhaps a bit more subtle. There are many 'theories' purporting to be "scientific." Especially in the realm of human behavior, it seems, is there a real danger of the charlatan. In the medical world we have institutional procedures for the discovery of the quack. The fraud in the classical sciences cannot claim easy success. The disciplines are so well established that the fake is readily seen. In the social sciences, however, this is not so true. One may, for example, offer a differential equation and perform an elementary operation claiming thereby some significant argument. Any reader could pick up a text in advanced calculus and, without knowing either mathematics or a subject matter area, perform in the way described. But is it science? How can one student be protected against the charlatan, the fraud? Only by a thorough knowledge of scientific method, only by knowing the agreed upon rules of the game may he protect himself.

Finally, one must be aware of the fact that some students of a discipline get addicted to it. Out of the undergraduate student body of today comes tomorrow's scientists. Their training requires a sophistication in the general methods of science, and in the specific methods of their discipline. This is not to say that there is a single 'scientific method.' It is highly improper for one area to copy the methods unique to other areas. That which distinguishes the areas in the social sciences is not the content of their data-statements. Rather, it is the kinds of problems which they choose to solve. The sociologist, the economist, the psychologist may all 'see' human beings; but the sociological problem is not that of economics nor is it a psychological one. Accordingly, the method which can solve one problem may not solve another. Copying a specific technique or mode or argument only imposes other problems on one's own. The apparent duplication of effort in teaching, say, statistics in each content area flows from this requirement. The duplication is only apparent: just as it would be inappropriate for an economist to define the sociologist's problem, so would it be inappropriate for a sociologist to tell an economist what statistics he must learn. Unless the student has a mastery of a wide range of methodological techniques, he is limited in his inquiry. He may never have to employ the devices he is capable of using; but that is of less waste than being unable to choose correctly the device capable of solving his problem. The erstwhile social scientist must have a good grasp of scientific method.

Teaching Social Science as "Social Action"

The ethical premises upon which inquiry rests do not necessarily include those necessary for the promulgation of the findings. In particular, they do not include the attempt deliberately to change the minds of "nonscientists." Teaching is a form of social action; and behind each teaching act lies an ethical decision. Some of the moral decisions which must be made can have far reaching consequences. This is not to say that all teachers face up to these ethical choices, or are even aware that they exist. Indeed, some teachers may act contrary to some of the moral judgments which are consistent with their behavior.

Consider the question: "Out of all we know about human behavior, which ought to be included in this course?" Let us imagine that the course relates to inter-group relations. Such a course may hinge around (a) 'racial minorities'; (b) 'religious minorities'; or (c) labor-management problems. Or bits of all three. In any event, it is quite clear that only some of the findings can be sensibly presented to a class, keeping fairly good contact with a general theoretical argument. That is, in any case, the points of theoretical interest may be well developed in each case. The specific content, however, may change. Which content? Or is 'content irrelevant'?

Suppose that the teacher were in a school in, say, Arkansas in 1958. He may present the results of the most rigorous research by social scientists. This is almost certain to be at odds with much of the 'common sense' formulations. To argue that it is "good" to destroy erroneous ideas is only part of the decision a teacher must make. For the destruction of erroneous ideas in this case will involve human relations. The teaching of correct sociological theory may well carry with it the implication that the students parents are either ignorant or deliberate in falsifying the facts of life. However 'good' it is to have correct knowledge, must it come at the price of destroying a student's relations with his family? The example could be continued in each case. In aspects of the 'Protestant-Catholic' relations we are certain to find incorrect information held by parents of both groups. To teach 'correct' ideas implies the inadequacy of one's parents or other relatives.

We should understand then that the teacher not only provides the student with a certain amount of information, he also has access to certain human relations. Anthropologists are, of course, aware of this difference.[5] Changing the orientation a student has to the world about him may alter the relations he has with others. The premise that the student ought to be "correctly oriented to this world" may be easily accepted. When the price of this orientation involves the destruction of other values—values which in many cases are more strongly held by the student and his family than those justifying science—the teacher has another decision to make. This decision is particularly difficult when those with whom the student relates are also 'related' to the teacher in the form of people who vote for school board members. The teacher's job may be at stake.

This is a special difficulty of the social sciences. As we noted earlier, students are more likely to have developed common sensical ideas about human behavior than about other disciplines. Accordingly, the social science teacher is in greatest jeopardy in this respect. A generation ago, perhaps, those in the biological sciences were held in equal distrust. But there is good socological reason why the social sciences are feared. We have, of course, alluded to some.

The social sciences deal with people, their hopes, their aspirations, their identities. We deal with the stuff that makes social life possible. The dissemination of social science knowledge changes the social world. People enter new relations, their identities change, their valuations are modified, as a result of 'knowing' social science information. In many schools, the "I.Q. test" became a new basis for stratification, and parents competed with each other to see whose child rated the highest on the various tests. This despite the overt attempt by teachers to be thoroughly democratic, removing any achievement comparisons made possible by a grading system.

It is one thing to talk about chemicals; ordinarily, few people are capable of expressing preferences. But in the social world, few people want to have "all the facts known." In a labor dispute, some laws may be violated by both parties. In political leadership studies, the existence of a real "king maker" may never be uncovered because such information would destroy the 'king maker' himself. And so it goes. The teaching of social science, and particularly, the teaching of a commitment to a scientific method goes deep. The student who is willing to learn may be required to sacrifice greatly to do so.

There are other problems to be faced. At what 'level' should the course be offered? Should it be "watered down" for the benfit of the least capable student, or should it be "made tough" to provide a challenge for the most capable ones? To what extent ought the teacher be allowed to assume motivation on the part of the students, and to what extent ought he provide the motivation?

Clearly, any answer to questions of this kind involve an ethical judgment.[6] To sacrifice the few for the many requires a different pedagogical philosophy than that justifying sacrificing the larger number for the smaller. To sacrifice both groups merely for the sake of avoiding a specific decision requires a particular moral philosophy of its own.

Few teachers are capable of making these decisions for themselves. For these decisions reach out farther than the classroom. They extend out into the community. In some areas, the pattern of education will be so well developed—one way or the other—that the decision will have been largely made by the community. The 'culture' of the student body carries with it some influence in achieving the answers to these questions just as the answers to these questions will affect the culture of the student body.

We are recurring to a point made earlier in this chapter. We live in a democratic society. Teachers, as experts in other fields, often feel themselves to be in that difficult position in which they are convinced that they know better than the citizenry. This feeling is mostly a difference in values, though sometimes it is not. Often a teacher will have made certain pedagogical assumptions about teaching and education in general only to find that such were not the motives of the public. For example, both the teacher and the citizen may work hard for a new school, the teacher out of a concern with improving teaching and the citizen out of a concern for being better than the neighboring town. After the school is built, say, the implications from the point of view of the teacher may contradict that of the citizen. In a democracy, the ultimate authority is the voter, and the teacher finds that a large part of teaching involves a new kind of political struggle. The decision to 'fight' or to 'surrender' is predicated upon some ethical premise. It may be that one takes an ethical stand to conform with the behavior in which he is engaged. The same may be true for teachers.

Some Remaining Problems of Social Science Teaching

The teaching of social science in a democracy is interesting, if one finds interest in challenges. Earlier we mentioned a concern for science as power and in particular who was the possessor of this power. The theory of the social sciences contains other sources of concern.

Consider the stuff of which much of social science is concerned: "culture," "social structure," "education," "social class" and other such concepts. Grant, for a moment, the possibility of developing a scientific account of human behavior from a conceptual scheme including these things as variables. Now consider the task of "engineering." Recall that the social engineer must take social theory as 'given'; he has no doubts about the correctness of the lawful statements of the field.

Ask, then, this question: who—what social unit—can manipulate variables such as these? Can you, or I, or any individual, manipulate 'culture,' or 'education'? *Undoubtedly not.* Many variables are of the order that a social unit greater than the individual must be employed in controlling its variation. It may well be argued that, in a democracy, the individual has some final, however weak, voice to say in the use of variables such as this. This raises the general question of the place of a citizen in a democracy. In view of the "professional politician," what importance is there in the vote of the citizen? Do they simply have the right of collective veto?

Our society, however, is predicated upon the assumption that the individual is, somehow, sacred. The individual is presumed to be "the master of his fate, the Captain of his soul." What de we do to our cultural heritage when we develop a theory which places the control of one's behavior in the hands of something other than the individual?

It is true, of course, that some social sciences begin "with the individual." In a way psychology may be said to be a science of individual behavior. Except for some major things. In psychological literature the 'individual' is a general category and does not refer to a unique entity. Moreover, an important branch of psychology has been that of "adjustment." This poses the problem: how can man, any man, adjust to a social system, any social system? Apparently, from a theory of adjustment one may learn how to be brain-washed most easily, without any severe moments of regret. Somehow, this is not the answer.

In the so-called "market research," a wedding of diverse motives for understanding human behavior, we find a curious consequence. How can one use mass media of communication as vehicles for the dissemination of those stimuli which will bring the greatest financial reward to the merchant? Here stimuli are presumably being presented in such a way as to *maximize* a "favorable response." If a rational comparison of alternatives would tend to minimize the purchase of a given commodity, X, and it almost certainly would, then a rational comparison must not be allowed. The ultimate of this kind of stimulation would require stimulation below the threshold of awareness, subliminal advertising. In this case, the object is completely unaware of being stimulated, he has a minimal opportunity to make comparisons.

When there is unanimity as to the end to be achieved, there could be little debate. But, behold: if there is unanimity, what need is there of subliminal stimulation? If we wish to be reminded of something, perhaps, but would prefer not to be consciously bothered, this may be a boon. But what if there is something less than unanimity? What moral position is required to warrant the use of a device such as this contrary to the moral position of the other?

If, in the case of subliminal advertising, we find grounds for moral apprehension, does not the argument extend to other possible uses of science? It probably does. Whatever answer one achieves will involve a moral commitment. In any case, the social scientist because of the subject matter which he studies is aware of the problems of value, and hence the *problem* of the position of science in society, in a way few other scientists are.

In the case of 'the problem of education,' the social scientist is aware, both from theory and from experience, that answers to the question, "what to do about our schools?" involve statements of value. Accordingly, they are aware that differences in value lead to different conclusions. While they are usually quite willing to take value positions themselves, they are not willing to impose their value judgments on others. In a democratic society, this seems to be a proper solution. In an authoritarian society, perhaps, the professional educator can get by with imposing his values upon his community. But this does not say that such ought to be the case.

Selected Bibliography

Max Weber, "Science as a Vocation" (From Max Weber: *Essays in Sociology,* translated and edited by Hans Gerth and C. Wright Mills (New York: Oxford University Press, 1946), chapter V. This essay is the basis for my philosophy of teaching science. While Weber did not fully explore the question of the kinds of ethical questions involved, nor did he make inquiry into the question of "who can manipulate these variables?" he states admirably the case for a "value-free" science.

Louis Schneider and Sanford M. Dornbusch, *Popular Religion* (University of Chicago Press, 1958). This small volume is not directly related to the question of teaching social science. However, it raises fundamental questions about the role of values in research of the social sciences. Since the thesis of my chapter has been that teaching involves value-decisions, any treatment of that subject is of interest to the problem at hand. Moreover, this book is valuable in itself.

H. Otto Dahlke, *Values in Culture and Classroom* (New York: Harper 1958). A systematic treatment, in a tradition of solid sociology, of the problems of schools in modern society. The subject matter ranges widely, and is based largely upon contemporary research. This book is a "must" for anyone who, accepting some notions about science, wishes to make sense out of the current debates about the value of our educational system. Those who are happy to argue without reference to acceptable research may choose to ignore this book.

R. Freeman Butts, *A Cultural History of Education* (New York: McGraw-Hill, 1947). Gives a historical perspective, and reminds one of the value assumptions behind much of the growth and changes in our school system.

John T. Doby, *Some Effects of Bias on Learning* (Madison, Wis.: University of Wisconsin Library, unpublished thesis, 1956). An empirical, and in part an experimental, assessment of the difficulties of learning subject matter which is contrary to one's value-biases. In specific, traces some difficulties in learning sociology when the sociological propositions contradict that of one's "religion."

G. Watts Cunningham, *Problems of Philosophy* (New York: Holt, 1924). An older volume, published as a textbook. Introduces the reader to the ideas of science and how science is related to philosophy. Various problems are discussed, of particular interest is the issue of values.

Lowry Nelson and Marvin J. Taves, "An Appraisal by Students of Rural Sociology Teaching," a paper delivered to the Rural Sociological Society's Convention in Seattle, Washington, August, 1958. This brief statement reports on student reactions to teaching techniques employed by various professors of Rural Sociology. The interest lies mainly in the fact that social scientists are beginning to use the methods of social science to evaluate the effectiveness of various techniques.

William Kolb, "Family Sociology, Marriage Education, and the Romantic Complex," *Social Forces* (October 1950). An excellent criticism of a traditional way of teaching "marriage and the family." The criticism is solidly based upon a theoretical position and is not simply a "carping" statement. Kolb has written a number of similar criticisms, each of which is valuable reading.

Charles S. Johnson (editor), *Education and the Cultural Process* (Reprinted from the *American Journal of Sociology*, Vol. XLVIII, No. 6, May 1943). Is made up of papers presented at the Symposium commemorating the 75th Aniversary of the founding of Fisk University, April 29-May 4, 1941. The authors contributing this volume read like a who's who: Margaret Mead, Robert Redfield, Robert E. Park, Ruth Benedict, Melville J. Herskovits, Bronislaw Malinowski, Scudder Mckeel, Louis Wirth and others. The relation between the educational process and its cultural setting is clearly shown through the use of a wide range of varying cultural groups.

M. H. Willing, et al., *Schools and Our Democratic Society* (New York: Harper, 1951). Designed as a textbook, this volume offers much material for thought. Especially interesting are Chapters 2 and 3. "Official Controls of the School," and "Unofficial Controls of the School." A fairly straight sociological account.

Arthur C. Bining and David H. Bining, *Teaching the Social Studies in Secondary Schools* (New York: McGraw-Hill, 1952.) Clearly a textbook in teaching the subject matter of 'social studies,' is good 'data' from which one may infer some of the problems and difficulties underlying the task.

Robert O. Blood, Jr., *A Teacher's Manual for use with Anticipating Your Marriage* (Glencoe: The Free Press, 1956). This pamphlet shows, primarily the self-conscious attempt to develop a sound practice of teaching one particular part of social science. Much of the difficulties are implicit, but stand out quite clearly when explicated.

Howard Becker, *Through Values to Social Interpretation* (Durham: The Duke University Press, 1950). A theoretical statement by an eminent sociological theorist regarding the implications of values, generally. It does not deal directly with the problem of social science education, but has provided an adequate frame of reference for many of the statements made.

ROY G. FRANCIS, Associate Professor of Sociology, University of Minnesota, was born in Portland, Oregon (1919), and educated in a one-room country grade school, and a rural union high school. He received his B.A. degree (Magna Cum Laude) from Linfield College (McMinnville, Oregon), in 1946, M.A. (with Honors), University of Oregon (1948), and Ph.D., University of Wisconsin (1950): was Departmental

Fellow, University of Wisconsin (1948-1949), SSRC Fellow (Mathematics) at Harvard (1952-53), and was appointed Acting Instructor in the University of Wisconsin (1949-50), Assistant Professor and Research Associate, Tulane University (1950-52), Assistant Professor, University of Minnesota (1952-56), and Associate Professor in 1956. Has authored numerous articles in professional journals, and *The Rhetoric of Science* (University of Minnesota Press) and has co-authored: *The Population Ahead* (editor and contributor), *Introduction to Social Research* (with Doby and others), *Research Methods in the Behavioral Sciences* (with McCormick), and *Service and Procedure in Bureaucracy* (with R. Stone).

NOTES FOR SOCIAL SCIENCE EDUCATION

1. For an extended treatment of this point of view, see Chapter 1, *An Introduction to Social Research*, John Doby, editor (Morningside, Pa.: Stackpole Press, 1954).

2. For an older, but highly readable, discussion of these and other issues, see G. W. Cunningham's *Problems of Philosophy* (New York: Henry Holt, 1924).

3. Felix Kaufman, *Methodology of the Social Sciences* (New York: Oxford University Press, 1944).

4. John Doby "Some Effects of Bias in Learning," to be published by the *Journal of Social Psychology*, and based on an unpublished dissertation of the same title (Madison, Wisconsin: University of Wisconsin Library, 1956).

5. See Charles S. Johnson, *Education and the Cultural Process*, reprinted from the *American Journal of Sociology*, vol. XLVII, May, 1943; Malinowski's discussion of "schooling" and "education" is highly insightful.

6. H. Otto Dahlke, *Values in Culture and Classroom* (New York: Harper, 1958). This volume could be variously cited.

Auxiliary Aspects

MATHEMATICS AND THE TRAINING OF SCIENTISTS

JOHN R. HILLS
Regent of the University System of Georgia
(Atlanta, Georgia)

There can be little doubt that our existence as a nation in these times is immediately dependent upon our achievement in science. Every day we are told that if the United States does not maintain leadership in fission, fusion, and rocketry, we are inviting invasion, destruction, and dissolution. It is equally doubtful whether any person can become eminent, or even highly competent, in the heavily-stressed physical sciences without a firm grasp of, and a facile manipulative skill with the concepts of mathematics. It is not difficult to document this statement; in fact, a few examples of the research findings may be worthwhile.

Engineering

Let's start with engineering, a close relative of the natural sciences. Siemens studied the relationship between grades in college mathematics and grades in upper-division engineering courses at the University of California.[1] He found that if all of the 1400 students he examined had been equal in their mathematical skills, the variation in their grades in engineering would have been reduced by one half. That is to say, if by some means all of the students had mathematics ability deserving of a grade of B instead of grades from A to C, then the engineering grades would range from something like B— to B+ instead of from A to C. In statistical terms, the correlation was .69. In one sense it was the differences in mathematical skills which permitted some of the students to excell in engineering and get A's and caused others of them to fail.

Higgins found a similar situation.[2] For his students the combined average grade in analytic geometry and calculus of the freshman engineering students correlated .84 with their four-year grade-point average. Over 70% of the variation in four year grade average of these engineers was associated with or determined by their mathematical skills.

Stuit and his colleagues summarize their extensive study of prediction of success in engineering and other professional schools by indicating that a measure of mathematical proficiency should be used in any battery of tests used for counseling engineering students.[3] They found that high mathematical achievement was the best single indicator that a student would successfully complete engineering training.

Physics

Next look at the situation in physics. In his very interesting chapter on the most modern approaches to the teaching of elementary physics, Michels has described mathematics and physics as being so intertwined that they are parts of the same intellectual activity. This idea is not new. Nearly twenty years ago Stuit and Lapp concluded that achievement in college physics was more closely related to ability in mathematics than to such things as high-school general achievement, ability to deal with spatial relationships, or even scores on measures of aptitude for physics![4]

The relationship holds at the very highest levels of attainment. Harmon found that the test scores of National Science Foundation Fellowship candidates in physics and in mathematics were the same.[5] That is, as far as test scores and undergraduate grades go, a candidate for a fellowship in mathematics has the same pattern of talents as a candidate in physics. And Anne Roe in her study of the most eminent of physicists and chemists found in interview after interview that these renowned specialists mentioned the importance of mathematical skill in their work.[6] Practically all of them indicated that they were not only good at mathematics but also enjoyed it. Among those who did not spontaneously volunteer their facility with mathematics were one or two who indicated that their lack of skill in that area was one of their major handicaps.

Other Sciences

There is more evidence from the field of chemistry, although it is perhaps not so extensive as in physics and engineering. In the fields of psychology and anthropology Dr. Roe's findings are at variance with the picture for the physical sciences.[7] In these fields and in biology it seems that the scientists who are now eminent did not need much mathematical background compared with such people as physicists and chemists.

Dr. Roe had a special mathematics test built for the purpose of measuring mathematical sophistication of all of her scientists. She was able to equate her test's scores to scores on the College Entrance Examination Board's Scholastic Aptitude Test mathematics score. This Scholastic Aptitude Test is taken each year by a great many high school seniors who must submit scores when they apply for admission to college. Scores on the Scholastic Aptitude Test are adjusted so that the average score of those who take the test is 500, the minimum is 200, and the maximum is 800. Since those who take the test are usually applying for admission to highly-selective colleges, only the better students are found in the test administration. Thus, the Scholastic Aptitude Test's mathematics section is rather difficult.

Dr. Roe's test was much more difficult, naturally, because she wanted to use it with eminent Ph.D.'s in science. In terms of scores equated to the Scholastic Aptitude Test, her biologists achieved levels as high as 1042. That score is equivalent to getting 27 of the 39 items in Dr. Roe's test correct. The most astounding comparison, revealing something of the relative skill in mathematics of top-level chemists and physicists, comes from the fact that Dr. Roe gives no data concerning their scores on her test. The test was so easy for them that it was no test at all! Their mathematical attainments leave those of the other scientists trailing far behind.

There are, however, signs that the pursuit of a career in social science may not much longer be maintained at a high level by people unschooled in the language and concepts of mathematics and its sister, logic. More and more it is becoming both necessary and fashionable for social scientists to work within the framework of matrix theory, to phrase problems in the structure of information theory, and to apply complicated multivariate statistical analyses to their data. It would be erroneous to conclude that because the eminent social scientists of today (who received their training some years ago in a very different social-scientific atmosphere) are not very good at mathematics, and do not find it important, that those newcomers to eminence twenty years hence will feel the same way.

It would be a mistake, too, for us to fail to make it explicit at this point that mathematics is but a tool, a prerequisite, for scientific productivity. It is a very precise language which provides a framework for posing problems and communicating findings. The attempts of the American Institute for Research to develop a test for selecting research personnel reveal many other qualities of a good scientist.[8] From quite another point of view James Newman, in reviewing Taton's book, "Reason and Chance in Scientific Discovery," presents in a fascinating way the background of some of the successes and failures in achieving or recognizing mathematical and scientific discoveries of great import.[9] It is indeed difficult to determine how and why such discoveries come about at certain times and places. A person skilled in mathematics has often played a role in scientific discoveries, but by no means always. In the history of mathematical invention there appears to be no discernible pattern.

With these reservations in mind, we find justified the conclusion that our country's attainment in science, and its leadership as a world power, are going to be closely associated with the degree to which our scientific laboratories are staffed with people who are competent in mathematics. And this is going to depend to a large extent on two things. First, it will depend on how successful we are at locating those people who have the talent for mathematics and inducing them to pursue scientific and engineering careers. Second, it will depend on how good a job we are able to do at teaching mathematics to them in such a way that they will appreciate and enjoy, as well as be fluent in, this highly abstract language.

Finding Mathematical Talent

Naturally, efforts are underway to provide the two ingredients which seem so necessary. Although there has been little research on the talents required for success in mathematics, a start has been made. In 1953 the writer commenced a study aimed at finding out what psychological "factors" are related to success in college-level mathematics.[10,11]

Role of Factor Analysis

During the early history of psychology, investigators were prone continually to coin new names for psychological characteristics of people. Gradually, the list of names grew to astronomical proportions, one listing including 18,000 different psychological "traits." Needless to say, a description of a person which required an estimate of the extent to which he possessed each of 18,000 different characteristics would be somewhat cumbersome for practical use. Ways were developed to simplify and bring order out of this chaos. (This is one endeavor in which mathematics has made a great contribution to psychology.) A mathematical technique called factor analysis has been developing over the last 30 years. Its function is to examine the interrelationships among measures of these many traits in order to see which ones of them are so similar that they need not be separately considered. The basic properties that account for most of the data in a collection of scores for a wide variety of "traits" are called factors.

Application of the technique of factor analysis has resulted in more useful organizations of our observations about the characteristics of people. Much remains to be done, and there is a great deal of contemporary study along these lines; what seems sound today may seem naive in the light of tomorrow's advances. But at present we feel confident that we have found some appreciable order in certain areas of the problem. In terms of basic abilities, there can be little question but that it is very useful to think of a Number factor as reflecting a person's ability to do simple numerical computations rapidly and accurately. This is quite independent of one's score on the verbal Comprehension factor, the possession of a large reading vocabulary. Some people are high on both of these abilities, others are low on both. Still others are high on one and low on the other. Where a person stands on one does not necessarily tell us anything about where he stands on the other. They are independent of each other.

As one advances to abilities which we tend to think of as on a higher level, we are treading on newly-broken ground, and the footing is not nearly so sound. The same holds true among the factors which we ordinarily call personality, as distinguished from aptitude or ability. Much more research will be required before we can feel confident that we have grasped such a useful organization of our observations that only a major discovery or interpretation could cause us to give up what we have at any moment for something newer. However, the concentrated efforts of several research groups are providing us with new information and new interpretations of this information. The factors from such recent research efforts played a large part in the writer's study of factors related to success in college mathematics.

Factors in Mathematics

Partly it was by design that we concentrated on the newly-isolated factors in high-level abilities; partly it was because they seemed to be the most promising variables available. A thorough search of the factor-analysis literature revealed about 16 likely prospects among personality, interest, and aptitude variables. These factors had been found in several different studies, were provided with tests that were of at least moderately good technical quality and suitable length, and the factors appeared related to the kind of things that seem to go on in mathematical thinking.

It would have been very desirable to measure personality factors and to study their relationships with success in mathematics. However, only four of the readily-measurable personality factors in the literature seemed to have much promise. This presented a difficulty which we did not try to surmount for the present research. The difficulty derives from the fact that personality traits are ordinarily measured by questionnaires. Generally, it seems to be unwise to ask a large number of questions about any one personality characteristic unless there are questions about other matters interspersed in the inventory. In order to have more than three items about other personality factors in between any two items about the same factor, it would have been necessary to insert miscellaneous questions as filler. But this wastes precious testing time. All in all, it was decided to forego the personality variables at this time and to concentrate on the aptitude and ability factors.

In the case of the ability factors which seemed appropriate, it was easy to pare down the number in the light of information about the quality of the tests available for each of the factors. Many of the interesting factors have been isolated so recently that adequate measuring instruments for them have not yet been developed. After the initial reduction of the number of factors by examination of the data on each, the remainder were evaluated by a group of psychologists and mathematicians in terms of the nature of each factor and how relevant it was likely to be for success in mathematics. The opinions of thirty-five different mathematicians were obtained, and a lesser number of philosophers and psychologists were consulted in this regard.

The battery which resulted from all of this deliberation contained nine tests to measure nine different ability factors in two hours of testing time. These factors were Numerical Facility and Verbal Comprehension, mentioned above, and the following: Spatial Orientation, the ability to perceive spatial arrangements accurately and to maintain orientation with respect to objects in space; Spatial Visualization, the ability to visualize mentally the movement of objects into new positions in space; General Reasoning, the ability to determine what is given in a problem and what is asked for by the problem; Eduction of Patterns, the ability to discover a principle, rule, or system; Originality, the ability or tendency to produce uncommon, clever, or remotely related ideas, i.e., the ability to go beyond conventional thinking; Adaptive Flexibility, the ability to change one's set or presupposition in order to meet new requirements, e.g., to discard an old hypothesis or method in order to notice or find and use a new one to solve a problem; and Logical Evaluation, the ability to evaluate the logical consistency of given deductions.

Tests to measure these factors were taken by 148 students from three institutions. These students were involved in learning college-level mathematics as they studied to become engineers, physicists, or mathematicians. For each of these students we obtained grades in mathematics courses. These cannot be questioned as relevant indicators of one's success in learning mathematics. However, grades are influenced by other things than proficiency in the subject matter being taught, and grades may not adequately reflect the promise a person shows for future development. So, wherever it was possible, we obtained ratings of potential or aptitude for mathematical work. Where that was not possible, we were able to obtain scores on a mathematics-proficiency test, an indicator which should not be heavily influenced by nonachievement contaminators.

Frequently, in studies such as this all of the people examined are thrown together and considered as a single group. Implicit in that procedure is the assumption that mathematics, wherever or for whatever purpose it is taught, is about the same, and that everyone achieves success in mathematics by means of the same ability or abilities. In our study we decided not to make that assumption. Basically, we considered each group by itself. We kept freshmen separate from sophomores and graduate students; we kept engineers separate from physicists and separate from mathematics students; also, we kept the students from the three different institutions separate. After examining the relationships found among the separate groups, we then combined the groups in various ways where this was appropriate. As will be seen, the decision to keep the groups separate may have been the most important decision in the entire study.

Results of the Study

Within each of these small groups of students we calculated the degree of relationship between scores on the measures for each of the factors and scores on each of the available criterion measures. For measures of four of the factors there were no important relationships. This finding is worthy of some comment. In particular, the measure of Originality did not appear to be associated with grades except in one of the very smallest groups of subjects. Now the tendency to be original or unconventional may actually be a handicap in our present pattern of education. However, the data from this study seem to point to a more conservative hypothesis, that mathematics as taught at the undergraduate level at these institutions does not stress Originality or give credit for it. Perhaps this is a necessary state of affairs where a very extensive background of knowledge must be laid before sound, new ideas are likely to occur. Maybe it is largely due to the practical or applied emphasis in many mathematics courses. Frequently, students are taught the correct formula for solving a particular type of applied problem rather than the basic ideas in mathematics, how they may be used to develop a logical system, and how they may be changed to permit the development of new systems.

Another possibility is that the present state of college education, with its large numbers of students, inadequate facilities, and overtaxed faculties, does not allow enough contact between students and faculty members for the trait of Originality to be noticed or encouraged. It seems unlikely that Originality on the part of a student will display itself in a class wherein the primary goal is for the teacher to lecture about several pages of the textbook and to provide the solutions to the problems assigned at the previous class meeting.

The one small group in which Originality was associated with grades occurred at an institution in which the members of the mathematics faculty considered Originality to be the most important factor in obtaining the doctoral degree in mathematics. This was the only mathematics faculty of those cooperating in that phase of the research that gave Originality such great importance. This finding may indicate that in a situation where faculty members feel that Originality is important, where there is sufficiently frequent and close contact between students and faculty members, and where the subject matter permits Originality on the part of the student, a sizable positive relationship will be found between Originality and success in course work.

In the case of another of the unrelated factors, that of Eduction of Patterns, it seems that the lack of relationship with grades may have been due to the test's being too easy for these college students.

The finding that Verbal Comprehension is unrelated to success in college mathematics is surprising, because this factor is usually weighted heavily in the kinds of intelligence tests that are often used for selecting college students. The results of our study indicate that verbal ability within the college-level range is not related to success in college mathematics. It may be that there is a level of verbal facility above which the relationship between this ability and success in college mathematics becomes negligible. This might account for the finding that intelligence tests generally are good predictors of general success in college while tests of verbal ability are not particularly helpful in predicting success in mathematics of those who specialize in fields where unusual skill in mathematics is important.

The finding that the measure of the general-reasoning factor is not associated with criteria of success in mathematics is somewhat baffling. It is difficult to find in this study any evidence which would help us understand this result.

The Importance of Criterion-Curriculum-Institution Context

The importance of the decision to examine relationships between factors and criteria within each of the smaller groups of subjects becomes apparent when we consider the remaining five factors. In each case, these factors were associated to an important degree within certain groups of subjects or for certain criteria, but not for other criteria or within other groups. For example, the measure of Numerical Facility was strongly associated with calculus grades of engineering students at one of the institutions. Although there is some relationship between Numerical Facility and calculus grades in the other institutions, it is not nearly as high. If all of the engineering students and physics students had been placed in a single group, we probably would have concluded that Numerical Facility was completely unrelated to success in mathematics. This would have been an error.

For the measure of Adaptive Flexibility, we find that the calculus grades and the ratings of potential or aptitude for mathematical work of physics students were strongly related to the test scores at one institution. The relationship between Adaptive Flexibility and grades for the engineering students at this institution is considerably smaller. The salient finding is that for engineering students at another institution there is a negative relationship between the measure of Adaptive Flexibility and tested mathematics proficiency. This negative relationship is not particularly large, but it is clear that a student who is adaptively flexible has a better chance of doing well in mathematics in one institution than a student who is not flexible, while in another situation (at another institution) one's Adaptive Flexibility might work against him in studying mathematics.

The same sort of finding occurs with the measure of Spatial Orientation. All things considered, our data seemed to indicate that Spatial Orientation is more generally associated with success in mathematics than any of the other eight variables studied. At one institution, the primary relationship seemed to be with calculus grades of engineering students. In another institution, Spatial Orientation is associated with ratings of mathematics graduate and undergraduate students. At the third institution, the scores are important predictors of grades and tested mathematics proficiency of the senior engineering students.

However, there are cases in which the measure of Spatial Orientation has essentially zero, or perhaps even a slight negative, relationship with the criteria of success in mathematics. For example, this test had a significant positive correlation of .68 with ratings of mathematical aptitude of mathematics students at one institution. This is an unusually high degree of relationship for a single, experimental test. The same test had a correlation of only .02 with the average grade in mathematics courses of these same students! Another example with this same test indicates that the problem is not that grades are just a poor criterion. In this case, the scores on the measure of Spatial Orientation correlate highly (.55) with the average mathematics grades of engineering students, but they correlate slightly negatively (−.16) with the grades in the same courses, at the same institution, of the physics students. Considered by themselves, these correlations are statistically significantly different, even though the physics and engineering students took the same courses at the same time, from the same instructors.

The same sorts of things, although not to quite the same extent, occur with the remaining measures. For the measure of Spatial Visualization, the high relationships are with the criterion of calculus grades of the physics and engineering students at one of the institutions. The measure of Logical Evaluation has the highest relationship with the criteria of any of the variables studied, this relationship being with the ratings of undergraduate and graduate mathematics students at one of the institutions. Logical Evaluation is also importantly associated with the calculus grades and the tested mathematics proficiency of engineering students at one of the institutions. However, the measure of Logical Evaluation has a moderate negative correlation with the calculus grades of sophomore physics students at one of the other institutions.

Is There a Mathematical Aptitude?

It is not to be expected with samples as small as those examined in this research, and with only moderately good measures, that relationships found in one sample will be identical with those found in another. If one believes that there is an ability, or several abilities or traits, that constitute mathematical aptitude, however, and if he is willing to accept ratings, grades in mathematics, or scores on a mathematics-proficiency test as being primarily measures of prowess in mathematics, then it seems likely that he would expect to find that an ability related to ratings of mathematical proficiency would be related to mathematics grades. He should expect also to find that an ability related to grades for one kind of student would be related to grades for another kind of student, especially if the grades are for the same course.

If this reasoning is correct then those who believe that there is an entity such as mathematical aptitude should be surprised at the significantly-different relationships found in this study. Although their number is not particularly impressive, considering the possible number, certain of the significant differences are quite striking. In addition, supportive evidence is accumulating from other sources. Kline conducted research into the nature of the psychological factors involved in high-school intermediate algebra.[12] He conducted two factor-analysis studies which were as nearly identical to each other as possible, and then he compared the psychological factors found in one study with those found in the other. The only major difference between the two studies is that they were done on successive classes of students. The first study was done on the class of 1950-51; the second study was done on the class of 1951-52. The relevant finding from Kline's research is that approximately half of the factors that were important for success for the students of 1950-51 were different from factors important for success in the class of 1951-52. Essentially, what this means is that during one school year certain factors or psychological abilities were relevant to achievement in intermediate algebra that were not important for the students at the school during the next year. Likewise, certain factors that were important for success in intermediate algebra during the second year had not been important during the first year. This occurred even though the studies were both done at the same school, with the same teachers, using the same textbooks and all of the psychological tests used in the factor analyses were identical from one year to the next.

Philip Vernon produces additional evidence indicating that in different situations a variety of approaches may lead to success in mathematics.[13] He feels that it is quite likely that variations between test validities in successive courses given by the same teacher to similar groups of students would be as large as the variations between validities for courses given by different teachers. This opinion is in line with Kline's results. Vernon, himself, has found in studying five closely parallel groups of naval apprentices that the degrees of relationship between the psychological variables and the criteria differ markedly from one group of apprentices to another.

Research Conclusions

Briefly then, the current research into the problem of locating the people who have the talent for mathematics appears to indicate that the psychometrist or counselor never will be able to find on his shelf a single test of mathematical aptitude. The data that we have examined in arriving at this conclusion have implicitly assumed that we are not interested in assessing aptitude by means of examining past achievement. Of course, if we use measures which depend to any extent on past experience with mathematics, we do not achieve our purpose. The students who do well on such measures probably already know through their experiences in mathematics courses that they are good at mathematics. For increasing the scientific manpower pool we need devices that will

locate the students who are not already studying mathematics and preparing themselves for careers which involve its use.

Now, if mathematical aptitude is not a fixed entity, if there are
many aptitudes or sets of them which lead to success in mathematics,
provided the person with such aptitudes is guided into the appropriate
curriculum-criterion-institution context, then there are probably many
people who could become skilled in the use of this basic tool of science.
There should be many more potential scientists and technicians, at least
as far as mathematical skill is concerned, than would be the case if
there were a unique mathematical ability or constellation of such abilities.

If we are to capitalize on findings that several constellations of traits
lead to success in this field, we must examine additional traits at other
institutions and study their relationships with success in any other curricula which involve mathematics as a basic tool. A fairly limited set of
traits may include those which are usefully related to success in this subject matter. Such a finding would be encouraging.

Future Study

To take advantage of a situation such as that, it would be necessary
for us to make a widespread and cooperative effort to collect local data
on the useful traits for the curricula and criteria which are locally meaningful. Then our habit of keeping these data secret must be abandoned;
the normative data from different institutions and curricula must be
made available to all persons involved in student selection and guidance.

Although even the local normative studies that would be required in
the above scheme would necessitate a rather formidable amount of
psychological investigation, the data so far are not convincing that it will
be even that easy to predict success in mathematics. If additional studies
reveal findings similar to those of Kline and Vernon, it would seem almost
impossible for us to do a very good job of predicting success in mathematics. That is, their data seem to indicate that the predictors which
work in a situation vary from year to year even though the situation remains essentially unchanged. If that is found to be the case, the variables
which predict this year may not predict next year, and it may be that
we won't know what variables to use during each successive year.

That picture is a grim one. But there may be some other evidence
accumulating which would brighten the situation. Maybe we have not
had the raw material which is needed for predicting success in mathematics. At the time the writer's study was commenced, the findings of
Guilford's research into high-level abilities had not yet been organized
into a "structure of intellect." Recently, Guilford has examined all of
his findings, and the findings of other people who participate in this sort
of investigation, and has synthesized them into a pattern which reminds

one of Mendeleev's chart. In this organization chart the intellect is divided into two major categories, memory and thinking.[14] In each of these categories of intellect there are three kinds of content. There are figural content, structural content, and conceptual content; that is, some aspects of the intellect seem to deal with figures, others seem to deal with structural relationships, and still others seem to deal with concepts expressed in words. It is Guilford's opinion that the predictors of success in mathematics, when they are found, will most likely concern structural content.[15] The measure of Adaptive Flexibility used in the writer's study, above, was a structural measure, as was the measure of Education of Patterns. We are only beginning to find measures of other structural abilities, and the future may reveal that had more variables of that nature been examined, our results would have been much more encouraging.

Developing Mathematical Talent

Since the research into the possibility of identifying mathematical talent is at present not particularly helpful or encouraging, we would do well to look into the second possible means of developing additional mathematical and scientific competence. Current developments in the area of teaching of mathematics in such a way that people will appreciate and enjoy it, as well as be fluent with it, promise to be very important. They may even involve us in a different pattern of traits from those we find predictive of success in learning traditional mathematics.

Mathematics Education

The reader of Dyer, Kalin, and Lord's brochure entitled, "Problems in Mathematical Education,"[16] will be aware of the great concern over the status of mathematical education in the United States. The general picture is that mathematics training is in a most unhappy state. The students don't like mathematics; they are afraid of it; they don't see any point to it; they like other subjects more; and they don't like mathematics teachers. Several studies suggest that mathematics has the dubious honor of being the least popular subject in the curriculum.

Mathematics teachers are also in a predicament. One study showed that of 211 prospective elementary school teachers, nearly 150 had a long-standing hatred of arithmetic. Numerous studies seem to indicate that many elementary school teachers have difficulty keeping ahead of the pupils. They are ignorant of the mathematical basis of arithmetic. This may not be surprising. Frequently, little knowledge of mathematics is expected, even officially, of prospective school teachers. Not all of the problem is with the teacher's background either. It takes more than a good background for a teacher to do an effective job in teaching mathematics to a class of 35 or 40 pupils of widely differing abilities.

New Mathematical Curricula

The third problem which Dyer, Kalin, and Lord stress is the mathematics curriculum. That curriculum has changed very little during the past century. Not only has the curriculum not kept up with modern mathematics, but it is suspected that it has not adjusted to modern knowledge about how people learn and how pupils develop. It is along the line of modifications of the high school mathematics curricula that the exciting new developments are taking place.

At present three major attempts to revise the curriculum are under way. These were discussed in some detail in Rosenbaum's earlier chapter. The College Entrance Examination Board, in 1955, set up a Commission on Mathematics. This group is preparing a revision of high school courses. In general, their approach involves three aspects. First of all, they would treat algebra and geometry as logical systems, rather than as tools for solving everyday problems. The reason for this is that once the basic logic of such systems is understood most of the principles can be derived. The person who is well acquainted with the basic notions of algebra and geometry should be able to develop the standard or even new techniques for solving problems. In the past, it has seemed that the basic nature of mathematics has been left out while mathematics has been taught solely with the idea of preparing a student to solve the mathematical problems of everyday life. The students, although perhaps learning to solve many practical problems, have failed to learn what mathematics is all about.

A second feature of the College Board Commission's program would be to remove from the mathematics curriculum much of the material in geometry and trigonometry which is standard in traditional courses. In place of some of the traditional material of plane geometry, the Commission would introduce material on analytic geometry and solid geometry.

The third aspect of the Commission's program would be to go into more advanced mathematics in high school than has traditionally been the custom. The Commission would introduce the ideas of the calculus and those of probability and statistical inference.

Another significant development of the mathematics curriculum comes from University High School, an adjunct of the University of Illinois College of Education. The Illinois group takes a very radically modern approach to mathematics training in the high school. They approach mathematics entirely by way of abstract generalizations. Algebra is taught by starting with a set of axioms from which the students proceed to derive the rest of algebra. The students are early introduced to set theory, and geometry is taught as a pure deductive theory without regard to content. It undoubtedly requires capable pupils and excellent teachers to start such a course, but the Illinois group feels that high school students are interested in ideas and enjoy working with abstractions. Many high school students are not much interested in the usefulness of mathematics for practical problems. They want to know how mathematics fits into their world, and they like the adventure of dealing with abstract mathematics.

The third development in the curriculum may come from the School Mathematics Study Committee, patterned after the Physical Science Study Committee which is revising the curriculum in physical science. It is too soon to know what this study committee will produce, but, so far, the parallel development in physical science seems likely to have a tremendous impact on physical science training in the secondary schools.

Changes such as these in the pattern of secondary school mathematics education would be expected, in the not too distant future, to require drastic changes in the college mathematics curriculum. It may be that in college the training can start in the freshman year at the level which is common to the middle of the sophomore year in the traditional program. The ramifications of such an event should fire the imaginations of college science teachers. It can be seen, however, that predictors of success in the traditional mathematics curriculum, if found, might prove useless as predictors of success in a modern mathematics curriculum. It may be that entirely different sorts of people will be most successful in the new style of mathematics. At present, we can only conjecture along these lines; future research holds the answers.

Selected Bibliography

H. S. Dyer, R. Kalin, & F. M. Lord, *Problems in Mathematical Education* (Princeton, N. J.: Educational Testing Service), 1956. This pamphlet discusses the problems of the learner, the teacher, and the curriculum, from the point of view of a survey of mathematics education conducted by Educational Testing Service. It stresses the importance of research in our efforts to solve the problem. A comprehensive bibliography of 250 items is included.

J. R. Hills, "The Relationship Between Certain Factor-analyzed Abilities and Success in College Mathematics." *Psychological Laboratory Report Number 15.* (Los Angeles: University of Southern California), 1955. This report is a condensation of a doctoral thesis of the same title presented to the University of Southern California. It contains a detailed account of the study described in this chapter.

A. E. Meder, Jr., "Mathematics for Today." *College Board Review,* Number 34, (Winter, 1958), 7-10. Dean Meder was on leave from Rutgers to be Executive Director of the College Board's Commission on Mathematics. This article is his description of the Commission's plan for modern mathematics to replace present "seventeenth-century" programs in the schools.

J. R. Newman, "Books: A Study on the Nature of Scientific Discovery," *Scientific American,* CXCVIII (April, 1958), 141-148. This is more than a review of Rene Taton's book, *Reason and Chance in Scientific Discovery.* It is a very thoughtful consideration of the whole problem of scientific discovery, and it is an excellent introduction to the complexities involved in attempting to find a pattern, or organized set of patterns, for understanding scientific productivity.

Anne Roe, "A Psychological Study of Physical Scientists," *Genetic Psychology Monographs,* XLIII (1951), 121-239. Dr. Roe's technical description of her study of eminent physicists and chemists. A large part of the volume is devoted to verbatim accounts of interviews, providing much information of interest to the reader not well versed in psychology or statistics.

Anne Roe, "A Psychological Study of Eminent Psychologists and Anthropologists, and a Comparison with Biological and Physical Scientists," *Psychological Monographs,* LXVII (1953), Whole Number 352. Dr. Roe's technical report of her findings in studying social scientists also includes much interesting verbatim interview material. Her discussion of the findings of her three studies is particularly worthwhile.

D. B. Stuit, G. S. Dickson, T. F. Jordan, & L. Schloerb, *Predicting Success in Professional Schools* (Washington: American Council on Education), 1949. Presents in a single volume a series of papers summarizing the essential facts (in 1949) concerning the use of tests for counseling and selection in professional colleges. Initially, the papers were prepared under the auspices of the Veterans Administration for the use of vocational advisement officers. The volume includes chapters on predicting success in engineering training, law training, medicine, dentistry, music, agricultural, teaching, and nursing training. Each chapter describes the training program, the admission requirements, the research findings, and the implications for counseling in one of these areas.

D. E. Super & P. B. Bachrach, *Scientific Careers* (New York: Teachers College, Columbia University), 1957. Super and Bachrach's monograph summarizes what research has so far shown to be the characteristics of natural scientists, mathematicians, and engineers. The research is evaluated, and studies and techniques which should be pursued are noted. In the final chapter, vocational development theory, a series of propositions which approach vocational choice as a continuing process, is presented. An excellent 229-item bibliography is included. This is the best overview of the research on scientific careers known to the writer.

JOHN R. HILLS, Director of Testing and Guidance, Regents of the University System of Georgia, was born in Ohio (1927), attended public schools in Minnesota, and graduated as Valedictorian of Wadena High School (1945); then attended St. Cloud Teachers College (Minnesota), San Bernardino Junior College (California), University of Redlands (California), and the University of Southern California. Received the Bachelor of Arts Degree, 1950, (Magna cum laude) the Master of Arts

Degree, 1951, and the Doctor of Philosophy (Psychology), 1955, from the University of Southern California. Has held psychological research positions with Sheridan Supply Company, Beverly Hills, California, Psychological Research Center for Business and Industry, Los Angeles, California, the University of Southern California, Los Angeles, California, and Educational Testing Service, Princeton, New Jersey. Became Director of Testing and Guidance for the Board of Regents, University System of Georgia, Atlanta, Georgia, in 1958. Member of Phi Beta Kappa, Phi Kappa Phi, Psi Chi, Sigma Xi, American Psychological Association, Eastern Psychological Association, American Statistical Association, and Psychometric Society. Numerous publications in the field of psychology.

NOTES FOR MATHEMATICS AND THE TRAINING OF SCIENTISTS

1. C. H. Siemens, "Forecasting the Academic Achievement of Engineering Students," *Journal of Engineering Education,* XXXVI (1942), 617-621.

2. T. J. Higgins, "Study of Mathematical Ability in Relation to Success in Engineering Studies," *Journal of Engineering Education,* XXIII (1933), 743-746.

3. D. B. Stuit, G. S. Dickson, T. F. Jordan, and L. Schloerb, *Predicting Success in Professional Schools* (Washington: American Council on Education).

4. D. B. Stuit and C. J. Lapp, "Some Factors in Physics Achievement at the College Level," *Journal of Experimental Education,* IX (March, 1941), 251-253.

5. L. R. Harmon, *Ability Patterns in Seven Science Fields: Technical Report Number Ten, Research on Fellowship Selection Techniques* (Washington: Office of Scientific Personnel, NAS-NRC), 10.

6. Anne Roe, "A Psychological Study of Physical Scientists," *Genetic Psychology Monographs,* XLIII (1951), 121-239.

7. Anne Roe, "A Psychological Study of Eminent Psychologists and Anthropologists and a Comparison with Biological and Physical Scientists," *Psychological Monographs,* LXVII (1953), 21, 22, 28, 30.

8. American Institute for Research, "The Development of a Test for Selecting Research Personnel," (Pittsburgh: Author).

9. J. R. Newman, "Review of 'Reason and Chance in Scientific Discovery,' by R. Taton in 'Books,'" *Scientific American,* CXCVIII, (April, 1958), 141-148.

10. J. R. Hills, *The Relationship Between Certain Factor-Analyzed Abilities and Success in College Mathematics: Psychological Laboratory Report Number Fifteen* (Los Angeles: The University of Southern California).

11. This study was partially supported by the Office of Naval Re-

search and by Educational Testing Service. The cooperation and help of Dr. J. P. Guilford and his staff is greatly appreciated.

12. W. E. Kline, *A Synthesis of Two Factor Analyses of Intermediate Algebra: ONR Technical Report* (Princeton: Princeton University and Educational Testing Service) .

13. P. E. Vernon, *Educational Testing and Test Form Factors: Research Bulletin 58-3* (Princeton: Educational Testing Service) , 30.

14. J. P. Guilford, *A Revised Structure of Intellect: Psychological Laboratory Report Number Nineteen* (Los Angeles: University of Southern California) , 14-15.

15. J. P. Guilford, *op. cit.,* 6, 11, 13, 16.

16. H. S. Dyer, R. Kalin, and F. M. Lord, *Problems in Mathematical Education* (Princeton: Educational Testing Service) .

ACTIVITIES OF THE FEDERAL AND STATE GOVERNMENTS

ORVAL L. ULRY
University of Maryland

Before plunging into the specific and somewhat more limited area of activities of the State and Federal Governments in science education it seems both fitting and proper to take a broader look into the activities of these in relation to the whole of the education program. Rooted in the culture of the Old World early American education quite naturally reflected the characteristics of its European predecessors. Thus, there evolved a dual system of schools steeped in the tradition of limited education for the masses with anything beyond that reserved for a small elite.

In the course of a relatively short time, however, American education developed many worthy characteristics uniquely its own. Among frontiersmen class differences resolved themselves and the school which emerged in the environment was a school which recognized no class differences—the so-called common school. The stage was set, however, and the actors, the local government, the state government, and the Federal government all began a smoldering struggle to determine the optimum level of coordination and cooperation necessary to accomplish a performance equal to the public necessity in a democratic society. This struggle, in fact, has never been resolved.

The State and Education

Surprisingly enough, no reference can be found in the United States Constitution indicating in any direct way that education should be a concern of the Federal Government. It is by virtue of default and omission, and through the vehicle of the 10th Amendment that the various states assume responsibility for public education. The Constitutions of the several states, consequently, do dedicate themselves to the maintenance and support of a system of free common schools wherein all of the children of the state may be educated. This is, for example, clearly discernable in the famed Massachusetts Law of 1647 which required towns of 50 families to provide primary schools for all boys and girls.

Although we do have, without any doubt, a decentralized school system in these United States a phenomenal similarity of pattern of growth and development is obvious among the states. This common organizational structure for education within a given state mandates a state board of education, a state superintendent of public instruction, and a state department of education. These key positions are supplemented by assistant and/or deputy superintendents as demanded by the size of the educational program and the recognized function of the state's role in public education.

The state departments of education have, with few exceptions, been the recipients of all federal assistance to educational programs. This has not generally been without regulatory involvement of the Federal Government. It has been, however, generally free of Federal control of program and curriculum.

The Federal Government and Education

While education in the United States is recognized as a state function and a state responsibility, there is actually nothing in the Federal Constitution that prohibits Congress from enacting legislation related to education. Actually, the Federal Government has been neither inactive nor without influence in the realm of education.

Federal participation in public education actually began with the passage of the Ordinance of 1785. Enacted by the Congress of the Confederation, this ordinance set aside section 16 of every township in the newly surveyed lands of the West for the "maintenance of public education." The ordinance of 1785 was strengthened and enhanced only two years later in the Ordinance of 1787. This ordinance declared religion, morality, and knowledge necessary to good government and proclaimed that "schools and the means of education shall forever be encouraged."

In keeping with these ordinances, regulations were laid down in the enabling acts covering land grants for educational purposes. The admission of Ohio to statehood in 1803 marked the first instance of this policy. This policy was altered until only a few years later found two sections of every township set aside for education in California and in Utah, Arizona, and New Mexico, four sections.

In general, up to now, little or no regulatory measures were prescribed regarding the use of these various aids. A slightly new dimension was added to Federal activities in education with the passage of the Morrill Act of 1862. The Morrill Act provided that large amounts of public land should be granted to each state for the establishment of an agricultural and mechanical arts college, more commonly known as a land-grant college. This act specified, for the first time, how the money was to be used and called for annual financial reports from each college specifying how the money was spent.

The same tendency toward greater control and more active involvement in educational programs was obvious in the Smith-Hughes Act of

1917. Herein was established an entirely new kind of Federal-State relationship with respect to the use of Federal money. Each state receiving money under the Smith-Hughes act was required to match Federal money, dollar for dollar. Furthermore, a Federal Board of Vocational Education was created by the Act to work with *similar boards in each state*. In this program also, state plans had to receive federal approval before financial aid was given.

The Federal Government began to assume much larger responsibilities in the Nation's education during Franklin D. Roosevelt's first term as President of the United States. Under the New Deal program, a number of emergency agencies were organized to combat the long shadows of depression and even though relief was the major objective of these agencies, several of them developed important and far-reaching educational programs. These include two programs designed specifically for the youth of the nation—the Civilian Conservation Corps (CCC) organized in 1933, for the purpose of providing employment and vocational training and the National Youth Administration (NYA) inaugurated in 1935, to assist needy students continue their education.

Two other emergency bodies created during the depression years left their mark on American education. Under the Works Progress Administration (WPA) the Federal Government provided funds for work-relief payments for unemployed teachers. This program of the Division of Education Projects under the WPA cost the taxpayers $93,180,790 and covered the following educational activities: literacy education, naturalization education, public affairs education, academic and cultural education, avocational and leisure-time education, nursery schools, parent education and homemaking, vocational education, teacher education and other educational programs.

Because of the emergency nature of these several programs, the Federal Government tended to by-pass, in many instances, the existing state departments of education. This was a bold new venture by the Federal Government and implications of this emergency move are aptly stated by Chamberlain in the following lines:

> The fact remains that the Federal Government extended its influence in education to a degree far beyond any it had previously exerted. Had this continued . . . there is little question but what we would have a national system of education paralleling state and local community systems of public education.[1]

The next several years are literally shattered with various and sundry activities of the Federal Government that directly or indirectly influence educational program and policies. None the least of these include the G. I. Bill of Rights, the Surplus Property Act, the School Lunch Program, the Surplus Commodities Program, and the Korean Bill for Veterans.

Each of these activities taken individually or collectively has extended immeasurably Federal activity in the total field of education.

Still another type of Federal Government involvement into education in general stems from the ever-growing United States Office of Education. As early as 1830 there was considerable discussion of the need for the collection and dissemination of educational information by a Federal agency. In 1866, at the annual meeting of the National Association of School Superintendents, it was agreed to present a memorial to Congress urging the establishment of a National Bureau of Education. Later James A. Garfield introduced such a bill in the House of Representatives. It was passed by a sizable majority in both the House and the Senate and was approved by the President on March 2, 1867. The Department of Education was not long continued for the annual appropriations act of July 20, 1868 reduced it to the status of a Bureau in the Department of the Interior. The present Department of Health, Education, and Welfare was created by organization plan 1 of 1953. This new department was charged with the tremendous task of improving the administration of those agencies of the government whose major responsibilities are to promote the general welfare in the fields of health, education, and welfare.

Thus, we have an exciting background of both direct and indirect activities of state and Federal Government involvement in educational activities. These activities have usually been carried on through a cooperative arrangement with the respective state departments of education. It has been shown, however, that in a few instances the Federal Government, under the banner of emergency status, actually operated educational programs other than or in addition to those carried on by the state departments of education. It is hoped that this brief background of activities of the Federal and state governments in education in general will serve as a firm backdrop for presenting activities of these major agencies in the area of science education.

The Dilemma of the 50's

Especially since World War II, the relation of science to the general welfare has received immeasurable recognition as a problem of profound national importance. The atomic bomb, the hydrogen bomb and Sputnik are only the most dramatic catalyses of a tremendous ferment.

Science has always been a proper concern of the United States Government. In a scholarly and detailed treatment of this hypothesis Dupree concludes that:

Through all the twists and turns of the . . . history of the United States, and through the immense changes wrought by 150 years of rapidly expanding scientific knowledge, the policies and activities of the Government in science make a single strand which connects the Constitutional Convention with the National Science Foundation.[2]

True as this may be, this "single strand" has not been successful in presenting a solid front in science and science education within the United States Government. As director of the United States Office of Research and Development and after an impeccable investigation of the situation Vannevar Bush exclaimed:

We have no national policy for science. The Government has only begun to utilize science in the Nation's welfare. There is no body within the Government charged with formulating or executing a national science policy.[3]

Bush is thoroughly convinced that it has always been basic United States policy that government should foster the opening of new frontiers. He further concludes that:

Without scientific progress the national health would deteriorate, without scientific progress we could not hope for improvement in our standard of living or for an increased number of jobs for our citizens; and without scientific progress we could not have maintained our liberties against tyranny.[4]

Thus, it is obvious that as late as 1945 all was not clear on the horizon of Federal activities in science. The years that follow abound in reports, plus suggestions and remedies. This general dilemma has many horns, none the least of which lie in the areas of education and technology.

A Dilemma in Public Education. It is indeed, ironic and quite pathetic that a severe shortage of qualified teachers should arise during the same general period of years (1950-54), when public school enrollments were hitting a new high each successive year and when the demand for highly trained personnel was at an all-time peak. In 1954, a low point 60 per cent below the 1950 number of new college graduates qualified to teach high school mathematics and science left our colleges. During this same period high school enrollments increased at the fastest rate in history.

Discouraging, also, is the fact that only about half of the teachers who graduate with certificates to teach science and mathematics enter teaching as an occupation. Science and mathematics instruction too often is being carried on by teachers not sufficiently grounded in these subjects. Teaching as an occupation is not attracting its fair share of the nation's most able persons. Many better college graduates are entering positions more attractive to them than teaching. Furthermore, there is a continuous drain-off of better qualified secondary school teachers moving to administrative and college level positions or leaving the profession.

The situation becomes even more exasperating when we recognize the hard facts concerning the probability of successful completion of high school and college by any one particular youngster. In the nation today, only slightly more than half of the youngsters that enter first grade in our public schools persist through high school graduation. Of this half, only a third enter college on a full-time basis and less than a fifth of the high school graduates attain a bachelor's degree. Less than two per cent of those judged to be mentally capable of studies at the doctorate level obtain the Ph.D. degree or its equivalent, the level of education required for highly qualified scientists.

Another area of grave concern today is directly focused upon the talented high school student and talented high school graduate. Recent data show that annually 200,000 able high school graduates do not go on to college. Over 60,000 of these do not continue because of financial reasons. Furthermore, able students while attending high school are thought to be substituting "easy" courses for the more difficult courses in science, mathematics and foreign languages. Three states, Idaho, Indiana, and Maryland have already made extensive surveys of courses actually taken during high school. The United States Office of Education has recently released pertinent data on subjects taken by students while attending high school. Among other pertinent data the following seem most apropos:

1. About 32 per cent of American high school pupils take Chemistry, and relatively few study the subject more than one year.
2. About 76 per cent of American high school pupils study *no* physics in high school.
3. Hundreds of high schools in rural areas, enrolling about 10 per cent of all pupils, provide inadequate education in mathematica and science to meet standard college entrance requirements for science and engineering.[5]

A Dilemma in Technology. It becomes immediately obvious that the dilemma we face in technological manpower is not mutually exclusive of the dilemma that we face in public education. As a matter of fact, the two are rather closely related. The Joint Committee on Atomic Energy takes the interesting position that the shortage in technology is really only a part of a much larger national problem—that of identifying and developing our best young minds in this country so as to reap the optimum harvest from our potential intellectual resources. This Committee continues:

It is the development of this brainpower and the upgrading of our educational system so as to make it more attuned to the needs of our economy which is at once the most challenging and most important job we have to do.[6]

Today the U. S. is in a rather precarious position where the demand for technical manpower has seriously out-run the supply. At every turn, there is evidence of America's increasing dependence on science and technology for its cultural, economic and defensive strength. Furthermore, this is part of a world-wide trend. Alan T. Waterman, Director of the National Science Foundation states his position in this dilemma in the following way:

It is obvious that this strength depends on high level education for more of our people—this education leading to some basic knowledge of the sciences on the part of everyone and also the continuing production of a corps of highly trained and creative scientists and engineers.[7]

Historically the United States has achieved its position of technical leadership due basically to the efforts of only a very small fraction of its people. In fact, progress has been largely due to the results of the splendid efforts of about one-half of one per cent of our population. Considering that technology is accelerating at an unprecedented pace and that this pace is being indulged in by all nations at a time when our supply of natural resources is dwindling and our population is at an all-time high, every effort must be expanded to increase both our quantity and quality of technically competent personnel. In November of 1957, shortly after Sputnik, Dunning, writing in the New York Times exclaims:

The future belongs to the nation with the best technical brains. It is scientific weaponry, not masses of brave men, that will win wars.[8]

Encouraging indeed, is the fact that within the last twenty years we have quadrupled the number of scientists and doubled the number of engineers. Consoling indeed, is the fact that what is military technology today is likely to be a higher and more dignified standard of living tomorrow.

Activities Prior to 1940

Dupree, in summarizing the importance of science to government down to 1940 states: "During a century and a half, science has not only contributed to the power of the government, but to the ability of the people to maintain their freedom."[9] This dual role is implied over and over again. President Eisenhower enhances this position in his charge to the National Committee for the Development of Scientists and Engineers. In it he states:

We in America have a unique technological ability to use science for the strengthening of our country's defense against aggression and for the application of our material resources to the improvement of human living.[10]

Even though Dupree[11] consumes four pages in listing a chronology of important events of Federal government activities in science, he certainly implies a need for some type of central organization to coordinate these activities. A few of the most important activities together with the year that the activity originated are listed below:

1787 Constitutional Convention
1790 First Patent law
1802 Army Corps of Engineers
1830 Authorization of work on weights and measures
1840 National Institute for the promotion of science
1847 David Dale Owen's and others' geological surveys of federally owned lands
1862 Morrill Act for land-grant colleges
1863 National Academy of Sciences
1874 U. S. Geological Survey
1881 Founding of the magazine *Science*
1901 National Bureau of Standards
1906 Pure Food and Drug Act
1914 Smith-Lever Act authorizing extension work, Department of Agriculture
1917 Smith-Hughes Act
1917 World War I
1918 Chemical Warfare Service
1930 National Institutes of Health
1933 Science Advisory Board
1934 Agriculture Research Center
1940 National Defense Research Committee

This partial list points up rather clearly the "topsy-like" growth of science activities in the federal government. The total four-page list is somewhat impressive but the individual entries taken separately are quite fragmental and unorganized.

Activities After World War II

The situation which existed before 1940 changed so rapidly in the following five years that increased interest was shown, both in the executive and the legislative branches of the Federal Government, as well as in the scientific world, in the proper relationship between government and scientific research. From the profusion of materials published in the 1945-50 period, a few selections have been made which relate, directly or indirectly, to proposals for a single centralized administrative agency comprising all, or most of, the scientific functions of the Federal Government.

Modern war uses (World War II) scientific knowledge as insatiably as it does material resources. These demands pressed the Government not only to enlarge its own capacity for scientific research and development, but also, of necessity, to seek out scientific talent and resources wherever these might exist or could be assembled. Working in close cooperation during the emergency were the National Defense Research Committee, the Office of Scientific Research and Development, and the military departments.

Early in 1945, the Subcommittee on War Mobilization of the Senate Committee on Military Affairs published a report entitled "The Government's Wartime Research and Development, 1940-44."[12] This subcommittee pointed out the temporary nature of the coordination between departments and agencies which had developed during the war and suggested that some mechanism for coordinating federal research activities must be worked out. After extensive hearings in October and November of 1945, the committee finally recommended the establishment of a *new coordinating science agency.*

About midsummer of 1945 Vannevar Bush, Director of the United States Office of Scientific Research and Development published his timely report to President Roosevelt. This report entitled "Science, the Endless Frontier"[13] is today, 13 years later, outstandingly important because of its profound analysis of the place of science in our contemporary democratic form of government. Bush states emphatically that the government should accept new responsibilities for promoting the flow of new scientific knowledge and the development of scientific talent in our youth. He further insists that the effective discharge of these new responsibilities will require the full attention of some over-all agency devoted to this purpose. The Committee, therefore, recommended, without reservation, the creation of a permanent overall Federal agency for the support of science.

On October 17, 1946, President Truman appointed John B. Steelman, Chairman of the President's Scientific Research Board and assigned to him the preparation of a report on the nation's position in scientific research and development with recommendations as to suitable action by the Federal Government. A five-part report "Science and Public Policy"[14] was published by this committee in the autumn of 1947. The report included 4 recommendations designed to produce an administrative organization which attempted to meet the charge President Truman assigned the committee. The 4th recommendation was for a *National Science Foundation on sound lines.*

The National Science Foundation, 1950

The National Science Foundation was created in 1950 by the 81st Congress to develop and encourage the pursuit of national policy for the promotion of education in the sciences. Although education in the United States is primarily the responsibility of the individual states, the Natonal Science Foundation as a federal agency encourages and lends support to the efforts of our colleges and universities as they attempt to improve education in the sciences. Details of the elaborate program of the National Science Foundation may be found in a later section of this same chapter.

The long discussion of an appropriate organization to coordinate the science activities of the Federal Government was ended only temporarily by the establishment of the National Science Foundation. Unfortunately, the scope and complexity of the problem increased at such a rapid rate that discussion began again within only a few years. In 1954 Prince points out:

But however hopeful we may be, we have to face the unhappy facts that democracy has been put on the defense in many parts of the world and that new weapons of destruction are a constant threat to civilization.[15]

The White House Conference on Education

In December, 1954, President Eisenhower appointed a 34-member committee for the White House Conference on Education and charged it with the responsibility in carrying out "the most thorough, widespread, and concentrated study the American people have ever made of their educational problems."[16] To the governors of 53 states and territories, he expressed a hope that each would call a citizens conference on education. Some $700,000 was appropriated by Congress to be distributed among the states and territories to defray the costs of these statewide meetings. In return for this money each state and territory was obligated to furnish a report on its conference program. The White House Conference was held in Washington, D. C., November 28-December 1, 1955. A full report was published in April, 1956.[17]

In 1956[18] and again in 1957[19] the National Science Foundation issued reports indicating the insurmountable task that had befallen their limited resources. Many government agencies broadened their scope of scientific activities. The number and variety of government laboratories were multiplying. The situation seemed little improved by the modest role played by such agencies as the National Science Foundation, the National Academy of Sciences and the National Research Council. As a matter of fact, considerable evidence is in hand to again suggest that stronger policy leadership would make a substantial contribution to the public interest, to the work of the scientific community, and to the efficient management of the government's scientific programs.

The Status—1958

Let us pause for a moment to take stock of our status in 1958. We have examined the activities of Federal and State Governments with respect to education in general and with respect to science education. We have followed these activities from the Constitutional Convention through

World Wars I and II, the creation of the National Science Foundation and Sputnik. What then is our status today?

The Technological Front. In late November and early December the establishment of a *Federal Department of Science* was advocated editorially by the Christian Science Monitor, the Memphis Appeal and other newspapers. On January 5, 1958 Senator Estes Kefauver, Senator Karl Mundt, Representative Kenneth Keating, and Republican Carl Albert appeared on the CBS television network in a discussion of the work of the 85th Congress. Although there was time for only a brief consideration of the proposed new Department, all four members expressed general approval of it. On January 27, 1958, Senator Hubert H. Humphrey, on behalf of himself and Senator John L. McClellan and Senator Ralph Yarborough, introduced a bill entitled *The Science Technology Act of 1958.*[20]

The Act attempted to add a bold new dimension to this whole confused area of relationship between science and the Federal Government. The Act did, in fact, propose a Cabinet-level department of science and technology, that is to create standing committees on science and technology in the Congress. Although unsuccessful, this Act gained considerable backing and will perhaps serve as a prospectus for action by the next Congress.

The fact remains that in 1958, 39 of the 80 agencies of the Federal Government are engaged to some extent in scientific activities. Many of these 39 agencies have established elaborate organizational structure to conduct their scientific activities.

The Educational Front. Critical shortages still exist in qualified elementary and secondary teachers. Degree granting institutions also show teacher shortages in all subject fields and more particularly in mathematics, the physical sciences, and education. Although the number of teachers prepared to day is slightly better than in 1954 the shortage is still extremely serious. The National Education Association continues to draw attention to low salaries as the principal contributing cause of the teacher shortage.

In the meantime, the federal government has expanded its educational programs and increased its support for education in the natural sciences including mathematics. Time and space will not permit a detailed description of all these programs, however, a few of these programs that are more directly concerned with science and science education follows. More detailed accounts of these programs may be found in a recent issue of Higher Education[21] and in Quattlebaum's report printed for the use of the Senate Committee on Government Operations.[22]

1. *Veterans' Administration.* The Veterans' Administration finances vocational rehabilitation training for disabled veterans and education and training for nondisabled veterans eligible to receive such benefits under act of Congress. Public law 346 for World War II veterans and Public

Law 550 for veterans of the Korean conflict provided a very large segment of our population with Federal assistance in educational pursuits.

2. *National Science Foundation.* Four types of institute programs are directly targeted toward the improvement of science and mathematics teaching in our public schools:[23] (1) Summer Institutes for High School Teachers of Science and Mathematics. The foundation supported 125 such institutes during the summer of 1958. A major increase in the number of institutes and participating teachers is planned for the summer of 1959. (2) Academic Year Institutes for High School Teachers of Science and Mathematics. During the academic year 1958-59, a total of 19 such institutes have been authorized, each of which will accommodate about 50 teachers. Future plans contemplate increasing the number to approximately 30 institutes. Full-time attendance is required for a complete academic year, and consideration is being given to lengthen this period by adding a summer session. (3) In-service Institutes for High School Teachers of Science and Mathematics. For the academic year 1958-59 the foundation will support 85 institutes, capable of enrolling 3,000 applicants. A sizeable increase in number is planned for the academic year 1959-60. Instruction for these in-service institutes is provided by the host institutions during non-teaching hours throughout the academic year. The usual schedule involves one two-hour or three-hour session each week during late afternoon, at night or on Saturday mornings. (4) Summer Institutes in Science and Mathematics for Elementary School Personnel. During the summer of 1959, the Foundation plans to support for the first time approximately 10 experimental institutes for elementary school personnel. The future of this program depends in large measure upon the success of the 1959 program.

In addition the foundation assists science and mathematics also on a broader basis but in a more indirect way. (1) Through a Supplementary Training Program, the National Science Foundation can support conferences, refresher courses, short institutes, and other gatherings to explore problems relating to science education, to develop methods of improving science teaching or to present short, intensive instruction in a specific area of science. During the academic year 1957-58 the Foundation supported eight conferences and two short institutes under the provisions of this program. Next year's plans envision about twenty conferences and as many institutes. (2) In an effort to bring the subject matter of science instruction up to date the Foundation is supporting eminent scientists and educators in the development of new course material and teaching aids applicable at the elementary, secondary, and college levels of instruction. Support is being given to the Physical Science Study Committee of the Massachusetts Institute of Technology and to the School Mathematics Study Group at Yale University. The Foundation is planning to support similar effort in other fields and at other educational levels. (3) An extensive Visiting Scientists Program is currently under support of the

N.S.F. Through this program eminent and distinguished scientists are enabled to visit small colleges and universities for periods of two to five days. During this period the scientist is available to give lectures, conduct classes and seminars, and to meet students and faculty members on an informal as well as a formal basis. Plans for the academic year 1958-59 envisage some 60 such visits. (4) The Foundation supports a Traveling Science Library program to high schools. Each library consists of 200 titles and a unit of 50 books may be borrowed by a school for a period of two months. After two months the unit is exchanged for another unit of 50 books, so that by the completion of the school year the teachers and students will have had access to the complete lists of 200 titles. Two hundred schools were reached during 1957-58 and next year the number of schools to be reached will be considerably greater. (5) In cooperation with the Atomic Energy Commission the Foundation sponsors a Traveling Science Demonstration Lecture program. After an intensive three-months of training at Oak Ridge a select group of science teachers visit high schools to present demonstration lectures on such topics as solar radiation, atomic structure, nuclear reactors, wave motion, space travel, etc. During the 1957-58 school year ten lecturers visited about 300 schools. Further expansion of this program is contemplated. The Foundation also assists with science clubs and student projects and with career information. A number of additional programs are under consideration by the Foundation as it continues to enlarge its program.

3. *Atomic Energy Commission.* The Atomic Energy Commission engages in a variety of activities directed toward increasing the number of scientists and engineers trained in nuclear science and technology. In addition, the Commission sponsors or helps to sponsor a number of activities that directly benefit public education. (1) In cooperation with the National Science Foundation and the American Society for Engineering Education, the Commission has sponsored summer institutes in nuclear energy technology for faculty members of engineering colleges. (2) The Commission provides, also in cooperation with the National Science Foundation radioisotope training for high school science teachers. As an adjunct to the training for high school teachers, the Commission furnishes each teacher a radioisotope demonstration kit for use in his home school. (3) The Commission offers colleges and universities assistance in establishing certain courses in the nuclear energy field by providing special equipment and materials. It makes grants toward the cost of acquiring equipment for courses and laboratory work dealing with nuclear energy technology.

4. *Department of Defense.* In addition to the assistance given enrollees in the regular Reserve Officers Training Corps the Department of Defense indirectly provides aid to graduate students through its research contracts awarded to colleges and universities. An estimated two-thirds of these research assistantships are in the physical sciences and engineering.

5. *Department of State.* Under the international educational exchange

program a sizable number (986 in 1956) of American citizens receive grants for graduate university study abroad and a group (1,984 in 1956) of citizens of foreign countries receive grants to come to the United States for university study during each fiscal year. As a whole, the program for foreign study by United States citizens contributes to the development of profesional manpower of this country. About one-fourth of these grantees from the United States pursue courses in the natural and physical sciences.

6. *Department of Health Education and Welfare.* Besides the usual function of gathering and disseminating important information, the Department of Health, Education and Welfare is engaged in at least two other kinds of programs directly aimed toward the objective of improving educational programs in the United States. The newest of these—the National Defense Education Act of 1958—will be discussed later. The other program is described in a Bulletin published in 1958 entitled *Co-operative Research Projects, Fiscal 1957.*[24] Public Law 531 authorized the Office of Education to enter into contracts or jointly financed cooperative arrangements with universities and colleges and state educational agencies for the conduct of research, surveys, and demonstrations in the field of education. The major purpose of this program is to develop new knowledge about major problems in education and to test new applications of existing knowledge.

7. *Student Trainee Program for United States Government Workers.* This expanded educational program of considerable importance throughout Federal establishments permits trainees to participate in special training programs consisting of on-the-job training in a Federal agency and regular scholastic training in an accredited college or university. While on the job, student trainees work under the direction of and assist professional personnel in one or more of the scientific and engineering, or allied fields, performing professional duties relating to the field for which they are training.

8. *National Bureau of Standards Graduate School.* This graduate school is sponsored by the Government primarily for employees in the area of physical science. Each year there is an enrollment of approximately 1,000 students in an average of 40 courses. The Bureau is also authorized to support up to 10 postdoctoral research associateships each year for young investigators of unusual ability and promise in chemistry, mathematics, physics, engineering, and allied fields.

9. *Department of Agriculture Graduate School.* The major objective of this graduate school has been to improve the Federal service by providing needed educational opportunities for Federal employees. The school offers courses in the biological sciences, mathematics, and statistics, the physical sciences, and technology.

There are many other Federal agencies and programs which contribute generally to professional manpower but not as directly or distinctly in the area of science education. Perhaps the above programs will suffice to indicate the sprawling program of activities in science education carried on today (1958) by agencies of the Federal Government.

It is axiomatic perhaps, to conclude that to meet the nation's needs for scientific, engineering and other professional manpower requires complete utilization of our entire organization and system of education. The statements of the more than 90 witnesses who appeared before the education committee of the Senate and the 185 different witnesses heard by the education committee of the House is summarized by Exton as follows:

(These statements) reflect the whole spectrum of opinion on what the Federal Government's role should be in strengthening America's educational system in general and in stimulating improvements in the teaching of science, mathematics, and foreign language in particular—the three subjects judged most closely related to our national security by many witnesses.[25]

The National Defense Education Act of 1958. On September 2, 1958, Public Law 85-864, known as the *National Defense Education Act of 1958* was enacted by the 85th Congress. The purpose of this act, as written in Title 1—General Provision is:

To provide substantial assistance in various forms to individuals, and to States and their subdivisions, in order to insure trained manpower of sufficient quality and quantity to meet the national defense needs of the United States.[26]

The Congress reaffims the principle that states and local communities have and must retain control over and primary responsibility for public education. Furthermore, section 102 of the Act expressly prohibits Federal control of education. The Act contains provisions for assistance through eight different programs:
1. Low-rate loans to students in institutions of higher learning.
2. Financial assistance for strengthening science, mathematics, and modern foreign language instruction.
3. National defense fellowships.
4. Guidance, counseling, and testing; identification and encouragement of able students.
5. Language development, (A) centers and research and studies (B) language institutes.
6. Research and experimentation in more effective utilization of television, radio, motion pictures, and related media for educational purposes.
7. Area vocational educational programs.
8. Science information service.

Of particular significance is the fact that the National Defense Education Act of 1958 is administered by the Commissioner of Education in the

Department of Health, Education and Welfare. The assistance goes directly from the Commissioner to State Departments of Education, and their subdivisions or in the case of low-rate loans and fellowships the assistance is arranged through institutions of higher education. This approach is outstandingly different from a decade of proposals for different forms of Federal aid directly to students to be alloted on a school-age-per-capita basis.

Trend of Action

Viewed generally and over the period of time since the Constitutional Convention of 1787, the trend of educational action has been away from an active federal policy—a policy by which the Government through the states or by-passing the states, seeks out, commends, and awards young people who have demonstrated superior intellectual ability. The trend has been rather, toward a more passive policy by which the Government makes available the machinery and the funds for those students who are qualified and who do need financial aid. In other words, rather than Government going to the student or the program, or the state, the student, the program or the state *may* go to the Government. If participation is desired then the participation must meet certain requirements and regulations inherent in the acts that provides the assistance. The decision to participate or not to participate is left to the participant whether it be an individual, an organization, or a state. This is as it should be in a democratic society.

Selected Bibliography

BOOKS

Vannevar Bush, *Science, the Endless Frontier* (Washington: United States Office of Scientific Research and Development, Government Printing Office), 184, 1945. Recognizing the great accomplishments of the O.S.R.D. during World War II and its possible application to peaceful living, President Roosevelt requested recommendations from Bush, Director of the Office, on several current issues. Bush suggested that the time has come when the Government should accept new responsibilities for promoting the flow of new scientific knowledge and the development of scientific talent in our youth. He, therefore, recommended a new agency, fully responsible to the President and through him to the Congress.

Leo Chamberlain and Leslie Kindred, *The Teacher and School Organization* (New York: Prentice Hall, 1949), 23-79. Chapters 2 and 3 of this text are most helpful in outlining the activities of the Federal Government (Chapter 2) and the activities of the State Governments (Chapter 3) in the field of general education. Each chapter includes a wealth of related reading.

A. Hunter Dupree, *Science in the Federal Government* (Cambridge:

Harvard University Press, 1957) . A scholarly and detailed treatment of the subject which is particularly significant in the present (1958) consideration of proper governmental organization and function in the field of scientific research. Dupree sees a single strand of policies and activities of the government in science connecting the Constitutional Convention with the National Science Foundation.

Don K. Price, *Government and Science* (New York: New York University Press, 1954) . In 1954 Price had the foresight to predict a new kind of rivalry with the Soviet Union. He advances the notion that the crucial advantage in the issue of power is likely to be with the nation whose scientific program can produce the next revolutionary advance in military tactics, following those already made by radar, jet propulsion, and nuclear fission. Price takes a bold broad look into this whole area of Government and science and concludes that the role of world leadership is an uncomfortable one; it requires a steadiness of purpose, an economy of our energies and a breath of philosophy that have never been characteristic of American temper.

John R. Steelman, *Science and Public Policy* (Washington: President's Scientific Research Board, Government Printing Office) . A five part report that includes four important and timely recommendations designed to produce an administrative organization for scientific activities of the Federal Government. The fourth recommendation was, in fact, for a National Science Foundation.

MAGAZINE ARTICLES

John R. Dunning, "If We are to Catch Up in Science," *New York Times Magazine* (November 10, 1957) . Implies that all is not lost—yet. The author suggests that by pooling the free world's resources and brains —not alone in science—we may partly overcome our numerical disadvantages. He declares that our main efforts must be in quality, not quantity. He closes the article with the suggestion that in winning the propaganda victory, the Russians may have given us the warning and the chance we needed to reestablish our technological supremacy and vindicate our traditional freedom.

BULLETINS

"National Committee for the Development of Scientists and Engineers," National Science Foundation (Washington: Government Printing Office) . Contains supporting statements by President Eisenhower and Alan T. Waterman, Director of the National Science Foundation. The bulletin proper presents a very convincing case for at least a limited amount of education in the sciences for all youth and excellence in the advanced training of a greater fraction of our most able and creative youth in the fields of science and engineering who have the requisite aptitude.

"National Defense Education Act of 1958," Public Law 85-864, 85th Congress, H.R. 13247 September 2, 1958. United States Congress (Washington: Government Printing Office), 26, 1958. This is an actual print of the National Defense Education Act of 1958. It includes general provisions of the Act together with all the various titles. This act embraces 8 different federal aid programs for the expansion and improvement of public education in the United States.

"National Science Foundation Programs of Interest to Science and Mathematics Teachers in Secondary and Elementary Schools." National Science Foundation (Washington, D. C.: The Foundation), 8, 1958. Contains essential facts about some of the Foundation's present and prospective programs for the encouragement of education in the sciences. The programs described are those of particular interest to teachers of science and mathematics in secondary and elementary schools.

"Offerings and Enrollments in Science and Mathematics in Public High Schools, 1956." United States Office of Education, (Washington: Government Printing Office), 44, 1956. By far the most comprehensive study of offerings and enrollments in the areas of science and mathematics to date. This bulletin contains a wealth of interesting and important data. In many instances the data presented refute staggering criticisms that have been leveled against public education in the United States today.

"Report to the President," Full Report, Committee for the White House Conference on Education, (Washington: Government Printing Office, 1956). Contains a complete rundown of the whole procedure of the White House Conference including participation by the states and territories. One section of this report contains recommendations from the various state and territory conferences while the other part carries the conclusions and recommendations of the White House Conference held in Washington, D. C.

"Science and Technology Act of 1958," Senate Committee on Government Operations on S.3126, Committee print, (Washington: Government Printing Office, 1958). Contains an abundance of facts and details for the support of a Cabinet-level department of science and technology. Even though the proposal was unsuccessful it does give a realistic preview of things to come in the 86th Congress.

ORVAL L. ULRY, Director, Summer Session, University of Maryland, received degrees from Ohio State University (B.S., 1938, M.A., 1944, and Ph.D., 1953). His experiences include five years of High School teaching in the physical science area and four years of school administration in public schools in Ohio as well as nearly ten years of first-hand experiences in the area of teacher education at the University level. He served as Assistant Coordinator of student field experiences at the Ohio State

University for two years, and spent the following five years as Director of Student Teaching, Miami University (Oxford, Ohio), where he initiated and administered a program of full-time, full-semester off campus student teaching for all students in thc teacher education curriculum both at the elemetary and secondary levels. Since September of 1956, he has been responsible for the organization and administration of the Summer Session program on the College Park Campus of the University of Maryland. His dissertation was entitled *The Program of Field Projects, Education 505, College of Education, The Ohio State University;* he also contributed Chapter V "Toward Improving Instruction in Secondary Education," in the 37th Yearbook of the Association for Student Teaching, *Improving Instruction in Professional Education.*

NOTES FOR ACTIVITIES OF THE FEDERAL AND STATE GOVERNMENTS

1. Leo Chamberlain and Leslie Kindred, *The Teacher and School Organization* (New York: Prentice-Hall Inc.), 37.

2. A. Hunter Dupree, *Science in the Federal Government* (Cambridge: Harvard University Press), 3.

3. Vannevar Bush, *Science, The Endless Frontier* (Washington, U. S. Office of Scientific Research and Development, Government Printing Office), 7.

4. *Ibid.,* 6.

5. United States Department of Health, Education and Welfare, Office of Education, *Offerings and Enrollments in Science and Mathematics in Public High Schools, 1956* (Washington, Government Printing Office), 44.

6. Joint Committee on Atomic Energy, *Shortage of Scientific and Engineering Manpower* (Washington, U. S. Government Printing Office), preface III.

7. National Science Foundation, *The National Committee for the Development of Scientists and Engineers* (Washington, D.C., The Foundation), preface V.

8. John R. Dunning, "If We are to Catch Up in Science," *New York Times Magazine* (November 10, 1957), page 19+.

9. A. Hunter Dupree, *op. cit.,* 381.

10. National Science Foundation, *op. cit.,* 383-386.

11. A. Hunter Dupree, *op. cit.,* 383-386.

12. Senate Committee on Military Affairs, Subcommittee on War Mobilization, *The Government's Wartime Research and Development, 1940-44* (Washington: Government Printing Office).

13. Vannevar Bush, *op. cit.*

14. John R. Steelman, *Science and Public Policy* (Washington: President's Scientific Research Board, Government Printing Office).

15. Don K. Price, *Government and Science* (New York: New York University Press), 191.

16. Committee for the White House Conference on Education, *A Report to the President,* Summary Statement (Washington: Government Printing Office).

17. ——, *A Report to the President, Full Report* (Washington: Government Printing Office).

18. National Science Foundation, *Organization of the Federal Government for Scientific Activities* (Washington: The Foundation).

19. ——, *Advisory and Coordinating Mechanisms for Federal Research and Development, 1956-57* (Washington: The Foundation).

20. Senate Committee on Government Operations on S.3126, *Science and Technology Act of 1958, Committee Print* (Washington: Government Printing Office).

21. Clarence B. Linquist, "Federal Civilian Education Programs in the Natural Sciences": Part I, Higher Education, 14 (April, 1958), 125-129.

22. Charles A. Quattlebaum, *Development of Scientific, Engineering, and Other Professional Manpower,* Committee Print (Washington: Government Printing Office).

23. National Science Foundation, *National Science Foundation Programs of Interest to Science and Mathematics Teachers in Secondary and Elementary Schools* (Washington, D. C., The Foundation), 2.

24. United States Department of Health, Education and Welfare, *Cooperative Research Projects, Fiscal 1957* (Washington: Government Printing Office).

25. Elaine Exton, "Grist from the Congressional Hearings," *American School Board Journal,* (August, 1958), 31-32.

26. United States Congress, Public Law 85-864, 85th Congress, H.R. 13247 September 2, 1958, *National Defense Education Act of 1958* (Washington, Government Printing Office).

LEARNED SOCIETIES

WILLIAM E. DRAKE
University of Texas

At a time when anti-intellectualism is so rampant in the United States, there is good reason for looking at a cross section of the intellectual's contribution to the American culture. Nowhere is this contribution more in evidence than in the work of the learned societies, for it is out of these societies that there has come much of the advancement in knowledge that has made the United States a better place in which to live, both culturally and materially.

Our history teaches us that the Romans were civilized by Greek literature, and that it was through the revival of learning that the mind of Europe was transformed. Thus, literature and the fine arts are an embodiment of man's efforts to reconstruct his past, to record those inner experiences which are the foundation of meaning and value structure. If the essence of education is to inspire then the problem of education can only be resolved through personal experience. In this sense the learned society provides bountiful proof of its educational significance.

Origins

Learned scientific societies, which now cover the globe, have been one of the foremost agencies for diffusing and increasing the world's knowledge. While they are traceable back to Plato's Academy, they are, in terms of structure and purpose, more properly identified with the Italian Renaissance. Here, in 1560, at Naples, there was established the first scientific society in the modern world, the Academia Secretorum Maturae. Soon other institutions of a similar pattern were to be found throughout the whole of Europe.

The big problem of transmitting scientific truth, in man's past struggles with nature and himself, was not resolved in the folkways or the mores, for each, by its nature, was subject to gross error; yet, nowhere does the old adage of two heads being better than one hold more conclusively than in the field of science. Individuality and universality are markedly characteristics of the scientific mind, but it was the learned society which provided the effective means of communication, a necessity for the advancement of knowledge and for high adventure in the field of learning.

It was the patronage of wealthy men, of state aid, and the pooling of resources that provided the early 17th century European scientific societies with the outlet for publication and for rewards in prestige. Sir Isaac Newton is more identified with the Royal Society of London than with Trinity College, and Linnaeus with the Academy of Sciences at Upsala rather than the University.[1] The history of the major contributions to the field of science during this period is to be found in the *Proceedings* and *Transactions* of the learned societies rather than in college curricula and professorships.

It is in the writings of Francis Bacon (particularly the *Novum Organum* and *New Atlantis*) that the modern scientific society finds its true spirit, for the search for a naturalistic philosophy is the foundation stone of the first era of scientific clubs in England and the United States.[2] The real mother of the American societies was the Royal Society of London (1662) which grew out of the weekly meetings of a group of "divers worthy persons, inquisitive into natural philosophy—; and particularly of what hath been called the . . . experimental philosophy."[3] The idea of an experimental philosophy was as revolutionary in the 17th century as the idea of evolution was in the 19th century.

The Royal Society was an immediate success, for it served as a medium of exchange between those of all branches of learning, churchmen, architects, philosophers, scientists, and men of letters. Among the American colonists who gained membership in the Society were the three Winthrops, Roger Williams, John Morgan, Benjamin Franklin, William Byrd II, and Alexander Garden. The most notable of these was Benjamin Franklin who was a regular contributor to *Philosophical Transactions*, the journal of the society.

Benjamin Franklin and the American Philosophical Society

Upon his arrival in Philadelphia (1727) at the age of nineteen, Franklin proceeded to the organization of the first learned society in the new world, the Junto Club, for the purpose of producing "one or more queries on any point of Morals, Politics, or Natural Philosophy . . ."[4] The success of the Club was such as to convince Franklin of the need for a society which was colonial wide in scope. Out of this new effort there was born, in 1743-44, the American Philosophical Society to "promote useful knowledge among the British Plantations in America." Meetings were to be held once a month, where members could discuss their observations and experiments; while, in between times, communication was to be by letter. Each member pledged his allegiance to the search for truth, the love of mankind in general, and the freedom of knowledge.

Over a period of a half century, the American Philosophical Society served as the intellectual voice of the new world. Questions discussed at the meetings of the Society included matters of moral science, history, politics, medicine, botany, mathematics, chemistry, mineralogy, mechanics, arts, trades, manufactures, geography, topography, and agriculture. In 1768, the Junto Club was reorganized as the "American Society Held at Philadelphia for Promoting Useful Knowledge," and was combined with the Philadelphia Medical Society which had been organized in 1765. In 1769, this society combined with the American Philosophical Society with Benjamin Franklin as its first President, the position to which he was elected annually until his death in 1790.

How shall we evaluate the work of the American Philosophical Society during this period? It would be difficult to overestimate the influence exerted by the Society upon the scientific movement, for the activities of the organization attracted attention both at home and abroad. One of the major activities of the Society was the collection of the first accurate data on the distance between the earth and the sun; and, for this achievement, David Rittenhouse and his associates received international recognition. A worthy activity of the Society was the awarding of a prize, following the adoption of the Federal Constitution, for the best plan of education "adapted to the genius of the government of the United States." Out of the eight plans submitted, the one which was finally chosen was written by Samuel Smith of Philadelphia.

Early Societies in Massachusetts and Connecticut

Early societies in Massachusetts and Connecticut were an outgrowth of the influence of the American Philosophical Society. In 1780, the American Academy of Arts and Sciences was established in Boston. At the same time, there was an effort on the part of a group of French scholars to promote an institution similar to L'Academie Française at Richmond, Virginia, but the attempt failed because of the French Revolution. The Chemical Society of America, formed in Philadelphia in 1792, was the first of such organizations in the world. The laboratory of the society was conducted by Adam Seybert who, along with Robert Hare and James Woodhouse, popularized the views of Lavoisier on combustion, and carried on an extended controversy with Joseph Priestley who was a defender of the phlogiston theory.

Two state societies came to prominence before 1800, the Massachusetts Historical Society, founded in 1791, and the Connecticut Academy of Arts and Sciences. In the case of the latter, the charter sought from the state, as early as 1783, was to give the Academy power to establish a botanical garden, and to purchase or erect a building with a meeting hall, library and other rooms for museum purposes. Only one paper was published in this early period (1787), that of Jonathan Edwards on "The Language of the Muhhekaneew Indians." Final success in securing a charter came in 1799 as a result of the leadership of President Stiles of Yale College.

Science in the Colonial College

At the same time that the societies were being formed, the teaching of science in the colonial college was taking root. John Winthrop was offering a course in Natural and Experimental Philosophy at Harvard College as early as 1746, covering such topics as mechanical powers, compound engines, and the laws of motion. As far as classwork was concerned, less than 9% of the time was allotted to the sciences. Provost Smith's plan at Pennsylvania (1756) was the first to mention chemistry, but the subject was more conspicuous in the medical schools. Toward the beginning of the Revolution of 1776, there was a distinct tendency to place more emphasis on the sciences, as is indicated by the curriculum of William and Mary College of 1779.

The development of the sciences in the colleges brought the learned societies and the college into a close working relationship. Local associations were formed and meetings held monthly. At first, there were only light refreshments, but this practice soon gave way to a sumptuous meal. This brought on adverse criticism from many of the members because it was considered a source of distraction, in spite of the "elegant and agreeable relaxation after the severer exercises of the meeting." It was held that the indulgence of the sensual appetites never made a philosopher, that animal pleasures were not suitable to the pursuits of scientific men for scientific purposes, and that late eating was injurious to the health.

Early Nineteenth Century Trends

During the period between 1800 and 1865, learned societies took on characteristics markedly different from those in Europe. As products of the American Culture, these societies took on many of those features which we have identified with the Era of Jacksonian Democracy. Membership in the societies, generally, was open to all interested parties, and emphasis was upon the local community rather than the nation. Nevertheless, highways and canals were being opened up, and railroads were being constructed.

It was the learned societites which took the initiative in spreading the new scientific knowledge, by way of an ever increasing number of local and specialized societies, and through more and more scientific journals. Of genuine significance in this respect, was the *American Journal of Science and the Arts,* published at New Haven, Connecticut, beginning in 1818, under the editorship of Benjamin Silliman. The journal attracted international attention as a leader in the field of mineralogy and as an agency for the promotion of the cause of learned societies.

The basic intellectual changes of the period, along with the industrial development, had a marked effect upon the learned societies. Around

1800, interest in the societies seemed to center around statistical reports on the cities and on state geological reports. Noah Webster, in appearing before the Connecticut Academy of Arts and Sciences, discussed the supposed moderation of temperature in modern times. Chronic debility of the stomach was to be cured by a little use of wine, for it was the only thing recommended by holy writ.

The shift toward a purer science is indicated by the use of the spectrum analysis theory of the conservation of energy, by the extension of the universe in space and time, by the unity of the universe of nature and process, and by the development of the idea of the history of the solar system. Also Laplace was contributing new methods dealing with gravitational attraction: Galvani, Volta, Ohm, Faraday, and Ampere were advancing the knowledge of electricity; Davy was inventing the mine safety lamp, Morse the telegraph, and Bunsen and Kirchhoff the spectroscope.

It was in the rapidly developing state and local academies that the scientific interest took deepest root. Interest was very high in the New England states, especially around the Boston area, where one of the most active of the local societies of natural history was established in 1830. The Essex Institute of Salem, Massachusetts, had developed a library of 22,000 volumes and a museum of more than 5,000 specimens by 1861, and had inaugurated field meetings to help carry on its educational program.

Developments in the South and the Ohio Valley

In contrast with the northeast, the southern states were very slow in the development of learned societies. There was a Literary and Philosophical Society formed in Charleston, South Carolina, in 1813, for the purpose of collecting, arraying, and preserving specimens in natural history, and for the encouragement of the arts, sciences, and literature, generally, but the society went to pieces in the 1840's. A Metropolitan Society was established in Washington, D. C., in 1816, for the purpose of creating a collection of minerals in the United States and from the rest of the world. Recognition of the society came from the United States Congress, in 1818, with the granting of a twenty year charter. While the organization folded up in the late 1830's (the name was changed to the Columbia Institute for the Promotion of Arts and Sciences shortly after establishment), it does deserve some recognition as the first society in the United States to unite the efforts of both science and government.

Interest displayed in the establishment of learned scientists in New England was carried over into the Ohio Valley region, along with the development of metropolitan centers. In 1818, the Western Museum Society was formed with a family membership fee of $50.00; and, while the society became extinct in the 1830's, the valuable mineral collection was turned over to the Western Academy of Natural Sciences which was organized in 1835. Societies were formed in St. Louis (1836), Cleveland

(1834 and 1844), Louisville, Kentucky (1851), New Orleans (1853), Grand Rapids (1854), and Chicago (1856).

Many of the societies, formed during this period did not last but a few years, but they did lay the foundations for the broad scientific development which took place after 1865. No societies were formed in Oklahoma, Texas, or the Rocky Mountain area during this period. A society was formed in San Francisco as early as 1853, which became the present California Academy of Sciences in 1868.

The Lyceum Movement

Along with the learned societies, no agency was more effective in the spread of scientific knowledge before 1860 than the Lyceum. Just as a local academy provided the means of diffusing knowledge from one man to a larger circle of intelligent men, so did the Lyceum provide the opportunity for that larger circle of community life. This outstanding adult educational society was first established in Massachusetts (1826) as Millbury Lyceum No. 1, Branch of the American Lyceum Movement. By 1834, there were upwards of 3000 Lyceum Centers in every part of the United States. The prominence and continuing significance of the movement throughout the 19th century is pointed up in the active participation of such men as Agassiz, Emerson, and Silliman in its program.

Specialized Learned Societies Before 1865

While the great period of specialization in the sciences did not occur until after 1865, the foundations for such were laid in the first half of the 19th century. The Columbia Chemical Society, of which Thomas Jefferson was a patron and James Cutbrush first President, was established in Philadelphia in 1811. A society, known as the Maryland Institute for the Promotion of the Mechanic Arts (1825), was active in promoting manufactures and the useful arts through popular lectures, the building of a library, and the establishment of a museum of minerals. The only mathematical association established during this period was the American Statistical Association, formed in 1839 for the purpose of studying the field of applied mathematics. Botanical, marine, entomological, ethnological, antiquarian, et al societies were formed during this early period. Also, standing committees were formed by many of the academies for the purpose of following through on many areas of specialization, but the setup was often too ambitious and little was accomplished.

Chemistry and geology attracted the greatest attention of the people to the sciences during this period, probably because of their cultivation for practical purposes. As long as chemistry was cultivated as an art, it belonged with astrology and magic, but when it became identified with medicine, and later industry, the genuine foundations of a true science were laid, the awe and mystery of the past centuries now being trans mitted into scientific curiosity and natural law.

Growing interest in geology was manifested by three attempts to establish a national geographic society in the early national period. There was first the short lived American Mineralogical Society, formed in New York in 1789. Again, in 1819, through the aid of Benjamin Silliman, the American Geological Society was established at New Haven, Connecticut, but it failed to survive over any significant period of time. Finally, in 1840, there was established the Association of American Geologists and Naturalists out of which there emerged (1848) the American Association for the Advancement of Science.

Geology was so much an applied science that it received more government aid during the 19th century than any other science. The first geological surveys were the work of the geological societies, and this continued to be true even after the states came into the picture. A number of the chemical and geological societies were general and national in character, but most of them were local and of a specialized nature. Their contributions included extending the knowledge of the given areas, creating a love for nature, and fostering the promotion of science.

The study of agriculture was a part of the intellectual culture of the 18th century, but in the early 19th century there was a tendency to apply the subject to economic objects. There was originated in South Carolina (1740) a society known as the Winyaw Indigo Society, but it would appear that the members were more interested in the convivial cup than in the science of indigo. While the American Philosophical Society was the first organization seeking to promote an interest in food production, there was a move toward the end of the 18th century, for setting up societies devoted exclusively to agricultural products. In 1785, a group of citizens in Philadelphia organized the Philadelphia Society for Promoting Agriculture, with Benjamin Franklin and Timothy Pickering among its first residents, and Robert Livingston and George Washington as honorary members. By 1789, there were honorary members in the society from all the thirteen states.

The economic revolution going on in the United States was very early identified with the development of the science of agriculture. In the early period of the societies, membership was confined to an elite—doctors, lawyers, and public men of high office, but the exhibiting of two Merino sheep in the public square of Pittsfield, Massachusetts (1804), by Elkanah Watson, marked the beginning of a new movement in the agricultural fair. Through Watson's efforts, cattle shows and state agricultural societies were established throughout all New England, and, in time, throughout the entire country. State agricultural societies were established in Pennsylvania (1823), New York (1832), Louisiana (1841), Wisconsin (1851), Illinois (1853), and California (1854). The first attempt to establish a national body came with the Columbia Agricultural Society, Washington, D. C. (1811), but the effort failed as a result of the War of 1812. In 1841, an effort was made to organize the United States Agricultural Society, but failure to secure adequate funds delayed activity until a charter was issued by the United State Congress in 1860.

Specialization tendencies in agriculture stand out especially in the field of horticulture which came of age in the organizing of the Pennsylvania Horticulture Society (1827), to be followed by Massachusetts (1829), Kentucky (1840), Rhode Island (1845), Delaware (1847), Illinois (1856), Missouri (1859), and Indiana (1860). There should be little doubt that it was these and other agricultural societies which created the public climate which brought about the establishment of state agricultural colleges, and the passage of the Morrill Act by the United States Congress in 1862.

Specialized Learned Societies, 1865-1917

The major characteristic and tendency of the learned societies during the period, 1865-1917, was that of specialization and the triumph of science over the classics in the field of education. Literally hundreds of science societies were established representing new fields of specialization and specializations within specialization. Because of the extensive nature of these developments it is possible to cite only a few of the outstanding creations.

(a) *Mathematics.* Although new discoveries in the field of mathematics in the United States lagged far behind Europe prior to World War I, there was much interest in the field due to the increasing importance of the subject in the development of the sciences. Many mathematics clubs were formed on university campuses consisting of small groups of university professors and students. Mathematics sections were formed in the various state teachers' associations, leading to the establishment, in 1907, of the American Federation of the Mathematical and Natural Sciences. Strong interest in mathematics in the midwestern states led to the forming of the Mathematical Association of America at Columbus, Ohio in 1915. The purpose of this association was more to promote effective teaching in the field than to discover new knowledge. One of the major contributions of the association was the sponsorship of the *American Mathematical Journal.*

(b) *Physics.* Evidence that physics was of increasing importance as one of the pillars of an industrialized societies was manifested in the large number of new societies which sprang up in this area. The study of radiant energy and of atomic theory attracted the best minds of Europe and America, and added new zest to the field. Interest in the area of fundamental constants brought on the organization of the American Meteorological Society at Columbia University in 1873. In 1899, the American Physical Society was organized at Columbia University for the purpose of the advancement and diffusion of physical knowledge. Evidence of specialization tendencies is indicated in the founding of the

American Radio Relay League (1914) for amateurs and professionals, the Optical Society of America (1916), the Metric Association (1916), and the American Institute of Weights and Measures (1917).

(c) *Geology*. Interest in the field of exploration, and an ever expanding industrial economy, continued the long time interest in the science of geology. Research on glacial geology and on the new sciences of petrography and paleontology marked significant advancement in the field. In 1888, the Geological Society of America was organized for the purpose of promoting "the science of geology in North America." Specialization tendencies are noted in the establishment of the American Climatological Society (1884), the Seismological Society of America (1906), and the Paleontological Society (1908).

(d) *Geography*. Geographers manifested a very active interest in the diffusion of geographic knowledge through the association medium. Although the American Geographical Society had been a real success since its founding in 1851, there was a felt need for an organization that would make a direct appeal to the public. As a result, the National Geographic Society was formed in 1888. This society, in addition to providing financial support for numerous expeditions, including that of Admiral Perry, has published the widely read *National Geographic Magazine,* and numerous monographs and maps. Public interest in the association is indicated by a growth in membership from 200 in 1888 to 1,600,000 in 1948. The Association of American Geographers, which is for those who have done original research in the field, was organized in 1904. In addition to the Peary Arctic Club, which was founded in 1899, numerous other exploration clubs grew up around the country.

(e) *Chemistry*. Slow developments in chemistry are indicated by the fact that it was not until 1881, at a meeting of the AAAS, that a subsection of this field was created. The American Chemical Society (1876) remained a local New York organization for some time, but the continued establishment of local societies, and their affiliation with this body, tended to build up a strong organization. The society has continued to publish the *Journal of the American Chemical Society* since 1879, as well as other journals and monographs. A national society of professional chemists, which was set up in New York in 1898 (the Chemists Club) made a significant contribution by building an outstanding library of chemical literature.

(f) *Biology*. It was the field of biology that created the greatest public furor, largely because of the publication of Charles Darwin's *Origin of Species,* and its attendant theory of evolution. Conflicts between science and theology on the nature of man tended to further stimulate study in such areas as genetics, anthropology, zoology, embryology, biometrics, horticulture, and bacteriology. There was a decided increase in the biological knowledge of the people due to the contributions of such men as Louis Agassiz, W. K. Brooks, E. D. Cope, and Asa Gray. Biological societies were further promoted by the establishment of aquariums, arboretums, zoological parks, and national and state experiment stations. The American Society of Naturalists (1883) was limited to those who were professionals—teachers, researchers, administrators, and affiliated technical and specialized biological groups. The American Nature Study Society was organized, in 1908, for the purpose of promoting the study of nature in the schools. Among the many other specialized societies were the American Microscopical Society (1878), Audubon Society (1886), Society of American Bacteriologists (1900), the American Association of Museums (1906), the American Society of Biological Chemists (1906), and the American Genetic Association (1913). Also there were many botanical and agricultural societies which were an outgrowth of the new knowledge.

(g) *Psychology*. It was during the latter half of the 19th century that psychology came into being as an experimental science. Significant contributions by German scholars, especially those of Wilhelm Wundt, who established the first psychological laboratory in Leipzig (1879), and of American scholars, such as William James and E. L. Thorndike, gave much impetus to the movement. With the emergence of the experimental approach, and the growth of a number of able psychologists, a meeting was called at Clark University (1892) which resulted in the formation of the American Psychological Association. Publication of the *Psychological Review* began in 1894, followed by *Psychological Monographs* (1895), the *Psychological Bulletin* (1904), and the *Journal of Abnormal and Social Psychology* (1916).[5] Numerous specialized and sectional associations were created beginning with the American Association for the Study of the Feebleminded (1876), and the American Society for Psychical Research (1885). One of the most significant of the developments was a result of the work of Clifford W. Beers who led the movement for creating the National Committee for Mental Hygiene. Developments in this area of recent decades have brought about a revolution in the treatment of the mentally diseased.

(h) *Philology*. Foundations for developments in the field of philology were brought over from Germany, with Francis Child of Harvard University being the first producing scholar. The American Philological Association was organized in New York (1868) to investigate the entire field, but the immediate interest was aboriginal American dialects. With the opening of Johns Hopkins University in 1876, and the establishment of the Modern Language Association in 1883, research in the field moved forward at a rapid pace. *Publications of the Modern Language Association* (1886), *Germanic Philology* (1897), and *Modern Philology* (1903) provided a solid publications foundation for the movement.

(i) *History and the Social Sciences*. Historians of the 19th century tended to move away from the idea that what they were doing involved philosophical assumptions to the idea of history as a science. This is understandable, since everything else, including religion, was moving in the same direction. The trend was away from the setting up of historical laws toward cold objectivity, a reliance upon facts which could speak for themselves. Before 1860, there was no profession of historians, historical studies being made by lawyers, gentlemen, and the clergy. With the founding of the American Historical Association and the launching of the *American Historical Review* (1895), the profession was well on its way. The establishment of professorships of history, in large numbers of colleges and universities throughout the country, provided for strong future developments.

Much of the expansion of the field of history since 1900 has been marked by a high degree of specialization. In the most recent listing of journals devoted to humanistic and social studies by the American Council of Learned Societies, 86 are classified under the division of history. This includes all of the state associations and some of the local societies. The World Congress of Arts and Sciences, held in connection with the St. Louis Exposition of 1904, indicated that the Americans were following the road of specialization as contrasted with that of Europe. In the airing of all new ideas, the pattern of the social sciences was laid down in its present form. Records of the American Historical Association point to the contributions of the learned societies in this respect.

The social science movement in the 19th century was marked by outstanding progress in professional organization. Among the most significant of the organizations were the American Social Science Association (1865), the American Economic Association (1885), the American Academy of Political and Social Science (1889), the American Political Science Association (1903), and the American Sociological Association (1905). Each in turn established its scientific journal which, in spite of many rivalries and jealousies, has carried on an intensive program of research.

(j) *Classical.* Significant and outstanding contributions in the fields of archaeology and papyrology, largely as a result of the able leadership of the Archaeological Institute of America, founded at Boston (1879) marked the closing decades of the 19th century. In addition to publishing the *American Journal of Archaeology,* the Institute figured significantly in the excavations carried out in Greece and in the Middle East. A similar program was followed in Italy, as a result of the work of the American Academy in Rome, which was combined with the American School of Classical Studies in 1911.

(k) *Philosophical.* The trend of science in the 19th century toward sheer description and administrative mechanism was not considered as progress in the philosophical sense of meaning and value. In this respect, there was a growing concern in the philosophical world over the widening gap between the humanities and the natural sciences. To bridge this gap, there was, first, "speculative idealism," which was borrowed from Kant and Hegel, and which found active expression in the United States in the work of Harris, Royce, and Howison. "Speculative idealists" held that the only way to maintain a civilization was on an ideal structure, for since knowledge was the mind's own work, the best that man could think or will had to be.[6] Second, there was "pragmatism" which insisted that we needed more relative grounds than those proposed by the speculative idealists. For John Dewey, the third in a line of pragmatists (Peirce and James preceded him), the determination of the foundations of meaning and value could be found only in experience, in the process of social evolution and the growth of the human mind.

These ideas and others were being wrought out individually and collectively through the continuing contributions of the American Philosophical Society and the Professors of Philosophy. Also, the American Philosophical Association was established (1900), for the purpose of promoting the study and teaching of philosophy in all of its branches. Today, there are some 1000 members in the Association which is divided into Eastern, Western, and Pacific Sections. There is a high degree of cooperation among the members and the philosophical societies throughout the world.

Learned Societies of Significance

Numerous other learned societies could be discussed, but limitations of space make it impossible to treat of more than four of the outstanding ones: the American Medical Association, the American Association for the Advancement of Science, the National Academy of Sciences, and the Smithsonian Institution.

(a) *The American Medical Association.* The setting aside of the medical man into a distinct profession is in large part a product of the organized effort of the American Medical Association. As early as 1735, William Douglas sent a letter to Cadwallader Colden stating that a medical society had been established in Boston, and, while the society only lasted a few years, it marked the beginning of a movement which resulted in the establishment of the Massachusetts Medical Society in 1781. This latter organization established a library in 1782, which it financed for over a hundred years, and began a series of "Medical Communications" in 1790. The oldest of the medical societies in name, however, is the New Jersey Medical Society, whose Constitution expresses a genuine interest in the cause of medical science, as well as the professional growth of its members. It was the constitution of this society that became a model for many of the future medical associations.

Medical societies were established in large numbers before 1865, in all of the states and in most of the large cities, but the most significant medical organization development came in 1847 with the formation of the American Medical Association. Delegates from the various state and local societies, along with those from hospitals and medical schools, made up the charter membership. In no other society do we find a more outstanding example of eminent success in raising the standards of the profession.

Numerous specialized medical societies have formed a part of the growth of the medical profession. Of these, some of the most outstanding have been the American Society of Dental Surgeons (1840), the American Psychiatric Association (1844), the American Institute of Homeopathy (1845), the American Veterinary Medical Association (1863), and the American Ophthalmological Society (1864). Possibly the most famous of all the specialized medical associations is the College of Physicians of Philadelphia which dates back to 1787. Among the charter members were Benjamin Rush, Samuel Duffield, and William Shippen Jr.

Since 1917, there have grown up many new kinds of medical societies. The significant ones would include the National Health Council (1920), the State Society of Neurological Surgeons (1924), American Society of Medical History (1924), American Federation of Clinical Research (1924), and the American Eugenics Society (1926). Rapid developments in the fields of psychology and psychiatry led to the Association for Research in Nervous and Mental Diseases (1920), the Orthopsychiatric Association (1924), and the American Foundation for Mental Hygiene (1928).

(b) *The American Association for the Advancement of Science.* Initiative for establishing the AAAS came from the members of the New York State Geological Survey who were experiencing difficulty in keeping up their professional contacts in other states. As a result, a meeting of American Geologists was called in Philadelphia (1840), but to include others covering a wide range of scientific interest. At the same time, the British Association for the Advancement of Science, formed in 1823, had been remarkably successful in bringing British scientists together. Finally, at the Boston meeting of the American geologists and naturalists (1847) it was decided to enlarge the purpose of the organization. A committee was appointed to inform scientists throughout the nation of the proposed change along with a copy of the draft of the proposed constitution. The result was the formation of the American Association for the Advancement of Science, which held its first meeting at Philadelphia (1848), with W. C. Redfield as the first president.

From the very first meeting, it was the policy of the AAAS to provide for the specialized interests of its members. The initial two sections included General Physics, Mathematics, Chemistry, Civil Engineering and the applied sciences, all in one section, and Natural History, Geology, Physiology, and Medicine in the other. In contrast to the American Philosophical Society and the American Academy of Science, membership was open to all who had a real scientific interest. Annual proceedings have been published from the first year, and continuously except for the Civil War Years. Men of eminence who participated in forming the organization were Benjamin Peirce, Alexander D. Bache, Asa Gray, Louis Agassiz, and Lewis H. Morgan.

The unitary character of the asociation was maintained until about 1900, when the real technical work was shifted to a number of "affiliated" and "associated" organizations. The policy of conducting the work of the society in two sections had been changed in 1882, when nine sections had been set up, with new sections being added in recent years. One of the most dramatic of the activities of the Association came with the program of entertainment and speeches put on in 1916, as a part of the Second Pan-American Scientific Congress held in Washington, D. C.

A new era for the Association followed the adoption of a new constitution in 1920. This is indicated by the great growth of the organization, for, in 1947, there were 201 affiliated or associated groups, and a membership of 33,000. An activity worth noting was the Declaration of Intellectual Freedom (1934), in which the suppression of independent thought and freedom of expression were declared to be a major crime against civilization, while the silencing of a scholar or teacher was held to be an intolerable act of tyranny. Another activity of significance has been the sponsorship of junior science academies among high school students.

(c) *The National Academy of Sciences.* While Jefferson, Washington, and others recognized the significance of a national academy of sciences promoted by the government, it was not possible to get any action on the matter during the first half of the 19th century. Jacksonian democrats were opposed to enlarging the activities of the federal government, and there was a general suspicion of the Northeast by the rest of the country. It was the dire need of the national government, for scientific technical advice in carrying on the war against the southern states, that led finally to the establishment of the academy. As a result, a bill providing for the establishment of a National Academy of Sciences was passed by the Congress and signed by President Lincoln in February, 1863. The work of the academy was initially divided into five committees, namely, those of Weights, Measures and Coinage; Protecting the Bottoms of Iron Vessels against Corrosion; Magnetic Deviation in Iron Ships; Wind and Current Charts and Sailing Directions; and the Question of Tests for the Purity of Whiskey. Other committees followed.

The National Academy of Sciences has been very active since World War I. In 1919, it was given the responsibility for putting the National Research Council on a firm basis, following a gift of $5,000,000 by the Carnegie Institution. The Academy has published the annual reports of the Councils in addition to its *Memoirs* and *Biographical Memoirs.* Much pioneering work has been done through the publication of *Science Service,* an activity carried on since 1921. In 1941, a national science fund was created for use in the physical and biological sciences.

The newly created National Research Council has served the purpose of stimulating scientific research in all branches. The Council is operated under an executive board of forty, and is divided into three areas of knowledge. By encouraging the support of research men throughout the entire country, and by having the various scientific societies appoint representatives to the Council, it has been possible to achieve a high degree of cooperation among scientists. One of the most significant contributions of the Council has been that of a survey of the laboratories, research funds, and scientific societies now available. Also, the National Academy of Sciences and the Council have been active in the work of the International Research Council, and in the work of the Council on International Cooperation, which was created in 1922.

(d) *The Smithsonian Institution* has been one of the most outstanding of the organizations for the spread of scientific knowledge during the past century. Foundations were laid for the establishment of the Smithsonian Institution by the work of the National Institution for the Promotion of Science, created in 1840, and chartered by Congress in 1842. Although having but a short life, this latter organization sponsored the first of the national scientific congresses ever held in the United States (1844), as well as aiding in several explorations and the reorganization of the Coast Survey. Following the expiration of the National Institute's charter (1862), the library and museum were turned over to the Smithsonian Institution.

The Smithsonian Institution owes its origin to James Smithson, an English scientist who, on his death (1829), left his property as a bequest to the United States for the purpose of founding in Washington an institution for the increase and diffusion of knowledge among men."[7] A long dispute, over what the nature of the reorganization should be, delayed its being chartered until 1846. Statutory members provided for in the charter included the President of the United States, the vice-president, the chief justice, and members of the President's cabinet. The Institution was to be operated by a board of regents who was responsible for selecting a secretary and the principal executive officer. It was the job of the secretary to act as custodian of the property, and to submit an annual report to the Congress.

It was the work and leadership of the first secretary, Joseph Henry, that made the Smithsonian Institution the great organization that it became. Holding that the work of the Institution should be both lasting and of world import, he was able to secure direct appropriations from the Congress, and to use the endowment for pure research. In time, its buildings came to house the National Museum and the National Academy of Sciences, as well as a great library of scientific documents. Through a system of informational exchange, learned scientific documents from all over the world became available to American scientists. *Annual Reports,* along with the *Smithsonian Contributions to Knowledge,* are among the most significant of the Institutions publications. The Institution has sponsored eminent lecturers such as James D. Dana and Louis Agassiz, as well as providing a daily meteorological service, the forerunner of the present Weather Bureau.

Through the years the work of the Smithsonian Institution has continued to stand out in the field of science. A new division of Radiation and Organisms has been set up to provide for the study of the effects of radiation upon living things; while outstanding work has been done on tropical vegetation in the Latin American countries. Today, the major problem of the Institution has become one of finance, since the endowment is a carry over from the 19th century.

Recent Developments

1917—to the Present. There has been much concern about the trends of American life and culture since World War I, and especially since World War II. This concern has been expressed by journalists, scientists, philosophers, social scientists, artists, and theologians. They see our culture being dominated by money getting, the desire for social betterment, and the power of militarization, all of which are lacking in imagination and that deep rooted sense of human betterment. Democracy, in its genuine sense, seems to terrify the individual, while industrialism seals his lips.[8] It has been said that the failure of rationality has caused practical men to turn to irrationality, substituting a tradition of words for an unjustified sense of reality.

From the standpoint of the learned societies these questions, involving the problem of increased control by business and government in determining the kind of research, have created a growing weakness in the realm of theory. While business has dictated a sense of practicality, a control for certain types of products, government has, of more recent years, been primarily concerned with the advancement of the military sciences. These controls have raised issues, involving competition for gifted men and women, along with the future of knowledge, of the country, and of the universities.

Marked Characteristics of the Learned Societies. Since the first World War, the American scientist has tended to function more and more through some kind of organization. In this respect, the learned societies have continued to serve as excellent agencies for the advancement of scientific knowledge. Well organized before 1917, these societies continued the trend toward specialization, with increasing importance for initiating research as well as disseminating the findings. It was this very high degree of specialization, however, which produced a sense of need for the integration reflected in the establishment of the National Research Council (1917), and the American Council of Learned Societies (1919). The Social Science Research Council, established in 1923, to counteract the growing parochialism among the various social sciences, has been especially effective in fostering unified research among the various specializations. In 1925, the sociologists, economists, and political scientists were joined by those in the American Statistical Association, the American Psychological Association, the American Anthropological Association, and the American Historical Association.

It is important to note the high ethical and moral standards practiced and cultivated by the learned societies. Here the brotherhood of man is more than words, for science knows no national, racial, religious, or class boundaries. Regardless of where a discovery originates or who originates it, the new knowledge is the property of all. The learned society discusses the new knowledge, and passes it on for what it is worth. Here is the kind of channel for world unity, sought by Comenius three centuries ago, for the scientist is a persistent wonderer, a questioner, a discoverer, who gives meaning to things that have no meaning for others.

Trends in Specialization. Developments in the field of the sciences since 1917 could almost be classified in the realm of the miraculous, especially those dealing with atomic power and automation. This has led to the establishment of many specialized types of societies.

In Astronomy, there was the Amateur Astronomers Association (1927) ; the Society for Research on Meteorites (1933) ; and the American Amateur Astronomical Association (1933) .

In Mathematics, there was the National Council of Teachers of Mathematics (1920) ; The Econometric Society (1930) ; and the Institute of Mathematical Statistics.

In Chemistry, the new societies are illustrated in the American Institute of Chemists (1923) ; the Association of Consulting Chemists and Chemical Engineers (1929) ; the Wood Chemical Institute (1929) ; and the Metropolitan Microchemical Society (1936) . The increased significance of applied chemistry is indicated in the Ohio Ceramics Industries Association (1924) and the Ceramics Association of New York (1933) .

In *geography,* additional state geographical societies were organized as well as pointing up a trend toward further specialization. These trends are indicated in the American Society of Photogrammetry (1934) and the American Polar Society (1934) .

In *geology,* numerous local geological societies were formed, along with new specialized societies such as the American Petroleum Institute (1919) , the Society of Economic Geologists (1920) , The American Shore and Beach Preservation Association (1927) , The Woods Hole Oceanographic Institution (1930) , the American Gem Society (1934) , and the National Speleological Society (1939) .

In *physics,* there was a marked tendency for the field to split up into sub-divisions, bringing about a number of new specialized societies such as the International Amateur Radio Union (1925) , the Society of Rheology (1929) , the American Rocket Society (1930) , the Photographic Society of America (1936) , and the Electron Microscope Society of America (1943). A number of these societies, and others, played a significant role in World War II, and have been responsible for much of the scientific advance in the post-war period.

In *biology,* there was still no one national society, but an organization known as the Union of American Biological Societies was created in 1921. Numerous specialized societies were set up, such as the American Society of Plant Physiologists (1924) , and the American Wildlife Institute (1935) . To get some idea of the larger pattern of development, there are today some 285 horticulture societies and more than 2000 garden clubs.

In *philology*, there have been continuing contributions to the field, but the general pattern in this area has been one of decay, due to the American culture situation, the decay of romanticism, and the worship of a pseudo-scientific ideal. The increasing separation of academic scholarship and the practice of literature has been paralleled by a growing anti-intellectualism.

In *political science*, there has been a tendency for the work to become more exclusively associated with public affairs, as illustrated in the study of public administration, rather than political theory of political psychology. The role of the expert has taken on added significance, as a result of the expansion of government employment and activities.

In *engineering*, the rapid industrial development, and the coming of the age of automation, resulted in much activity among the engineers. The long time interest of the engineers in organized professional activity has continued into the present decade, with new areas of specialization constantly opening up. Good illustrations are found in the American Engineering Council (1920), the American Society of Safety Engineers (1924), the Institute of Aeronautical Sciences (1932), the National Society of Professional Engineers (1934), and the Television Engineers of America (1940). At the same time, there have come into being numerous technical societies, such as the National Technical Association (1926), the Aluminum Research Institute (1929), the Industrial Research Institute (1938), the National Inventors Council (1940), and the Refrigeration Research Foundation (1943).

In *history*, the most significant of the recent developments was the growing interest in the history and philosophy of science, an indication of the maturing of scientific research. Also, lack of interest in the fundamental propositions upon which historical research rested, brought on a heated controversy in the fourth decade of the 20th century, which resulted in a shift toward historical relativism. As early as 1919, a history of science section was organized in the American Historical Association, and, in 1920, a similar section was organized in the AAAS. The flight of many scholars from Europe, beginning with 1917, made American scientists more sensitive to the cause of freedom. Stimulated by the growing realization of the unity of all science, there was established in Paris (1928) the International Academy of the History of Science. Out of 150 members of the organization in 1933, ten were Americans.

In *philosophy*, the American Philosophical Society, and the American Philosophical Association, continued to function at a high level of scholarship, even if often ignored by the American people. In the early 1920's, it was evident that there were many philosophers who were still operating on the assumption that genuine knowledge was a discovery of things as they exist, this in spite of the fact that, since the days of Descartes, the philosophical tradition was against them.[9] This rebirth of realism was a product of an era of scientific mechanism, an era in which the models were being provided by physics and biology, and by the merging of logic and pure mathematics.

The need for a philosophy, where meaning and value in life could be explained in a pattern of thought consistent with the findings of science, was becoming increasingly evident. In 1932, there was organized the Philosophy of Science Association, which has continued to bring together men of eminence in philosophy and science. Its official organ, *Philosophy of Science*, has published papers dealing with varied interpretations of the nature of the universe, both logical and mechanistic. In 1936, the organization of the Association for Symbolic Logic gave evidence of a further trend toward specialization. The social and educational implications have been reflected in the organization of the American Association of Scientific Workers (1930), the Philosophy of Education Society (1940), and, of more recent date, in the organization of the Federation of American Scientists (1945).

Basic Contributions

By way of a summary, there is justification in pointing to the basic contributions of the learned societies to the American culture. These can be stated briefly in terms of: (1) *Meetings*. Over a period of three hundred years, the meetings of the learned societies have been the true university for the scholars of the modern age. (2) *Journals, Proceedings, et. al.* These have been the true textbooks of the scholars, for no substitute has as yet been found for them. (3) *Libraries and Museums*. These numbered more than 20,000 in 1932, and have served both the public and the scholar in the advancement of scientific knowledge. (4) *Aid to the Universities, Governments, Philanthropic Agencies, Industrial Corporations, and Press Syndicates*. There are few examples in American life where men have contributed so much to the general welfare at their own expense. (5) *High Specialization*. There is no doubt that the learned societies have been a major factor in the rapid advancement of knowledge in all areas. (6) *Low Cost Membership*. The democratic nature of the learned society is pointed up in the universality of its membership. (7) *Professional Unity*. The societies have served the major purpose of protecting their membership, professionally and economically, against the enemy of free knowledge.

The Problem of Humanizing Knowledge

Man's civilization and his culture have been founded on two languages, those of the intellect and the emotions, otherwise expressed in terms of science and the arts. Ours is an age of science in the sense that it is marked by genius in the technical and applied science fields, and in the fact that we have come to worship the material products of the scientific mind; but, in no sense is our age scientific in those areas of deepest human need and value, in man's relation to his fellow man and to himself. Faith is still posited against reason, and assumed to be possessed of divine insight. A good example in point is that few people in the United States have any scientific knowledge of the *Bible,* or even desire it. The continued advancement of scientific knowledge along present lines, that of fracturing, and a failure to make application to the more personal and social needs of man, presents a major threat to further human existence. Specialization, by tearing knowledge into fragments over the past century, dehumanized it. Also, while specialization is significant for research, it, at the same time, acts as a barrier to the scientific frame of mind, for the science of a subject is not at its center, but at its periphery.[10] The overall tendency in our present situation has been to weld into our educational system the technical divisions of scientific research, causing us to lose sight of the trunk of the tree. Vocational education has been substituted for intellectual education, and psychological truth for philosophical truth. Scholars, writing for themselves in the learned societies, while serving the purpose of advancing knowledge, at the same time produced an unwarranted prejudice against the popularizer of knowledge, yet, the problem of science and democracy is a continuing one.

We need to recognize that man's emotional life is primary, a fact that has been verified over and over in human experience. Man's beliefs originated in what was in him, and in what he was able to grasp with ease. It was not the truth of an idea that brought its acceptance by man, but its appeal. Today, man's fate rests on what he does with his knowledge of the physical universe, for all knowledge is interrelated and of one piece. That scholarship which has been defined as high competence in a delineated field has undermined one of the pillars on which western culture was built, that of the separation of man and nature.[11] Now that it has been clearly demonstrated that man is a part of the natural order, there must be, not only a new concept of culture, but the humanizing of knowledge itself. To this end, the learned societies must now dedicate themselves.

Selected Bibliography

Ralph H. Bates, *Scientific Societies in the United States* (New York: John Wiley & Sons, Inc., 1945). This is the most thorough and scholarly treatment of the subject now available.

H. Butterfield, *The Origins of Modern Science* (New York: The Macmillan Company, ed. 2, 1957), 71-76. An excellent, brief, and pointed analysis of the role of the learned society in the origins of science.

James M. Cattell, *Scientific Societies and Associations* (Albany, N. Y.: Lyon Company Printers, 1904). This is one of the pioneering studies in the field, written by an eminent and well known scholar.

Merle Curti, *American Scholarship in the Twentieth Century* (Cambridge, Massachusetts: Harvard University Press, ed., 1953). The book is made up of significant contributions by leading scholars in the several fields of present day scholarship endeavor.

G. B. Goode, *The Smithsonian Institute, 1846-1896* (Washington, D. C., 1896). A fascinating account of the first half century of a great institution.

Martha Ornstein, *The Learned Societies and the Universities* (Chicago: The University of Chicago Press, 1928). The most significant contribution of this study is an insight into the close working relationship between the learned societies and the universities.

G. S. Robinson, *Adventures in American Medicine* (Cambridge, Massachusetts: Harvard University Press, 1957). There is no better way to get an appreciation of the contributions of science education than through a study of this kind.

F. W. True, *A History of the First Half Century of the National Academy of Sciences, 1883-1913* (Washington, D. C., 1913). An excellent example of how the scientific societies have helped to mould the American culture.

Bulletins, Proceedings, et. al.:

American Association for the Advancement of Science, Proceedings (Washington, D. C., 1848-   ). One of the most authoritative accounts available on the basic contributions of a great learned society.

Simson E. Baldwin, "The First century of the Connecticut Academy of Arts and Sciences," *Transactions of the Connecticut Academy of Arts and Sciences,* XI (New Haven: By the Academy, 1901-1903). A good, model example and first hand account of the origin of the learned societies in the United States.

Bulletin of the National Research Council, 115, April, 1948, *Handbook of Scientific and Technical Societies and Institutions of the United States and Canada,* Fifth Edition, (Washington, D. C.: National Research Council, 1948). This is the best and most thorough of the handbooks on the learned societies.

"(A) History of the National Research Council, 1913-1933," *Reprint and Circular Series of the National Research Council,* 106 (Washington, D. C., 1933). A good digest of the contributions of the National Research Council during the first two decades of its existence.

Periodicals:

"(The) American Anthropological Association" *American Anthropologist,* V (1903), 178-192. A brief but excellent statement of the work of the anthropologist.

C. A. Brown, "A Half Century of Chemistry in America, 1876-1926,"

American Chemical Society Journal, XLVIII (1926). The revolution in chemical knowledge is clearly indicated in this contribution.

H. L. Fairchild, "The History of the American Association for the Advancement of Science," *Science,* N.S. LIX (1926), 365-369, 385-390, 410-415. The evolution of scientific knowledge in the United States is nowhere better illustrated than in the growth of this learned society.

Thomas Lange, "A Function of Regional Scientific Societies," *Science,* N.S. LXVII (1928), 272-273. There is a need for more research in the area of the regional societies, a point which this paper clearly indicates.

E. L. Roberts, "Scientific Societies and the Government," *Science,* N.S. XLIV (19160, 312-314). This article points to one of the crucial issues in the present era, that of the relation of the government to the learned society.

WILLIAM E. DRAKE (A.B., M.A., Ph.D., University of North Carolina) is Professor of the History and Philosophy of Education at the University of Texas. Prior to his Texas appointment, he was Professor and Head of the same department at the University of Missouri, a member of the faculty of the Pennsylvania State University (1930-1939), and has served as a Visiting Lecturer at the University of Illinois (1953) and the University of North Carolina (1936). In addition, his teaching career includes four years of public school work in the state of North Carolina, and one year at Shrivenham American University (Shrivenham, England). In addition to the publication of more than 30 articles in scholarly journals (such as: the *Educational Theory,* the *AAUP Bulletin, Phi Delta Kappan*), he is author of *The American School in Transition* (New York: Prentice-Hall, 1955), and co-author of *The Sociological Foundations of Education* (edited by Joseph S. Roucek, 1942), *Significant Aspects of American Life and Postwar Education* (1944), *Teaching World Affairs in American Schools* (1956), and *American Education,* Vol. I (1958). Has served also as a member of the Executive Committee of the National Society of College Teachers of Education, and as Executive Secretary of the National Philosophy of Education Society. He is, at the present time, Chairman of the Graduate Studies Division of the College of Education, University of Texas.

NOTES FOR LEARNED SOCIETIES

1. Theodore Hornberger, *Scientific Thought in the American Colleges, 1638-1800* (Austin, Texas: The University of Texas Press, 1945), 3.
2. Harrison Ross Steeves, *Learned Societies and English Literary Scholarship* (New York: Columbia University Press, 1913), 43.
3. Ralph H. Bates, *Scientific Societies in the United States* (New York: John Wiley & Sons, Inc., 1945), 2.
4. *Ibid.,* 5.
5. S. W. Fernberger, "The American Psychological Association: A His-

torical Summary, 1892-1930," *Psychological Bulletin*, XXIX (1932) ,1-89.

6. Merle Curti, *American Scholarship in the Twentieth Century* (Cambridge, Massachusetts: Harvard University Press, 1953) , 173.

7. G. B. Goode, *The Smithsonian Institution, 1846-1896* (Washington, 1897), 20.

8. John Jay Chapman, *New Horizons in American Life* (New York: Columbia University Press, 1932), 46.

9. Curti, *op. cit.*, 190.

10. James Harvey Robinson, *The Humanizing of Knowledge* (New York: George H. Doran Company, 1923) , 78.

11. Curti, *op. cit.*, 20.

SCIENCE AND ADULT EDUCATION

JOHN M. BECK
Chicago Teachers College

Adult education has played an important role throughout the history of our civilization. This role, however, has never been as crucial as it is in today's world which is increasingly dominated by scientific and technological forces.

The phenomenal scientific advances of recent years accompanied by the rapid technological transformation of society necessitate an expansion of educational opportunities for adults. Living in an atomic age, survival itself, depends on wise, social perceptions of science and technology. If American society is to be equal to the great challenge of this dynamic technology, not only must the universal schooling of the young be evaluated critically but substantial improvement must be sought in the education of adults. Furthermore, our educational institutions which have been traditionally slow in adapting their curricula to a changing society have lacked qualified staffs and school facilities to produce the kind of scientific literacy required for community life. Although the corrective or remedial function could hardly be imposed upon adult education, the emerging personal responsibilities of adults in today's society argue strongly for organized continuing education which may even become established as the fourth level in the American educational system.

It may well be that some sort of universal education is necessary to realize the kind of understanding needed in modern society, as C. Scott Fletcher, President of the Fund for Adult Education, expressed it:

The world cannot wait until the new generations take charge. It is *adults* who make the homes, the churches, the schools, the communities. It is *adults* who control the mass media, which assail us every waking hour. It is *adults* who determine policy in our political, economic and social lives. It is adults who must cope with the dangers and opportunities of every pressing day. Men and women must act while their children prepare. Their decisions will create the world their children inherit.[1]

A continuing program of lifelong education is as essential today as formal elementary schooling was in the nineteenth-century America.

Even in the twentieth century the increasing adult responsibilities have widened tremendously the required educational attainments of the individual. Knowles points out that our grandfathers could learn by age twenty all they needed to know during a lifetime. In contrast a life-long program of education is desirable to keep abreast of a rapidly changing world in the second half of the twentieth century.[2] The need for continuing education is becoming critical due to the ever-shortening lag period between discovery and the application of scientific forces. More and more, immediate controls must be established to safeguard the welfare of mankind.

Man's everyday life has been surrounded by numerous problems with scientific implications. Modern life has become completely involved with the world of science. The major role of institutions or agencies concerned with adult education is the initiation of effective educational processes to enable the average man to better comprehend the scientific force that shapes his life, so that rather than fear science through ignorance, he may through educational exposure discover a useful servant in science.

Science curriculum experiences to explain the world to man have been designed in recognition of the need for the diffusion of scientific knowledge, scientific appreciations, and scientific attitudes among the general public. Proposed objectives of science education for adults have ranged from a commonplace understanding of everyday science to the more complex human and social applications in a technical age.

Dimensions of Adult Education

Adult education has been typically defined as a "way of learning, voluntarily selected by an individual or group of people, usually 18 years of age or over, in which they carry on sustained inquiry," directed toward the "improvement of themselves or their environment."[3] Today some seven hundred public and private agencies are providing almost unlimited opportunities for continuing learning, formal programs of education and an extensive field of informal activities. Like the American academy of the eighteenth and nineteenth centuries, adult education institutions have been catering to the particular needs of the participating learners.

Early in the 1950's the Adult Education Association of the U.S.A. was founded to coordinate plans for a program of adult education which may develop into life long learning to meet the complex problems of a changing world. This step can provide the basis for a nationwide, well organized system of adult education.

One of the observable trends in adult education has been the increasing emphasis on learning to control and direct environmental and cultural changes. Although the "free time" available in our culture permits a liberal sampling of random recreational activities and essential remedial endeavors, contemporary life as it increases in complexity has made it more expedient for more adults to participate more in an organized curriculum with common objectives in continuing education.

Knowles realistically foresees an expanding curriculum adapted to the problems of society:

The curriculum for adults will concentrate less on vocational, recreational, and remedial subjects and place more stress on advanced programs of liberal, civic, and creative education. . . .

Adults will become more deeply involved than ever before in the schools' welfare, with far-reaching effects on financing, policy-making, and community relations.[4]

Principal Sources

The heart of the adult education movement is set in the vast ongoing activities of educational institutions. A wide range of educational interest is served by colleges and universities, public schools, and public libraries. These institutions join forces in providing a continuing education for adults comprising mainly of degree programs, non-degree programs and community action programs designed for the improvement of civic welfare.

The nation's colleges and universities have shared the major responsibility for adult education for many years. In the thirty-year period, 1924-1954, the number of adults participating in college programs has grown from about 5,000,000 to 15,000,000. These figures which include university extension, correspondence, evening enrollments, and an estimate of numbers served by radio and television represent about one third of the total number of adults engaged in formal and informal programs.[5]

Millions more will be able to share in educational services as colleges and universities exploit the potentialities of television. It is not an idle prediction that in another generation through commercial and educational television, utilizing live, film, and kinescope programs, colleges and universities will carry instruction to the most remote hamlet. Opportunities in adult education, formerly largely centered in the vicinity of educational institutions, can be extended equally to distant rural areas. Pioneer telecast educational programs have already drawn large viewing audiences. A program, *Shakespeare on TV*, first presented on a commercial station in Los Angeles, California, one morning each week, drew a local viewing audience well over 100,000. Western Reserve University in Cleveland, Ohio, during its initial telecasting in the early 1950's presented courses to an average audience ranging between 10,000 and 30,000 persons.[6]

An important phase of adult learning is also carried on under the auspices of the public schools, particularly in the evening divisions of high schools and junior colleges. By 1955 approximately 3,500,000 adults were enrolled in public school programs which offered extension and enrichment of opportunitie: in the popularly accepted fields of adult education. The functions of these institutions have been enlarged to emphasize the interdependent relationships between school and community. The cooperative endeavors of local agencies and the school make it possible for a greater utilization of various resources to prepare individuals and groups to live in a complex and dynamic society.

This interrelatedness in junior college has been expressed in the community college function. The comprehensive pattern of education adopted by the junior colleges includes numerous non-college preparatory courses, credit and non-credit, suitable to adult programs in continuing education. One survey of 144 public junior colleges revealed that nearly sixty per cent had adult education programs. In addition almost ninety-seven per cent offered non-preparatory courses.[7]

During the school year, 1956-57, over 300,000 adult students were enrolled in junior colleges. The increasing numbers of adult students in the past twenty years indicate that the two-year colleges are becoming more sensitive to community needs outside the regular collegiate degree offerings. The past growth of the two-year colleges, particularly public institutions, and the plans being made now in many states for junior colleges point to the prediction that in another generation the junior or community college may be generally available to the public throughout the country. It appears destined to play a significant role in the continuing education of adults.

Status of Science

Although there has been substantial agreement among educators that sciences should occupy a prominent place in the education of adults, the proportion of science classes in adult courses during the past twenty-five years has been small. This is true despite the tremendous growth of scientific literature, both technical and popular, and the ever-increasing impact of mass media communication. The diffusion of easily available information has popularized science through the exploitation of startling achievements but has failed to develop a public understanding of how science produces its results, how it affects the modern ways of living, what can be done with it and what public policy should be formulated to realize the greatest good for mankind.

Dixon characterizes the average adult at the present time as "living in a world of impressive material achievements and construction, made possible by science, of complex human relationships greatly influenced by science, and of terrifying dangers arising from the misuse of science."[8] At the same time the adult who is ignorant of the subject matter and method of science is called upon to make decisions affecting scientific policy. Obviously the manifold purposes, deemed desirable, can be attained only in systematic courses of instruction and require a far greater extension of adult educational effort than currently evident in the sciences.

The fact that the average citizen is "under-educated" in the ways of science can hardly be attributed to a dearth of potential learners among the adult lay groups. In 1957, a sample survey by the Census Bureau revealed an enrollment of over 9,000,000 adults in formal programs of education. Adult enrollment in the broader field of informal activities has been estimated as high as 35,000,000. This widespread interest, though commendable, covered a great variety of subject matter, mostly non-academic and non-scientific.

Science education failed to keep pace with the rapid growth of adult education during the recent decades. Excluding the yet undetermined influence of the *sputniks* and *jupiters,* educational statistics from various sources have indicated that the science curriculum has been at a standstill if not on a decline. Surveys of the field of education in the 1930's showed with considerable consistency that the attention given to science in organized courses amounted to five or six per cent of the total. For the school year 1931-32, the ratio of science courses to total offerings in sixteen university extension courses ranged from 0.6 to 22.2 per cent; and the ratio of enrollments in science to the total enrollments ranged from 0.2 to 15 per cent. Comparably low ratios were reported for evening high schools, correspondence courses and other miscellaneous adult agencies.[9]

Another study of adult education in the independent liberal arts colleges of the United States in 1953 found that little attention was given to science and mathematics. Of the 404 responding colleges, 233 reported programs in adult education. Courses in the humanities, social and behaviorial sciences, and natural sciences and mathematics accounted for about two-thirds of the adult offerings. Science and mathematics placed a poor third among the three major divisions in the adult curricula. The courses in biology, chemistry, mathematics, physics, and related subjects were offered less than one-fourth as often as were courses in the humanities. More credit courses were offered in the field of business than in any one academic discipline, and almost half as many as in all of the liberal arts and sciences combined.[10]

Mid-century reports on adult education in two densely populated regions in the United States also revealed a very limited science program. A survey of 651 cities with a minimum population of 2500 in New York, New Jersey, Pennsylvania, Maryland and Delaware indicated the greatest annual increases in practical courses, such as hobbies, arts, crafts, and homemaking; the smallest gains were registered in academic courses which included science.[11] In California, the total of 389 different courses offered by 118 centers included only 17 courses in science. The small proportion of science in adult education is more striking when the number of classes is compared. The California centers offered a total of 6326 classes and only 123 of these were science subjects.[12]

The Promise of Television

Television, a new force on the educational horizon, appears to be the one means to bring about mass education for adults, augmenting the many millions already in the program. And in view of the growing concern over scientific advancement, this creation of modern science will undoubtedly serve as the means to improved and widespread education of adults in the sciences. The number of television courses has increased steadily since 1950. Many of the college television courses were open to the general public. An inventory of telecourse instruction for the academic year 1955-56 listed a total of 162 non-credit or adult education courses being offered by 39 institutions. The kinds of subjects taught as adult education subjects were similar to college credit courses. Twenty-nine of the 162 courses available to viewers of non-credit courses were in the field of science, eleven of these in general science.[13] The growth of television instruction suggested by these figures represents only a small part of the field recognized as educational television. Numerous network programs, e.g., the excellent Johns Hopkins Science Review, were not included in the inventory.

The rather modest representation of science telecourses for adults in the *pre-sputnik era* is certain to expand with the increased national concern over science education. Following the thrust of the *sputnik* into its orbit, noncommercial education television stations have nearly doubled the production of programs in science and technology, according to a survey made by the Educational Television and Radio Center. During a sample week in 1958, the 27 non commercial educational television stations surveyed telecast a total of 89 hours of programs in scientific subjects; compared to 47 hours for a similar period in 1957. The scientific and technological programs now head the list in terms of volume.[14]

Probably the outstanding televised science project since the *sputniks* has been the University of Chicago's *Science 58* which was telecast over a commercial station. Thousands viewed the early morning program daily for thirteen weeks. Primarily designed for adults, the basic purpose of the program was to develop an awareness of the methods and achievements of science.[15] It may well be that with the aid of television and the pressure of international rivalry, science education will be firmly established on all fronts in adult education.

Many plausible reasons have been suggested for the apparent indifference of the lay public to science education. Often the indifference has its origin in the dreary, inadequate instruction given in secondary schools where the memorization of isolated facts and the lack of scientific apparatus stultified student growth in the appreciation and understanding of science and ultimately precluded a possible adult venture in the same field. In adult education, too, the lack of scientific apparatus has had a restricting effect. Other factors which are said to account for the slow progress of science in adult education include the shortage of science teachers, the indifference of classically bent promoters of adult programs, and the popular adult preoccupation with the vast number of recreational activities and commercialized forms of entertainment during the daily free hours.

New Directions in Science Education

While the *sputniks* have aroused concern over our scientific capabilities and have created unprecedented activity in the field of science, it does not suffice to have the young and old between 6 and 60 devoting their educational efforts to produce more scientists to launch more *jupiters*. Nevertheless, the spectacular events have thrust into the limelight the urgent need for a continuing education to enable the adult learner to integrate the scientific knowledge as science advances to greater pinnacles of discovery in every field of endeavor. Adult education is called upon for greater efforts to produce well-informed laymen concerning the various scientific and technological aspects of community policy and action. As the importance of science increases in the United States, a broader understanding of its impact upon today's culture must be acquired by the vast number of laymen whose scientific interest stems from the numberless gadgets, useful as well as non-essential, that materially influence living in a modern age.

Unified science curriculum. The dominant position of science in contemporary life situations calls for new designs and new directions in adult science education. Adult agencies must devote more time and resources to capitalize on the rapid growth of interest in teaching science and expand the curriculum offerings which have been relatively meager in the past. The primary purpose would be to give a wider degree of understanding of science by a close study of basic scientific concepts in a unified curriculum. To meet the crisis in human experience, there seem to be several directions in which science courses might be developed. In his analysis of essential science elements in adult education, Dixon suggests courses in subject matter, methodology, history of science, and science as it relates to social problems. Dixon proposes a subject matter course for instruction in what is commonly called known science. Such a course in adult education would be less analytic in character and much wider in scope than are conventional school courses. A second course in scientific method is the most difficult to administer effectively since the usual classroom procedures are not adaptable to the kinds of learning experiences required to attain the course objectives. Lectures on the scientific method are of little value. Group investigations of social and psychological phenomena in the lives of students would be of some value in gaining knowledge of and insight into the scientific method. The history course would present a careful study of scientific theory and the application of scientific method as well as to show the interaction of society and evolving scientific concepts. In addition the historical approach would include a resume of the major technological achievements. The concluding course of the program would center on the important part played by science in society, its potentialities and the possible consequences of present and future technology.[16]

What is intended in the foregoing courses is the cultivation of a basic understanding of science for citizenship in a technological society. It represents a goal which challenges adult education in an unparalleled manner in view of frenetic demands pressing on all educational fronts that priority be given to the sciences.

Science and the liberal arts. Apart from the present expansion of an organized science curriculum in adult education, a far more promising trend developed in adult education after World War II. New directions were charted for the liberal education of adults. New designs in the liberal arts were related to the integrative needs of different specialist groups.

The Center for the Study of Liberal Education for Adults, was established in 1951 by a grant from the Fund for Adult Education to work with university agencies providing liberal education for adults. The aim of the Fund is the expansion of opportunities for all adults to continue their education throughout life. In addition to colleges and universities, the Fund has helped to broaden and improve the area of liberal adult education under public school auspices, public libraries, and a number of national organizations. Both credit and non-credit programs in liberal education have been sponsored, ranging from an institute of a few weeks to a four year course. To further extend liberal education on a nationwide basis, large sums have been allocated by the Fund to improve the utilization of mass media, particularly television and radio.

Although the satellite fears brought science and technology to the forefront, the science curriculum was held in balance. The approach to science was adapted to the specific program in liberal education. In the Dartmouth College Program, one of several college programs for the A. T. and T. executives, the course, *Science and Man,* was presented in two main parts. The first was philosophical—how modern man has had to adjust his thinking on religious and philosophical issues in confronting the changing world revealed by science. The second topic, social aspects, considered the effect of science and technology and the direction of change now taking place. Readings for courses were selected from works of past and modern writers.

Science in the executive program at Clark University carried the course title *Man and Nature.* The four divisions of the course were *Science and the Modern World, The Evolution of Man and Culture, Man's Place in Nature,* and *Nature and Human Values.*

The second year of the Memphis Plan for the education of executives linked the humanistic studies with the technologies of industrial society in a three-unit course entitled *Man and His Life As Affected by Science and Technology.* The content was related to how science and technology created new materials and shaped new patterns and new modes of living, how applied science has brought economic changes, and how science and technology have shaped the life of men.

At Wabash College the Personal Development Program for young men in business and industry consisted of a combination of seminars and classroom sessions. Sciences were treated in more traditional fashion. During the first year sessions the course presented a general survey of all forms of science in the development from the first phases to modern applications. In the second year, courses were conducted in physics and chemistry. The plan for the final three years includes the further development of the knowledge and use of the sciences.

Thirteen programs of liberal education for executives were in progress in 1958. All programs were non-credit and the sessions, mostly short in duration, ranged from a period of a few weeks to a full academic year. Similar programs were instituted by eight universities for labor organizations. Other programs, thus far, have involved teachers, private secretaries and college presidents.[17]

More extensive science courses in adult liberal education are found in the older institutionally-centered credit programs. An example of a successful venture of this kind is the *Basic Program of Liberal Education for Adults* which has been in operation for some ten years at the University of Chicago. In the Natural Science Tutorial, physics is offered the first year, chemistry in the second, biology in the third and mathematics in the fourth. No scientific or mathematical background is presupposed. Included among the course experiences is the analysis of the methods of investigation utilized in solving problems in the various fields of science.[18]

The Fund is also engaged in exploring the potentialities of'educational television as a means to aid colleges and universities in the development of programs in liberal adult education and also to serve organized community discussion groups in a variety of subject-matter fields.

The pioneer programs in continuing liberal education represent a much desired trend today when independent liberal thinking based on informed adult experiences is urgently needed to shape national policy in an era of cold war and possible atomic warfare. Scientific achievements and world tensions only accentuate the importance of high quality of education in all areas of endeavor. To meet the exigencies of contemporary life, education must meet certain basic requirements. In discussing the essentials of liberal education for adults, the Fund's president, C. Scott Fletcher, emphasizes five requirements for effective education. These are:

1. Balance. The nation needs broadly educated men and women in all fields of endeavor. Scientists and technicians need the vision and

breadth that a well-rounded education gives. Laymen need familiarity with the role of science and technology in modern life.
2. Concert. The sciences can be liberating disciplines. The liberal arts and humanities have relevance to all human activities, including science & technology.
3. Essentialness. This calls, not for a choice between scientific and technical education on the one hand and liberal and humane education on the other, but for a choice between that which is essential to both and that which is essential to neither.
4. Quality—excellence in both scientific and humane education together.
5. Continuation. This means, not only more attention to all higher education, but also more attention to liberal education continuing throughout life.[19]

These requirements represent a basis for establishing needed educational goals for adult education as well as a basis for the complete revision of curricula in high schools and colleges. Progress in this direction is essential if our educational institutions are to provide a more effective means for the solution of basic problems of contemporary American life. Man's self-realization in a machine civilization is dependent upon a wise utilization of a lifelong accumulation of knowledge. Neither science alone nor any other single discipline can provide the magic formula for the kind of life which is good for man to live.

Conclusion

Adult education has become a major force to insure quality of policy decisions in a society which is being transformed rapidly by technological advances. Until World War II, relatively little attention was given to science in adult education. Post-war conditions, in particular the cold-war tensions with the stress on war technology, produced a trend marked by excessive specialization in science education. In each instance the imbalance was an obstacle to intellectual synthesis.

The initiation of new liberal education programs in adult education may be expected to improve balance and quality. These new directions point to prolonged, continued adult education for the heterogeneous lay groups, an education strong in a qualitative sense rather than in a quantitative sense. However, increasing support must be available for continued and sustained liberal education if adult education is to serve the nation well and help meet the problems of science and technology and their offsprings, automation and leisure time.

Selected Bibliography

James E. Crimi, "Adult Education in the Liberal Arts Colleges," *Notes and Essays on Education for Adults*, No. 17 (Chicago: The Center for the Study of Liberal Education for Adults, 1957). Investigates the nature and extent of adult education in four hundred and four American independent liberal arts colleges.

Cecil Dixon, "Science in Adult Education," *Adult Education*, XXI (March, 1949), 123-27. Considers the plight of the ordinary citizen who is grossly ignorant of science subject matter and the method of science. Suggests essential aspects of science education for adults.

William Y. Elliot, *Television's Impact on American Culture* (East Lansing, Mich.: Michigan State University Press, 1956). Includes a chapter on the tremendous potential of television in adult education. Traces the development of educational television.

Paul L. Essert, "Adult Education in the United States," *The Annals of the American Academy of Political and Social Science*, CCLXV (September, 1949), 122-29. Gives the motives for continued learning and reveals the influences of the changing conditions of culture upon adult education programs.

J. W. Getsinger, "What Do the California Adult Schools Teach?", *Adult Education Journal*, VIII (October, 1949), 231-36. Analyzes course offerings in 118 public adult schools in California. Indicates the small proportion of science classes.

Benjamin C. Gruenberg, *Science and the Public Mind* (New York: McGraw-Hill Book Co., Inc., 1935). Discusses the place of science in relation to adult education. Emphasizes the importance of science to the laymen and describes the types of programs in adult education. This is one of the best sources on the subject.

Herbert C. Hunsaker, "What Are the Responsibilities of Higher Education for the Continuing Education of Adults?", *Proceedings of the Tenth Annual National Conference on Higher Education, 1955*. (Washington, D. C.: Association for Higher Education, 1955), 85-90. Shows how the colleges and universities have assumed a major responsibility for adult education.

Malcolm S. Knowles, "What Should You Know About Adult Education?," *The School Executive*, LXXVII (August, 1958), 19-21. Discusses briefly the different meanings of adult education, the trends in adult education, and predicts that adult education shall become an "integrated, articulated, and perhaps a preponderant part" of our educational system.

Hideya Kumata, *An Inventory of Instructional Television Research* (Ann Arbor, Mich.: Educational Television and Radio, 1956). Surveys research literature on educational television and includes a summary of credit and non credit courses in adult education programs.

S. V. Martorana, "Status of Adult Education in Junior Colleges," *Junior College Journal*, XVIII (February, 1948), 322-31. Presents an analysis of kinds of adult education programs in 144 public and private junior colleges.

Richard A. Mumma, "Trends in Adult Education Offerings in Region II," *Adult Education Bulletin*, XIV (August, 1950), 181-86. Gives detailed information on available classes for adults in 651 cities with a minimum population of 2500 in five eastern states.

Peter E. Siegle, *New Directions in Liberal Education for Executives* (Chicago: The Center for the Study of Liberal Education for Adults, 1958). Describes various successful programs in liberal education to which business, industry, and academia have committed themselves.

Peter E. Siegle and James B. Whipple, *New Directions in Programming for University Adult Education* (Chicago: The Center for the Study of Liberal Education for Adults, 1957). Discusses liberal arts programming for businessmen, professionals, technicians, alumni, aging, and community groups. Evaluate the use of mass media in adult education.

The Fund for Adult Education, *Continuing Liberal Education.* Report for 1955-1957. (White Plains, N. Y.: The Fund for Adult Education, 1957). Summarizes the advances made in the Fund's liberal adult education through educational institutions.

JOHN M. BECK, Chairman, Department of Education, Chicago Teachers College, received his Ph.D. from the University of Chicago (1953). He taught at De Paul University (1948-1953) and, in addition to having written articles for such periodicals as the *Junior College Journal, School Review, Review of Education Research, Encyclopaedia Britannica,* and others, is the periodicals editor of the *Chicago Schools Journal.*

NOTES FOR SCIENCE AND ADULT EDUCATION

1. *Continuing Liberal Education* (White Plains, N. Y.: The Fund for Adult Education, 1957), 11.

2. Malcolm S. Knowles, "What Should You Know About Adult Education?", *The School Executive* (August, 1958), 19-21.

3. Paul L. Essert, "Adult Education in the United States," *The Annals of the American Academy of Political and Social Science,* CCLXV (September, 1949), 122.

4. Knowles, *op. cit.,* 21.

5. Herbert C. Hunsaker, "What Are the Responsibilities of Higher Education for the Continuing Education of Adults?" *Proceedings of the Tenth Annual National Conference on Higher Education* (Washington, D. C.: Association for Higher Education, 1955), 85-86.

6. William Y. Elliot, *Television's Impact on American Culture* (East Lansing, Mich.: Michigan State University Press, 1956), 292-93.

7. S. V. Martorana, "Status of Adult Education in Junior College," *Junior College Journal,* XVIII (February, 1948), 323.

8. Cecil Dixon, "Science in Adult Education," *Adult Education,* XXI (March, 1949), 123.

9. Benjamin C. Gruenberg, *Science and The Public Mind* (New York: McGraw-Hill Book Company, Inc., 1935), 141-149.

10. James E. Crimi, *Adult Education in the Liberal Arts Colleges* (Chicago: The Center for the Study of Liberal Education for Adults, 1957), 16-17.

11. Richard A. Mumma, "Trends in Adult Education Offerings in Region II," *Adult Education Bulletin,* XIV (August, 1950), 181.

12. J. W. Getsinger, "What Do the California Adult Schools Teach," *Adult Education Journal,* VIII (October, 1949), 231.

13. Hideya Kumata, *An Inventory of Instructional Television Research* (Ann Arbor, Mich.: Educational Television and Radio Center, 1956), B-10-B-11.

14. *News of National Educational Television* (Ann Arbor, Mich.: Educational Television and Radio Center, 1958), 30-31.

15. Alex Sutherland, "Science 58," *NAEB Journal,* XVII (May, 1958), 11.

16. Dixon, *op. cit.,* 125-127.

17. Peter E. Siegle, *New Directions in Liberal Education for Executives* (Chicago: The Center for the Study of Liberal Education for Adults, 1958), 9-43.

18. Peter E. Siegle and James B. Whipple, *New Directions in Programming for University Adult Education* (Chicago: The Center for the Study of Liberal Education for Adults, 1957), 65-66.

19. Continuing Liberal Education, *op. cit.,* 12-13.

SCIENTISTS AND ENGINEERS FOR THE ARMED FORCES

COL. CHARLES W. HOSTLER, U. S. AIR FORCE
and
COL. DONALD J. DECKER, U. S. MARINE CORPS

In the fiscal year 1957 the United States Department of Defense and the Atomic Energy Commission obligated nearly $3 billion of the taxpayers' money for scientific research and development. In these days of long-range missiles, weapons of mass destruction, nuclear propulsion, and satellites, there can be no question of the dependence of the security of the United States on scientific education. In the bipolar world struggle, even this great effort by the United States may not be enough to meet the challenge of the Soviet Union.

Before considering the problems in obtaining adequate scientifically qualified personnel for our defense structure, let us look at the broad duties of the armed forces. They are responsible for the military security of the United States today, and for the development of weapons for the future. Defense has an annual budget of more than $40 billion. 3.6 million civilian and military personnel are employed directly by the Department of Defense and the military services. The tasks range from development of weapons systems for the future to the constant maintenance of large military forces at a high degree of readiness in many parts of the world. The defense of the country five or ten years hence is dependent upon the decisions and developments of today. The Department of Defense must also be prepared today, and every day, to defend the United States by employing all-out nuclear retaliation or by more conventional operations in numerous strategic areas of the world. This requirement for constant readiness during what is a transitional phase, due to the tremendous advances in weapons technology, requires a delicate balance in the allocation of our defense effort. The voice of science must be adequately represented in securing that balance. Scientific education is essential at the policy level, and also at every level in the development and maintenance of our armed forces. One of the most urgent needs in the defense organization is for men, in and out of uniform, who understand the scientific, strategic and operational aspects of weapons systems in these days of rapidly advancing weapons technology. We will discuss below just who these scientifically trained men of our defense organization are and what are Defense's problems in relation to them.

The specific requirements of the Army, Navy, Air Force, and Marine Corps for scientifically trained personnel vary with their operational missions, but the procurement, classification and training of these personnel follow a generally similar pattern in each service. They are conveniently grouped as military and civilian (including contractual and foreign scientists).

Military Personnel

The military personnel are primarily operators. They man and maintain the operating forces—the divisions, fleets and air wings—which are deployed overseas and in the United States. It must be borne in mind that the end product of all efforts of the Department of Defense is to produce and maintain these operating combat units with adequate weapons at a state of readiness to meet any threat to our security. As the equipment of these units becomes increasingly more complex, engineering and technical training becomes more essential.

Many officers receive a basic engineering education at the Military, Naval, or Air Force Academies, or receive their commissions through ROTC programs at various universities. After graduation and commissioning, officers are normally assigned to a military combat or combat support unit in order to acquire proficiency in their service. This initial indoctrination as well as screening process is most desirable before specialization in a scientific field. To meet the requirements for scientifically trained personnel, officers are selected for graduate level education in science and engineering after the period with an operational unit. This education is sought either at civilian graduate schools or at service-conducted postgraduate or advanced technical schools. The Army depends almost exclusively on civilian graduate schools for this advanced training. The Navy and Air Force conduct, in addition, their own service postgraduate level schools to provide the large number of officers with the advanced technical education required for the operation and maintenance of ships, aircraft, and space vehicles.

This advanced education provides technically qualified officers for the operating forces and a source from which those who will receive further scientific training may be chosen. Selected graduates from these schools are given a more advanced scientific and engineering education with the aim of providing officers who are qualified for the supervision of research and development and design of weapons. These officers must work closely with civilian scientists and translate military requirements into scientific terms. They must also be able to identify the military usefulness of new scientific developments revealed by the scientists. This interchange of information and ideas between the military officer and the civilian scientist will determine the success with which scientific developments will be employed for military means.

There are, of course, many individual variations from the pattern outlined here. Reserve officers with special qualifications in scientific fields have been employed when they could be obtained. During emergencies when large numbers of reserves are on duty this has been a valuable source of talent. Some of these reserve officers have been attracted to the regular service, but most of them return to civilian life and more lucrative employment when the immediate emergency has passed.

This system of selecting a few professional officers for advanced education in scientific and engineering fields to provide the supervision of research and development by civilian scientists has met the service requirements. The number of these officers is such a small percentage of the total military officers required for operational tasks, that the process should be quite selective. It is essential that they maintain their familiarity with combat requirements and operating conditions and that they possess the imagination to see the combat potential of scientific discoveries. They must know the problems of the scientists and the operating combat units and must understand the strategy of our defense plans.

Educational Programs for Military Personnel

The increasing complexity of modern weapons systems and communications has made it desirable for Defense to encourage and assist Military personnel to continue their formal education to the baccalaureate level or higher.

Approximately one-half of all the officers on active duty at the present time have a college degree (an estimated 164,000 with degrees). Of these, about 15% are qualified in the natural sciences and engineering, but less than 2% of the officer force have a graduate degree in these fields.[1] At least a college degree in science or engineering is required for officers who are assigned to supervisory-type positions in the research and development field. The need for officers with a technical background in the sciences or engineering is not, however, confined to the research and development field. Such fields as intelligence, combat engineers, air defense, maintenance and production and procurement have many jobs for which a technical background in engineering is required.

Current Service programs to improve the quality and increase the supply of scientific and technically qualified officers include regular tours of duty at civilian universities, in-service education, tuition assistance programs, and off-duty study.

An educational program at civilian universities has been established at the graduate level: (1) to provide qualified officers for duty in the scientific and engineering fields in which a graduate education is required and (2) to improve the educational level of officers serving in these specialties. Under this program, the Services contract for specified education with selected civilian universities. Courses include—among others—engineering, nuclear physics, applied mathematics, ordnance, fuel technology, seismology and metallurgy. From 1946 to 1958, the program produced 7,500 officers with a master's degree or higher in the scientific and engineering fields. There are now approximately 1,250 officers doing graduate work at civilian universities such as MIT, California Tech., University of Michigan, Iowa State, Lehigh, University of Maryland, Stanford, etc.

Principal in-service graduate education programs in science and engineering are conducted by the Air Force Institute of Technology (AFIT) School of Engineering at Wright-Patterson Air Force Base and the Naval Postgraduate Schools at Monterey. The program at AFIT, designed primarily for graduate engineering work, includes courses in nuclear, aeronautical, electrical and air weapons engineering, as well as graduate courses in management. It has an average annual enrollment of 270 students. The Institute began a graduate astronautics program in September 1958 with an enrollment of 30 students and will have 10 students at the Massachusetts Institute of Technology.

The Naval Postgraduate School, with an enrollment of 85 in the physical sciences and 448 in engineering for FY 1958, is expanding its science program for FY 1959 to 215 students and its engineering course to 527. Expansions in the former program will include Advanced Science Ordnance and Aerology. The latter program will install a new general engineering curriculum for the fleet officer.

Under the Tuition Assistance Program, military personnel receive tuition assistance for voluntary courses taken off-duty, or are placed on temporary duty at a college or university for a final semester to complete college degree requirements. The Service pays up to three-fourths (not to exceed $7.50 per semester hour) of the tuition costs for each course. Subsidized courses, generally, are restricted to those leading to a baccalaureate degree or, at the graduate level, to the fields of mathematics, physical sciences, international relations and management. Army figures show that, from 1948 to 1958 1,293 degrees have been earned, and 176,540 single college courses have been completed, by its officers. Army enrollment in 1958 in college courses is 17,149. Of that number 21% are in mathematics and the physical sciences. Under the Air Force program ("Bootstrap"), approximately 6,000 officer students participate annually in off-duty study with tuition aid. Final semester temporary duty participants in the tuition assistance program are limited to not more than 200 at any one time. Approximately 3,500 degrees have been earned by Air Force officers during their off-duty time, of which 8% have been in the scientific and engineering fields.

Officers and enlisted personnel of all Services participate in the Armed Forces Institute program, monitored by the Department of Defense Armed Forces Information and Education Service, and providing a wide variety of voluntary off-duty educational programs for military personnel. This program is carried on through correspondence and self-teaching courses up to the second year college level.

In addition to the foregoing, the Armed Forces have developed extensive programs for the training of enlisted personnel, ranging from formal school education to on-the-job training. New weapons and technical equipment have increased the skill requirements as well as the number of skilled jobs. In FY '57, it was estimated that about 250,000 persons were trained in technical and mechanical skills in Service schools. The length of the courses varied with the subject matter from a few months to a year.

Recognizing the need for advanced education for selected enlisted personnel, the Navy inaugurated an Enlisted Advanced School Program in 1956 and the Army in 1958. Under the Navy program, selected enlisted personnel of proven technical skill are sent to engineering colleges where they study for two years, return to the operating forces for two years, and then return to college for the final two years of engineering education leading to a baccalaureate degree. These men compete for officer status along with their contemporaries. In 1957 there were 90 such students enrolled at Purdue University and 47 students at the University of Washington under this program. The planned input is 100 men per year.

In January 1958, in recognition of the increased emphasis on science, the Navy inaugurated its Enlisted Scientific Education Program. Selected enlisted personnel will be enrolled in civilian universities to pursue regular college catalog courses in scientific fields. Upon graduation they will be offered a commission if otherwise qualified. They will be required to serve four years of obligated service in return for their four years of college training. About 200 students entered colleges in September 1958. It is planned ultimately to enter 400-500 students a year in this program.

Civilian Scientists

The major requirements for professional scientific and engineering personnel in defense are met not by military personnel but by civilian scientists. They are either employed directly by the Department of Defense or their services are secured through contractual arrangements.

The civilian force employed directly in the Department of Defense are largely civil service employees. The professional grades alone number approximately 43,000, nearly half of whom are in Research and Development. They are assisted by subprofessional workers, many of whom through training will be advanced to professional grade during their careers. It is this large force that carries on the work at such installations as Cape Canaveral, Redstone Arsenal, the Office of Naval Research, and White Sands. In addition, there are numerous other Army, Navy, Air Force, and Marine Corps maintenance and repair installations throughout the country and overseas where these civilians carry on much of the technical work done to support the military operating units.

This civil service force is recruited in competition with industry. The recruitment of persons with engineering and scientific skills for these positions normally must conform to the competitive examination, pay, and appointment requirements of the civil service system. Since 1950 the Services have not been able to recruit and retain a completely adequate force of professional engineering and scientific personnel.

Various work-study and advanced education programs are used to alleviate this situation. These begin at the undergraduate level for college students. Students of engineering, science, and mathematics who are interested in a civilian career with the Department of Defense are given part-time employment largely during the summer in the field of their interest, and receive some financial assistance toward their educational expenses. Other co-operative educational programs provide work experience with military establishments which lead in some cases to interest in civilian careers with Defense. The Military Departments also operate programs under the principle of alternate work-study periods for each candidate, with students being hired in pairs. One works at a military installation while the other is at school. They change places at the end of each college quarter. Current enrollment under this program is 1,860.

There is also considerable emphasis on continuous training of scientists and engineers of the civilian professional force. Technological advances alone require this, but there is also the objective of widening the scope of individual competency. Accordingly, there is continual assistance offered to employees to induce them to participate in educational activities at both graduate and undergraduate training level. This education is primarily for advanced study, and in new areas of endeavor. It assumes many different patterns and varies from full-time graduate study at universities to university sponsored evening classes at the place of employment. These evening courses are suited to the needs and schedules of the student's work. They may be on government time and at government expense, and have the added incentive of giving credit toward advanced degrees.

Contractual Scientists

The Department of Defense found that it was neither feasible nor efficient to exclusively meet their scientific needs within their own organizations. They contract with advanced educational institutions and with private industrial organizations to carry out many of their research and development projects. Of approximately $3 billion budgeted by the Federal Government for the conduct of research and development in fiscal year 1958, about $1.5 billion was used in Federal laboratories, and $1.5 billion for contracts with educational and other organizations.[2] Precise data are not available but a reasonable estimate of present Federal research and development funds going to institutions of higher education is $500 million per year. The predominant proportion of these funds goes to defense or defense-oriented programs. This method of obtaining the scientific assistance required by defense was an outgrowth of World War II. The Korean conflict brought it to a new peak. During those and succeeding emergencies, many university and industrial scientists have worked upon the development of new weapons and other military applications of scientific discoveries.

The Government's reasons for contracting for such work instead of expanding its own laboratories were originally related to the urgency of wartime. The establishment of new government laboratories is a slower process than is setting up an organization outside the Government. The Air Force, founded in 1947, lacked the Army's and Navy's long-established military laboratories and contracted to meet its scientific needs quickly. Contractual operation avoids to a considerable extent civil service regulations with respect to hiring and firing, salary scales, promotions, and vacations. For some scientists these regulations tended to make Government duties less attractive than industrial or academic employment. The contractual means made possible quick utilization of scientific talent, experience, and facilities. Many research centers had their start in an existing nucleus of talent. Project "Lincoln" was placed at MIT where a highly qualified group of scientists were located and where excellent facilities were available for engineering and electronic research and high speed computation. The importance of contractual scientists to the Air Force is indicated by the fact that 95% of Air Force research and development is performed by contract; less than 5% is done directly in Air Force organizations.

There are three distinct kinds of Government research and development programs managed by educational institutions. *First* are those in which university scientists are financed to conduct research of their own choice—generally, but not invariably, basic research with no immediately foreseeable utility. Examples of these are the basic research contracts of the Office of Naval Research, the Army's Office of Ordnance Research, and the Air Force's Office of Scientific Research. All have one common objective—the increase of scientific knowledge. Auxiliary objectives of greater or less importance to different agencies are those of maintaining contact with university scientists and of increasing the supply of scientists. The defense agencies regard student training as a valuable byproduct of these research programs. For example, ONR has specific interest in increasing the number of scientists in so-called "naval sciences," like oceanography, hydrodynamics and hydrobiology. The Air Force interest in developing scientific skills in hypersonics and meteorology is comparable.

A *second* category of programs are those in which the Government departments, on their own initiative, contract with universities for applied research necessary to the development of products, systems and processes, as well as for fundamental research related to defense. The principal programs in this category are: (a) the applied and operational research contracts going to universities from the ONR, (b) the contract programs of Air Research and Developments Command administered through centers like the Wright Air Development Center, the Rome Air Development Center and the Cambridge Research Center, (c) the contracts of the Army technical services, and (d) some important military research projects handled on a multi-service basis. The Defense agency may contract for a university's staff, services, and facilities much as it would contract for the services of an industrial firm. In either case the objective is the solution of particular problems arising in the performance of the contracting agency's defense mission.

The *third* category of government sponsored Research and Development programs is concerned primarily with broad, important military and nuclear energy problems. Some 40% of Government contracts and grants for research and development at educational institutions are with large government-owned and university-managed laboratories. The distinguishing features of university-managed research centers are size and organizational independence from the contractor university. The utilization of research centers has been a most significant development in the efforts of the Armed Services to solve their scientific problems. Some of these centers are managed by nonprofit organizations such as RAND and the Woods Hole Oceanographic Institution; others by profit organizations such as General Electric and the Thiokol Corp.; but the large majority are managed by educational institutions. Examples of these are the Operations Research Office managed for the Army by Johns Hopkins University; the Naval Biological Laboratory managed by the University of California; and Project Lincoln managed for the Air Force by MIT.

These research centers have substantial budgets and considerable freedom to tackle the job in whatever way seems best. In some cases the creation of completely separate organizations by the university was dictated by the size of the undertakings, their specialized nature, the requirements for unusual facilities and personnel, or the demands of military security. The research center has met the organizational needs for research and development on new weapons and material with marked success. However, some industrial organizations have been more critical of their value. The research center device has advantages to the universities, in terms of organizational separateness; and to the defense agencies, in terms of freedom from governmental procedural rigidities. Most important is great ease in recruiting and compensating scientific personnel since research center personnel are employed by the contractor and not by the defense agency itself.

It is through these means of contracting with educational institutions and with private industry that the military services have been able to supplement their own scientific agencies and thus meet their skyrocketing requirements for scientific talent. All of these facilities are, however, in competition with each other and with private industry for the personnel who have the scientific education required in the problems of defense.

Foreign Sources of Scientists for Defense

The Defense Department has shown imagination in augmenting the numbers of United States personnel who have a scientific education. Prior to World War II, the majority of the new concepts in purely scientific knowledge came from Europe. America was the practical land of applying this knowledge in industry. In 1945, it was realized that the Germans and Austrians had many scientific and technical concepts which were advanced far beyond any similar ones possessed by the Allies. Preeminent foreign scientists were brought to the United States on a temporary contract basis through what was known as Operation "Paperclip."

In 1947, it became evident that there was a critical shortage of certain specialists in civilian industry. Thus the program was enlarged to allow certain aliens to immigrate to this country where their special research skills could be used in defense programs. The Defense Scientists Immigration Program was designed to assist in all phases of defense research and development through utilization of selected foreign scientists. Individuals must volunteer and undergo extensive background and professional investigation before they are accepted for immigration. They are free to accept any employment offers made by the military services, government research agencies, industry and universities. Today the names of Oberth (propulsion), Kober (ballistics), Stuhlinger, Lusser, Zeigler, Strughold (aero medicine), Friedrich, von Braun and many others are well-known in research circles. These scientists have become United States citizens and have made outstanding contributions to the United States. The Defense Scientists Immigration Program processes any person for immigration whose qualifications are in short supply and has biographical data on a large number of foreign specialists. This information is available to any institution associated with the defense effort who seeks to employ scientific and technical specialists not otherwise available in the United States. The number of scientific personnel obtained through this program to augment our scientific manpower has been substantial.

Problem Areas

The limitations in the nation's supply of scientists and engineers will affect the resources on which defense will draw for its military and civilian scientists. There are only predictions as to how successful the United States educational system will be in reducing the shortages in professional manpower; but the predictions generally express doubt as to the possibility of making a large-scale quantity improvement in the next ten years. The unfortunate facts are that too many young men fail to acquire the elements of a fundamental education. The United States Department of Labor stated in a study entitled, "The Skilled Work Force of the United States" (1955), that roughly 50% of all men of military age are not high school graduates. Further, many high school graduates do not have the basic background for college training in the fields in which Defense is short.

The effect of this situation on Defense is and will continue to be twofold: The resource of scientists and engineers will be at an austere level for the forseeable future; and Defense will remain in active competition for the most skilled college graduates.

A good deal of the difficulty experienced by the Government in attracting qualified scientists and engineers would exist regardless of the overall market conditions. Even in a period of surplus, those employers restricted by competitive handicaps experience difficulties in attracting and holding the type of individual needed for creative technical work.

Pay

The Government, in terms of pay particularly, is neither competitive with its own contractors nor with other employers of scientific manpower.[3] There are significant differences between the level of pay established by the current Federal pay structure and that of other employers of engineers and scientists. The compensation for top engineering and scientific positions in Government does not even come close to the compensation for comparable type positions in industry.

In the past many budgeted Federal vacancies remained unfilled for long periods and the turnover rate among civilian engineers and scientists in the Federal service is at a high level. Several studies indicate that the turnover is due primarily to movement to jobs paying more in private industry. On the other hand, pay is not the most common reason for turnover in private employment. Almost 70% of Defense's scientists in Grades GS-11 and above who leave, do so because of pay. Defense cannot match allowances offered by industry, such as payment of travel expenses for interview, transportation to first duty station, and relocation allowances.

The Federal Government, with over 84,000 scientists and engineers (roughly 10% of the nation's supply), is not able to attract the same proportion of the college graduates in these fields. Furthermore, at least the experience of one agency indicates that during recent years, the bulk of the recruitment has been from students with lower and lower standing in their graduating class. A survey of attitudes indicates that while 42% of the Government supervisors feel the quality of replacements is about the same as those being replaced, 44% believe the quality of replacements is lower.

Problem of Military Retention

While ROTC and Service Academy sources provide college trained military personnel in large numbers, an insufficient percentage remain in service beyond the initial tour. This leaves a gap between the new officer input and the senior level groups. For most of these junior scientific officers, the incentives of a career with civilian industry outweigh those of a military career. Many ROTC officers are committed to join industry immediately upon completion of their military tour. For example, a July 1956 survey of Air Force ROTC graduates on active duty, by school, indicated that none of the graduates of the following schools intended to stay in the Air Force: California Institute of Technology, Stevens Institute of Technology, and Lowell Technical Institute.

While it is an oversimplification to say that "the best get out of the service," there appears to be a direct relation between graduate education and attrition.

Effects of Military Service Obligation

It has been felt by some that military service is a drain on the supply of newly trained scientists and engineers. About 8,000 of a recent year's 27,000 engineering graduates were in ROTC programs and committed to active duty after graduation.

Dr. A. W. Davison, Chairman of the Engineering Manpower Commission of the Engineers Joint Council, has said[4] that "in most cases no attempt is made by the Armed Services to assign these young officers to duties for which their engineering education specifically prepared them. They are not only withheld from industry and education for two years but also are not utilized in defense programs requiring more engineers and research scientists."

Mr. Gus C. Lee, Staff Director, Manpower Utilization, of the Office of the Secretary of Defense, indirectly answered this and similar criticisms in a statement before the House Armed Services Committee.[5]

Lee stated that Defense fully supports the student deferment policies under Selective Service. As of March 1958, 145,000 students were deferred to complete their education. In addition, 34,000 persons with critical civilian occupations were deferred because they were employed in defense-supporting industries and research.

The 260,000 ROTC students enrolled in our colleges and universities are also deferred. These deferments serve to keep open our educational pipeline. Defense policy allows the ROTC graduate qualified in a critical occupation to delay his active duty long enough to obtain a master's degree in his specialty. Many potential scientists and engineers enrolled in ROTC programs may elect six months active duty for training rather than two years. In the event it becomes necessary to meet active duty requirements by ordering ROTC officers to active duty for two years involuntarily, an officer trained in the sciences or engineering may be so ordered only if the Service has a requirement for his critical civilian skill.

Defense has acted to remove from the Ready Reserve those who possess critical civilian skills which are not positively required by the

Services. These reservists are subject to recall only upon determination by Selective Service that the individual's skills are more needed in the military than in the civilian economy. Selective Service records show that approximately 90,000 reservists have been screened in to the Standby Reserve because they possessed critical civilian occupations. An estimated 40% of these were scientists and engineers.

In 1956 Defense initiated its critical skills program, under which qualified young scientists or engineers working in research or defense industry may discharge their military obligation by volunteering for three months of active duty training, with the remaining time spent in reserve status. About 5,100, of whom 3,400 are scientists and engineers, have been found qualified, given their basic training, and promptly returned to civilian economy after three months in order to minimize their loss to defense-supporting industries.

Utilization

Maximum utilization involves making the best possible use of the scientists available to Defense. Opportunities for professional growth and the working environment are particularly important to scientists and engineers, and they look at it in terms of the following:
1. The management atmosphere in which they work.
2. Prestige through recognition—by their organization and the public.
3. Standing in their profession.
4. Growth or professional development through formal education and broad work experiences.
5. Excellence of physical facilities.

Industry has recognized the importance of each of these items in terms of keeping up with technological changes, and of the impact on both recruitment and retention.

Among the positive aspects of Federal employment, for example, is that Government laboratories have among the best facilities and equipment. The President's Committee on Scientists and Engineers for Federal Government Programs felt that through better personnel management of our scientists and engineers the Government can improve recruitment, utilization and retention.

A 1957 survey of attitudes of scientists and engineers[6] indicated in part that: (a) the Government apparently utilizes the skills of scientists and engineers as well as but no better than industry, (b) Government scientists and engineers by and large personally look favorably upon the Government as an employer. However, only half of the Government respondents, contrasted with four-fifths of the industrial scientists, believe their working with their Government organization is looked upon favorably by professional colleagues and friends, (c) Nine out of ten Federal scientists and engineers are satisfied with such aspects of their employment as geographical location, working hours, safety, health and comfort.

Results of attitude survey and turnover studies in the Air Force indicate that:

1. Civilian scientists and engineers in Air Force and industry feel the same about how their employers are utilizing their skills. Forty-eight per cent of the Air Force civilian scientists and engineers and 54% of industrial scientists and engineers feel that their jobs require their full skill.

2. Sixty-four per cent of Air Force scientists would like to spend more time on technical duties as opposed to administrative and other nontechnical duties, although they have said that only about 10% of their work can be done by technicians or administrative personnel.

3. Nine-one per cent of the college graduates in the Air Force in engineering and scientific positions are in jobs related to their academic training.

4. Only 1% of the scientists who resigned in 1956 indicated they left because of their skills not being fully utilized.

Actions to Improve Utilization

Defense shares the difficulties of all employers who must draw from skills which are in short supply. Defense's shortage of scientific manpower is substantial, but by no means critical in terms of bulk requirements. Many civilian scientists stay with Defense despite pay disparities and offers from the outside. Defense is generally regarded as a good place to work since the most effective source of recruitment is the existing workforce. Nearly one-third of the scientists and engineers come to Defense through personal contact with an individual already working for Defense. These factors afford no basis for complacency. Many of the program efforts are in the stage of infancy and the West has probably only just entered the contest for world scientific supremacy.

Defense has taken a variety of actions to improve the utilization of the scientists and engineers, both military and civilian, which it already possesses.

Each service has now created research and development career fields for their military officers qualified in the sciences and engineering. Moreover, new regulations permit more continued assignment of qualified officers in these career fields.

The Military Departments, under Department of Defense guidance, are stepping up their programs for improving the management and utilization of their civilian scientists and engineers.

The Military Departments conducted extensive surveys of the attitudes of approximately 16,000 of their scientists and engineers during fiscal year 1957. These surveys revealed several areas where remedial action by the laboratory or Service would improve utilization, raise morale, increase retention and improve productivity of scientists and engineers.

Action has been taken to improve the physical facilities in which scientists and engineers perform research and development duties. It is believed that a better working climate will permit better utilization of their talents, encourage creative thinking, and maximize their research and development production. The Secretary of Defense has directed that projects to improve the physical working conditions for technical personnel be given appropriate priority in construction and installation.

The Military Departments have taken action to relieve their scientists and engineers of much of the administrative details and nonprofessional work. Such work is often assigned to an Administrative Assistant, or "Man Friday," who works with a team of scientists and engineers, doing a portion of the detailed work which would otherwise consume much of their technical time.

Conclusions

1. One of the most urgent needs in the Defense organization is for broadly educated men who understand the scientific, strategic, and operational aspects of modern weapons systems.

2. The limited number of professional officers who provide the supervision of research and development for the Armed Forces should be carefully selected from those who receive advanced scientific education. They must possess the imagination to see the combat potential of scientific discoveries, comprehend the problems of scientists and the operating combat units, and understand the strategy of our defense plans.

3. Since the major requirements for professional scientific and engineering personnel in Defense are met by civilians, steps must be taken to insure the recruitment, training, and proper compensation of an adequate civilian force. Civilian recruiting and retention problems can be largely attributed to the fact that over the years the gap between industrial salaries and Government compensation for scientists and engineers has been widening.

4. Defense should continue to augment, wherever politically feasible, on a worldwide basis, its Defense Scientists Immigration Program.

5. The marked success of Defense's use of contractual organizations and research centers, with their advantages of freedom from Governmental procedural problems and pay limitations, recommends this method as a means of "getting the job done."

6. Defense should continue to improve the utilization and to expand the capabilities of the scientists and engineers which it already possesses.

7. The expanding space age will continue to make military technology increasingly more complex. This will require Defense to increase its qualitative and quantitative demands for the services of scientists and engineers. The future defense of the United States,

in the final analysis, will depend to a large degree on the adequacy of the United States educational system in meeting this challenge.

Selected Bibliography

U. S. National Science Foundation, *Scientific Manpower in the Federal Government 1954,* NSF 57-32 (Washington, D. C.: May 1957). A comprehensive survey of federally employed scientists and engineers engaged in research and development and certain related scientific activities.

U. S. National Science Foundation, *Government-University Relationships in Federally Sponsored Scientific Research and Development,* NSF 58-10. U.S. GPO (Washington, D. C.: April 1958). The evolution of Federally sponsored Research and Development at universities and the nature, trends and magnitude of the financial support. Problem areas in government-university relationships and recommendations are included.

Report to the President of Committee on Scientists and Engineers for Federal Government Programs (29 April 1957). A summary of the findings and recommendations of the President's Committee to survey and develop methods to improve personnel management of scientific and technical personnel in the Federal service.

Committee on Engineers and Scientists for Federal Government Programs, *Survey of Attitudes of Scientists and Engineers in Government and Industry.* U. S. GPO (Washington, D. C.: April 1957). The findings of a survey based on questionnaires of what scientists and engineers in government and in industry think about their work and careers.

Herbert S. Parmes, *Effective Utilization of Engineering Manpower—A Survey of the literature.* Prepared for the President's Committee on Scientists and engineers (Washington, D. C. Undated 1957). A review and synthesis of recent literature on methods of improving the use of civilian engineering personnel. An analysis of the suggestions advanced and an appraisal of their effectiveness.

Unpublished mimeographed summary entitled *Officer Education in Scientific, Engineering, and Foreign Language Fields.* Prepared by Assistant Secretary of Defense (Manpower, Personnel and Reserve) (dated 1 April 1958). A compilation of data with condensed summaries on the requirements, sources and educational programs of the armed services in these fields.

Unpublished statements to Subcommittee 6 of the House Armed Services investigating into Phase II of Defense-Posture Hearings, March, 1958:

    a. Robert E. Cronin, Rear Admiral, USN, Chief of Industrial Relations, Department of the Navy. The Navy's requirements for civilian engineering and scientific personnel, the problems in meeting those requirements and programs to improve training and retention.

    b. Gus C. Lee, Director, Manpower Utilization, Office of Secretary of Defense, (MP&R). Defense requirements for scientists and engineers, actions taken to lessen the impact of military obligation on the supply of scientists and engineers; and actions to improve utilization.

c. C. W. G. Rich, Brigadier General, USA, Office of Deputy Chief of Staff Personnel, Department of the Army. The Army programs for the procurement, utilization, and development of military personnel in scientific and engineering fields.

d. John A. Watts, Director of Civilian Personnel, Department of the Air Force. Requirement, recruitment, utilization, and retention of civilian scientists and engineers for the Air Force and plans for improved development and utilization.

e. Bureau of Naval Personnel report of 17 March. Shortage of Naval technological, scientific and engineering personnel with particular reference to new development areas such as guided missiles, sources, training programs, retention experience, evaluation.

Theodore H. White, *Where Do We Fit In? The Scientists Ask.* New York *Times Magazine* (May 18, 1958). The problems of scientists in government who are lost in a bureaucratic labyrinth. The need for organization in which their talents may be used more efficiently by the nation.

COLONEL CHARLES WARREN HOSTLER is a regular Air Force officer who has had an outstanding career as a military planner and observer in responsible assignments with the U. S. Government in Turkey, the Arab States, the Balkans, and Europe. Following his recent duty as a Strategic Planner with the Joint Chiefs of Staff in the Pentagon, Colonel Hostler has lately been assigned as U. S. Air Attache in the American Embassy at Beirut, Lebanon. He is also an author, linguist, and scholar with four university degrees in International Relations and Political Science, including a Ph.D. from Georgetown University. Doctor Hostler is a Professorial Lecturer in International Relations at the American University, Washington, D. C. He is the author of the recent book *Turkism and the Soviets* (The Turks of the World and their Political Objectives), (New York: Frederick Praeger, 1957) and numerous articles in periodicals such as *Middle East Journal* and *Middle Eastern Affairs*.

COLONEL DONALD J. DECKER, U. S. Marine Corps, is currently a member of the faculty of the National War College, Washington, D. C. He was born in Baldwinsville, New York, on September 3, 1908; graduated from Cornell University in 1931; and has since served in many parts of the world with the Marine Corps. In his assignments in recent years, he has worked on top level problems of maintaining Western defense. After study at the Naval War College in 1951, his duties in the Navy's Strategic Plans Division concerned NATO problems. From 1953 to 1955, on the joint staff of the U. S. Naval Commander in Europe in London, he was associated with the problems of European and Middle East defense. Early in 1956, he was assigned to Baghdad, Iraq as observer for the Joint Chiefs of Staff at the initial military planning of the Baghdad Pact. During 1956 and 1957, he was assigned to the Joint Chiefs of Staff in the Pentagon. He is continuing his academic studies with graduate work at The American University in Washington.

NOTES FOR SCIENTISTS AND ENGINEERS FOR THE ARMED FORCES

1. Mimeographed summary entitled "Officer Education in Scientific, Engineering and Foreign Language Fields" prepared by Asst. Secretary of Defense (Manpower, Personnel and Reserve), dated 1 April 1958, p. 2.

2. U. S. National Science Foundation, *Government-University Relationships in Federally Sponsored Scientific Research and Development* (U. S. GPO, Washington, D. C., April, 1958), 4.

3. *Report of Committee on Scientists and Engineers for Federal Government Programs,* 29 April 1957, 2.

4. Quoted from *The Shortage of Scientists and Engineers,* a booklet prepared by the McGraw-Hill Publishing Co., Inc. (undated).

5. Statement to Subcommittee No. 6, House Armed Services Committee, by Gus C. Lee, OASD (MP&P) on March 31, 1958.

6. *Survey of Attitudes of Scientists and Engineers in Government and Industry,* April 1957 (U. S. GPO, Washington, D. C.: 1957).

SCIENCE IN FICTION AND BELLES-LETTRES

SHERWOOD CUMMINGS
State University of South Dakota

Literature records, interprets, and evaluates human experience. In Zola's words, "Literature constantly encroaches on life; if it has been lived or thought, it will one day become literature." And since Western life from the time of the Renaissance has been profoundly and pervasively changed by science, literature can only recognize the change. In subject matter, attitude, and style, writers of the last three centuries have made obeisance to science. Sometimes, especially in the last century and a half, writers have deplored the dominance of science, have tried to forget it, have pointed in anti-science directions, mostly backwards or inwards; but their revolt does not change the fact that science rules. Science has caused the writers and readers of human experience many agonizing re-appraisals. It is only fair that the scientist should use literature to apprise himself of how a great many people have felt about living in a world that is continually being remade by science.

The New Astronomy and Seventeenth-Century Literature

In 1609 Galileo began constructing telescopes, each better than the last, with which he surveyed the heavens. What he saw he reported in *Sidereus Nuncius,* published in 1610. The moon, he said, is not a flat, smooth disk; its surface is rough and uneven with towering mountains and deep valleys. Around Jupiter he discovered four moons, and his telescope resolved the haze of the Milky Way into myriad separate stars, stars not recognized by astrology. In 1632 his *A Dialogue on the Two Principal Systems of the World* implied endorsement of the proscribed Copernican cosmology. The Inquisition forced him to renounce his belief in the heliocentric theory, but from that time the idea of a geocentric universe was moribund.

In his own country of Italy, poets like Giambattista Marino and Piero de' Bardi responded quickly to Galileo's earliest discoveries, hailing him as a latter-day Columbus voyaging fearlessly into sidereal wilderness. English writers were similarly stimulated, and as they ruminated the implications of the New Astronomy they expressed many new ideas, some playful, some profound. Romantic references to the moon (except in present-day popular songs, the sanctuary of the unscientific mind) as a luminous disc or as the residence of Diana became quaint. Poets like John Dryden conceived of the moon as a new "world" and speculated whether it and the planets might be inhabited. The effect of the New Astronomy on many poets was disturbing. The neat universe of Ptolemy had given way to the vast, perhaps limitless, universe of Galileo. There were not several hundred stars but thousands upon thousands. The larger the universe, the smaller man's part in it: the idea was painful. John Donne expressed the anguish in "First Anniversarie," lines 205-214:

> And new Philosophy call all in doubt,
> The Element of fire is quite put out;
> The Sun is lost, and th'earth, and no mans wit
> Can well direct him where to looke for it.
>
> *   *   *
>
> 'Tis all in peeces, all cohaerence gone;
> All just supply, and all Relation.

Newton and Eighteenth Century Literature

If Galileo's New Astronomy left Donne's world concept "all in peeces," Newton's exposition of the immutable laws of nature organized and harmonized that concept for the leading thinkers and writers of the eighteenth century—among many others, Voltaire in France, Alexander Pope in England, Benjamin Franklin and Thomas Paine in America. No period has been intellectually more comfortable to man than the eighteenth century to the deists. Newton demonstrated that the planets circled the sun with exquisite precision, perfectly regulated by the laws of momentum and gravity. Not only was the universe orderly, it was obviously conceived and created by a consummate Intellect who since its creation had stood aside and let it run like a well regulated clock. Man had only to discern the universal principles through observation of nature and reason and then to accommodate himself to those principles.

Order, precision, balance, harmony, reason were the key concepts of the century, the Age of Reason. Working under their influence English writers like Dryden (who died at the beginning of the century but who lived in Newton's time), Jonathan Swift, Joseph Addison, and Richard Steele developed a prose style unmatched for its clarity, elegance, and precision of expression. With them modern English prose was born; to appreciate their accomplishment one has only to read the baroque and tangled prose of such Renaissance writers as Thomas Lodge and Sir Philip Sidney. Following the same principles of precision, Pope developed the balanced, measured heroic couplet, the dominant English poetic form of the century.

Not only was the style of eighteenth century writing affected by

Newtonian science, but its philosophy as well. Pope spoke for a large segment of his age in his "An Essay of Man" where he perceived in the orderly universe a natural hierarchy, a great chain of being, descending from God down through the angels, man, and the lesser creatures. He differed from the radicals in seeing in human society a microcosm of the universe. Social classes and institutions were a part of the great chain of being, not to be tampered with:

> From Nature's chain whatever link you strike,
> Tenth or ten thousandth, breaks the chain alike.[1]

One implication of his poem is that a man had jolly well be satisfied with his lot in life, with his little link in the great chain.

But the Newtonian concept of natural order was a two-edged sword and it was used as effectively by the American revolutionaries as by the conservative Alexander Pope. The revolutionaries explained themselves this way: The universe is orderly and harmonious; society is not. Why is there injustice and disharmony in society? Because tyrants, with their unreasonable laws, and society, with its artificial institutions, have disrupted the natural order of things in the sphere of man. It behooves us, then, to defy the tyrant's law, to allow society to take its natural course, and so achieve justice and peace among men. The first sentence of Thomas Jefferson's *Declaration of Independence*, it will be remembered, justifies the colonies' break with Britain by the authority of "the laws of nature and of nature's God." The political tie with the mother country had become an outrage to the order of the universe. Jefferson's idea of the less government the better, that men (who were naturally good) should be untrammeled so that they could discover their orbit in the harmonious machinery of nature, achieved wide currency in literature. It finds early development in the works of Jean Jacques Rousseau in France and a sort of apotheosis in William Godwin's *An Enquiry Concerning Political Justice* and the poetry of Percy Bysshe Shelley in England.

The century's appreciation of Newtonianism suggested the proper choice of subject for literature as well as its style and ideas. In one word, that subject was nature, not just bucolic nature, but all observable phenomena. Actually, the eighteenth century writers did not poke into so many corners of the world nor record what they observed with such particularity as did the realists of the next century, but they did respect the scientist's injunction to learn truth by studying nature. In *Rasselas*, Samuel Johnson, a figure as monumental of his age as Voltaire, tells how his poet, Imlac, learned the proper subject matter for poetry:

> I soon found that no man was ever great by imitation. My desire of excellence impelled me to transfer my attention to nature and to life. Nature was to be my subject, and men to be my auditors: I could never

describe what I had not seen; I could not hope to move those with delight, whose interests and opinions I did not understand.

Being now resolved to be a poet, I saw everything with a new purpose; my sphere of attention was suddenly magnified; no kind of knowledge was to be overlooked; I ranged mountains and deserts for images and resemblances, and pictured upon my mind every tree of the forest and flower of the valley. I observed with equal care and crags of the rock and the pinnacles of the palace. Sometimes I wandered along the mazes of the rivulet, and sometimes watched the changes of the summer clouds. To a poet nothing can be useless.[2]

In principle Johnson's statement is revolutionary. It is a declaration of independence from classical authority and example and a translation of the scientific method to poetry. In actuality there had always been examples of natural description in writing, but Johnson's formulation concerns a pervasive philosophy derived from the scientific spirit of the times.

The Rise of the Detective Story

Not only had science put its mark on the manner and matter of writing in the eighteenth century, it was also responsible, in a perverse way, for the emergence of a new type of literature—the mystery story. The first mystery novel, *The Castle of Otranto,* was written in 1764 by Horace Walpole in England, and it is rightly called the "ancestor of the modern detective novel,"[3] though there is little detection in it. In the novel for the first time mystery was cultivated for mystery's sake by an artistic manipulation of plot, a manipulation that piled terror on terror with no let-up until the denouement. After Walpole set the mode, there was a succession of mysteries, or Gothic novels, as they were called, throughout the rest of the century and, indeed, up to the present.

One might well ask how the mystery story, and especially *The Castle of Otranto,* which is full of the flummery of secret passages, ominously fluttering drapes, corpses appearing and disappearing, and the like, could have anything to do with science. The answer is that the mystery story filled a cultural vacuum that science had created. By the middle of the eighteenth century superstition was out of fashion, if not completely dead. Voltaire's cry, "Écrason l'infâme," roughly translated as "Let us crush religious superstition," had been heeded. The Enlightenment had burned away the mists of superstition. John Wesley remarked in 1768, "The English in general, and indeed most men of learning in Europe, have given up all account of witches and apparitions, as mere old wives tales."[4] Uncomfortable as witches and apparitions are to live with, when they depart their absence is felt. Man had been living in fear of the inscrutable for thousands of centuries. The ancients did not need to manufacture mystery; it was the air they breathed. From the time of primitive man's peering into the darkness from the small, safe circle of his campfire to the time of the Middle Ages and its supernaturalism, man was surrounded by things which he could not explain and which held him in awe. After the searchlight of science illuminated the dark corners of the world, life must have seemed monotonously matter-of-fact to many people. They hankered to be scared again. Thus the contrived mystery of the Gothic novel became popular.

But as the public continued to be educated in science, it became more critical about its literature of suspense. Readers still wanted to be scared, but they could not believe in ghosts. The writers of the Gothic novel beginning with Mrs. Radcliffe in 1794 were generally forced to explain their terror-impelling circumstances from natural causes. Charles Brockden Brown, the American novelist, prefaced his Gothic novel, *Wieland,* in 1798 with this statement:

> The incidents related are extraordinary and rare. Some of them, perhaps, approach nearly to the nature of miracles as can be done by that which is not truly miraculous. It is hoped that intelligent readers will not disapprove of the manner in which appearances are solved, but that the solution will be found to correspond with the known principles of human nature.[5]

Under the exigency of science the Gothic novel was evolving into the present most popular form of mystery—the detective story. Two of the three necessary ingredients of the detective story—mystery and scientific solution—had been supplied. The third ingredient—crime, or more exactly, a society that deplored crime and efficiently coped with the criminal—was ready to be added. The handling of criminals had throughout the history of the world (except, perhaps, in Greece and Rome) been a haphazard and often cruel and unjust process. Moreover, public sympathy had often been on the side of the criminal, and crime had, indeed, sometimes been regarded as a profession in its own right. But in mid-eighteenth century, as a facet of the Enlightenment, Henry Fielding (the novelist) and his brother John organized in London a police force called the Bow Street Runners. In his *Plan for Preventing Robberies* Fielding describes his Runners as handpicked for integrity, daring, and intelligence. They inaugurated the tradition of a police force which is not only efficient but respected. In France crime was not effectively coped with until 1817 when Vidocq, a former criminal, became chief of the Paris police. Thus, when Edgar Allan Poe wrote his first "tale of ratiocination" or detective story, "The Murders in the Rue Morgue," in 1841, the ingredients had been put in his hands and the public was ready.

The detective story resolved the paradox of the co-existence of mystery and science. In the detective story the modern reader may satisfy his craving for mystery and not only not do violence to his enlightenment but further confirm it. The success of the genre is evidenced in the fact that in modern times one out of every four works of fiction published in the United States is a detective novel.[6]

Nineteenth Century Science and Realism

The scientists of the eighteenth century were, relatively speaking, concerned with gross phenomena. Newton's discovery and the preoccupation of the age was with universal law. Even Samuel Johnson, after exhorting the poet to study nature, cautioned him that it is the "business of a poet . . . to examine, not the individual, but the species; to remark general properties and large appearances . . ."[7]

It was left for nineteenth century scientists to discover the workings of universal law in the minutiae of nature and to study microscopic and sub-microscopic phenomena. The existence of the atom in its modern concept was predicated by John Dalton in 1804. Since Dalton's time physicists have been splitting the atom, both in theory and in terrifying reality, into smaller and smaller components, until now there are some twenty atomic particles identified and named. The trend in science has been from the large view to the fine, from universals to particulars. The trend in literature is the same.

The early nineteenth century poets like Wordsworth and Keats described nature with unclassical particularity, but it was not until the advent of realistic fiction just past mid-century, with Gustave Flaubert in France and William Dean Howells in America, that the relation of science and literature again becomes marked. Realism in literature is the recording of life with photographic fidelity. Settings are described in detail; dialogue is recorded at such length and with such exactness that it sounds like a transcription from a tape recording; the motivations, activities, gestures, and appearances of fictional characters are described with great particularity. The following typical excerpt from Henry James's *The American* illustrates the realistic technique:

On the threshold stood an old woman whom he remembered to have met several times in entering and leaving the house. She was tall and straight and dressed in black, and she wore a cap which, if Newman had been initiated into such mysteries, would have been a sufficient assurance that she was not a Frenchwoman; a cap of pure British composition. She had a pale, decent, depressed-looking face, and a clear, dull English eye. She looked at Newman a moment, both intently and timidly, and then she dropped a short, straight English curtsey.[8]

Photographic writing like this can hardly be found before the nineteenth century. Sustained realism, the steady focus on the minutiae of life, is a modern phenomenon. Its causes are complex, but all are more or less closely related with nineteenth century science. One obvious aid to realism is the invention of the camera itself; before Daguerre writers could witness life only in flux or as it was congealed on canvas in the shape of another artist's conception. The photograph stopped life and presented it for perusal exactly as it was. Another cause was the flood of the products of nineteenth century technology. Invention and manufacture changed the face of the Western world; people's attention was diverted from ideas to things. The railroad, telegraph, telephone, phonograph, electric light; the cheap manufacture of clothing, furniture, and other articles of creature comfort and conspicuous consumption made material accomplishment seem more exciting than philosophy. Consequently, what a large segment of the public expected in its fiction was an accurate reflection of the engrossing material world, not abstractions and romances. But behind the new shape of the material world and behind literary realism was the scientific method: natural phenomena must be studied empirically and with close attention to detail. Essentially, the realistic technique is an application of scientific method to literature.

Darwin and Naturalism in Literature

From 1632, when Galileo was sentenced by the Church for the Copernican implications in his astronomy, to the present, there have been elements of conflict between science and traditional Christianity. That conflict is voluminously recorded in literature. Voltaire's agnosticism, Goethe's unorthodoxy, and Paine's deism were typical expressions of the Enlightenment. In the nineteenth century secularism was reinforced by Darwin's exposition of evolution through natural selection (1859). The popular interpretation of Darwinism was that humans descended from monkeys, an idea that seemed to contradict the Biblical statement that man was made in God's image. From that contradiction, and from the evidence of nineteenth century geologists and astronomers that showed the universe to be more ancient and vast than religious fundamentalists liked to think, arose the Great Debate between science and theology. So many Victorian writers were disturbed by the revelations of science that their period has been called the Age of Doubt.

There was, however, a group of novelists who accepted the implications of Darwinism and who used them to develop a new literary genre called naturalism. Emile Zola in France was the first novelist to be called a naturalist, though he and his colleagues owed much to the French critic and historian, Hippolyte Taine, who in the 1870's applied the evolutionist's theories of the formative influences of heredity and environment to his analysis of national cultures. In America Stephen Crane, Frank Norris, Jack London, and Theodore Dreiser wrote naturalistic novels. The vogue of naturalism lasted roughly from 1880 to 1930, though naturalism, like realism, is still evident among the varied forms and approaches of contemporary writing.

The basic idea that the literary naturalists took from Darwin is that man is a creature of nature. They then added the proposition that man should be studied as a scientist studies any natural phenomenon—dispassionately and objectively. As literary scientists they felt obliged to refrain from making moral judgments on their characters' actions or to assume the existence of undemonstrable moral and spiritual forces.

Since man in the naturalists' view is a creature of nature, it follows that he, like the other creatures, is the product of heredity and environment. He is born with certain potentialities which are developed or inhibited by the environment he moves through. As a consequence, delineation of environment is an important part of the naturalistic novel. In Zola's *Germinal*, for example, the resentful attitudes and impoverished existences of the miners can be fully appreciated only in reference to the carefully detailed setting of the repressive mining community. Moreover, the emphasis on the influence of environment has inclined much naturalism toward determinism. That is, if man is simply a product of the complementing forces of heredity and environment, then he has no will of his own. He reacts to stimuli in a mechanical way. For example, Henry Fleming, the soldier boy in Crane's *The Red Badge of Courage*, has no idea if he will be brave or cowardly in his first battle:

He finally concluded that the only way to prove himself was to go into the blaze, and then figuratively to watch his legs to discover their merits and faults. He reluctantly admitted that he could not sit still with a mental slate and pencil and derive an answer. To gain it, he must have blaze, blood, danger, even as a chemist requires this, that, and the other.[9]

As a naturalist Crane declines to endow his "hero" with courage; the boy must be tested in his environment as a chemist would test an unknown compound.

Incidentally, Henry turns coward and runs away from battle. Refusing to assume undemonstrated virtues in their characters, the naturalists are inclined to credit man with few of his traditional dignities. The "natural" actions of man are identified with the instincts and urges of other animals. Harried, man fights back in atavistic fury, as in the novels of Jack London. Amorous, he makes love with the handiest female, as in rather too many of the paperback "breast sellers." This uncomplimentary portrait of man has given naturalism a bad name. As an unidentified critic has remarked, "The realist calls a spade a spade; the naturalist calls it a dirty shovel." To give the naturalists their due, however, it must be said that they have enterprisingly explored the possibilities implicit in Darwin's revelation of man's animal heritage.

Science Fiction

Perhaps the earliest science-fiction story is the legend of Dedalus's constructing wings to fly out of the Labyrinth, but science fiction as a type is, like the detective story, barely a century old, and it has been widely popular only in the last thirty years. Several nineteenth and early twentieth-century writers laid the foundation for modern science fiction: Mary Shelley, Poe, Jules Verne, H. G. Wells, and the pseudonymous author or authors of the Tom Swift series. It was not, however, until 1926, when Hugo Gernsback published the first issue of *Amazing Stories*, that the science-fiction magazine appeared. Thereafter such magazines enjoyed a rapid rise until a few years ago. In 1931 there were nine science-fiction magazines published in America; in 1953 there were thirty-one; but by 1956 the number had been reduced to fifteen.[10]

Much science fiction is neither good science nor good fiction. Several magazines and two comic strips (*Buck Rogers* and *Flash Gordon*) specialize in "space operas"—melodramatic heroics by men in space suits. The theme of space opera, if such fiction can be dignified with the term, is Earth chauvinism; the good guys from Earth are threatened by and eventually overcome villainies that erupt in every sector of space. The theme is based on the fear of the unknown which dictates the premise that whatever is outside one's experience is probably dangerous and should be destroyed. Another brand of science fiction, not totally distinct from space opera in plot or theme, is the "Bug-Eyed Monster" story, a term used scornfully by sophisticated science-fiction fans for the horror stories about weird creatures constructed by "mad scientists" or confronted on other planets. Unfortunately for the reputation of science fiction, the space opera and the Bug-Eyed Monster have been used far more in television and the motion pictures than more significant science-fiction material.

The best science fiction is written and read to explore the apparently limitless possibilities of science and their effect on human beings. As fast as physicists, mathematicians, astronomers, psychologists, and social scientists have opened new territory in the last century, men of imagination have leaped to the frontiers to extrapolate future trends, and in so doing have anticipated many actual developments. Jules Verne's *Nautilus* traveled twenty thousand leagues under the sea before Simon Lake developed the submarine; he "sent" a nine-foot caliber cannonball to the moon from the sands of Florida nearly a hundred years before earth satellites were launched from Cape Canaveral. In 1904 Mark Twain wrote "A Visit from Satan" in which he envisioned the power of radium as capable of destroying the earth in a gigantic explosion. The theme of atomic explosion intrigued other writers before 1945; indeed, during the second World War a story of the still-undeveloped atomic bomb in *Astounding Science Fiction* prompted an investigation by the FBI to see if there had been an information leak from atomic laboratories. The existence of planets outside the solar system had been a cliche of science fiction long before Harlow Shapley and other astronomers used statistical probabilities to assert that there must be numerous such planets. The accuracy of such fictional prognostications, the solid scientific background of many writers (some of whom are practising scientists), and science's perennial accomplishing of the incredible combine to make serious science fiction the bridge between the present and the future. If inter-galactic space travel, visitations from other planets, robot workers, backwards or sideways time travel, man-controlled evolution, and physical immortality become realities, readers of science fiction will have been prepared.

Science Versus Humanity?

The picture of the future in science fiction is exciting to the imagination but not always comforting. Many science-fiction writers and other literary people do not regard the present and future of humanity in a science-dominated world with cheerfulness. Speculative science fiction seldom predicts utopias. On the contrary, much of it offers only a variety of human catastrophes. One may take his choice, for example, among Aldous Huxley's *Brave New World* where people have lost their souls in exchange for pleasure and security, George Orwell's *1984* with its ultimate in fascism, Karel Capek's *R. U. R.* and its revolt of the robots, or the many stories which show remnants of humanity living as cavemen after an atomic war. If the society of the next centuries is shown as orderly, the order is usually brought about at the sacrifice of individualism. The behavioral sciences are utilized to regulate men's motives and channel their activities. The last privacy, the mind, is invaded by telepathic devices or, as in Harry Kuttner's "Private Eye," any past activity and utterance of a suspect can be recreated by a machine and projected for a court of law to witness.

So many writers have found the present out of joint and the future ominous that the present literary period has been called the Age of Anxiety. H. G. Wells at the end of a career distinguished for its enterprising vision wrote, "The human story has come to its end . . . and Homo Sapiens in his present form is played out. . . ."[11] T. S. Eliot, repelled by the shallow materialism of the twentieth century, predicted the world's ending with a whimper. In "Tyranny over the Mind" in the May 31, 1958, issue of *Newsday*, Aldous Huxley claims that social engineering has brought us to the edge of his Brave New World. Anxious writers are worried about the survival of humanity, but they do not blame science alone for what seems to them man's precarious situation; science, like fire, can bless or burn. They recognize that if civilization fails it will do so because man's moral and social progress has not kept pace with his ingenuity. But writers, like all responsible human beings, know that the world is for people and that people are now threatened by their own inventions. They have done what they could to reflect and evaluate the impact of science on society. Any one interested may read.

Selected Bibliography

For a more complete literature and science bibliography see: *Symposium*, VI (May, 1952), 241-245; VIII (Summer, 1954), 208-213; X (Spring, 1956), 182-187.

Pearl S. Buck, "The Artist in a World of Science," *Saturday Review,* XLI (September 20, 1958), 15-16, 42-44. Mrs. Buck's essay is notable for its challenging optimism. The artist's role, she writes, is "to use the findings of scientists and to illuminate them . . . in order that human beings will no longer be afraid."

Douglas Bush, *Science and English Poetry* (New York: Oxford, 1950). In his broadly documented study of the effect of science on English poetry from the Elizabethan age to the present, this leading scholar of literature sees that the poet has been increasingly alienated from and hampered by the mechanized world and positivistic philosophy of science.

Everett Carter, *Howells and the Age of Realism* (Philadelphia: Lippincott, 1954). Carter analyzes the origin and growth of American literary realism in the last part of the nineteenth century, with special reference to William Dean Howells. His study includes all important aspects of realism, and among other things he demonstrates that Howells, Garland, and Eggleston formulated their literary theories partly in response to the new emphases of Hippolyte Taine and Charles Darwin.

Harry H. Clark, "The Influence of Science on American Literary Criticism, 1860-1910, Including the Vogue of Taine," *Wisconsin Academy of Sciences, Arts and Letters,* XLIV (1955), 109-164. Not a few, but virtually all American writers and critics of 1860-1910 responded in one way or another to the idea of evolution. Mr. Clark's thoroughly documented work presents the most inclusive view of science's effect on the American literature of the period.

Basil Davenport, *Inquiry into Science Fiction* (New York: Longmans, Green, 1955). Surveys science-fiction from Verne to the present. He classifies some of it as insignificant, but concludes that the best of it, written by men of enterprising imagination and solid scientific backgrounds, significantly explores the future "extrapolations" of science; the appended bibliography is helpful.

Frederick Hoffman, *Freudianism and the Literary Mind* (Baton Rouge: Louisiana State University Press, 1945). In following the tangled trails of Freudianism in literature, Hoffman avoids speculative pitfalls that catch amateur psychoanalysts. He pays particular attention to Lawrence, Kafka, Mann, and Sherwood Anderson.

William Irvine, *Apes, Angels, and Victorians* (New York: McGraw-Hill, 1955). Irvine's "story of Darwin, Huxley, and evolution" not only brings those two scientists vividly to life but details the intellectual and philosophical ferment resulting from the evolutionary theory, including the reaction of English writers.

S. Mandel and P. Fingesten, "The Myth of Science Fiction," *Saturday Review,* XXXVIII (August 27, 1955), 7, 8, 24-25, 28. Sees science fiction as an unhealthy attempt to escape the difficulties of the present by constructing mythical worlds of the future made orderly by science but unsatisfactory to humans.

Marjorie Nicolson, *Science and Imagination* (Ithaca, N. Y.: Great

Seal, 1956). Nicolson's concern is principally with the impact of science on the English writers and philosophers of the seventeenth and eighteenth centuries. Drawing from a broad and solid background of information in both science and the history of ideas, she pays particular attention to Donne, Milton, and Swift.

J. K. Robertson, "Science in Literature," *Queen's Quarterly*, LVIII (Spring, 1951), 36-55. Surveys references to science in English literature of the last three centuries with special attention paid to the trip-to-the-moon theme.

Karl Shapiro, "Poets and Psychologists," *Poetry*, LXXX (June, 1952), 166-184. A leading poet declares his antagonism not only to psychology but to science in general because of "its absolute authority for limiting the areas of truth."

Antonina Vallentin, *H. G. Wells, Prophet of Our Day* (New York: Day, 1950). The biography as a whole capably explores the complex and mercurial personality of Wells. The last two-thirds deals more specifically with Wells' early fascination with science fantasy and his ultimate despair that men would be defeated by their technology.

George H. Waltz, Jr., *Jules Verne, The Biography of an Imagination* (New York: Holt, 1934). The biography entertainingly directs the reader along Verne's career in science fiction to reveal an imagination that was always exciting if not always profound.

SHERWOOD CUMMINGS was born in Weehawken, New Jersey (1916), and raised in Rockford, Illinois. He attended the University of Illinois (B.S., 1938) and the University of Wisconsin (M.A., 1946; Ph.D., 1951) and has taught for the last ten years in the English Department of the University of South Dakota. For seven years, 1939-1946, he was a metallurgist in the laboratories of the Barber-Colman Company, Rockford, Illinois. His interests derive from his dual career. A student of literature and science, he has published articles and read papers on Mark Twain and science and the development of the detective story. He is a member of the Literature and Science Group of the Modern Language Association, past president of the American Studies Association of Minnesota and the Dakotas, and member of the Bibliography Committee of the American Literature Group.

NOTES FOR SCIENCE IN FICTION AND BELLES-LETTRES

1. "An Essay on Man," I, lines 245-246.
2. *Rasselas*, Chapter X.
3. M. H. Needleman and W. B. Otis, *Outline History of English Literature* (New York: Barnes and Noble), II, 404.
4. Quoted in Edith Birkhead, *The Tale of Terror* (London: Constable), 6.

5. C. B. Brown, *Wieland* (New York: Harcourt, Brace), 3.

6. Howard Haycraft, *Murder for Pleasure* (New York: Appleton-Century), VIII.

7. *Rasselas,* Chapter X.

8. Henry James, *The American* (New York: Rinehart), 168.

9. Stephen Crane, *The Red Badge of Courage* (New York: Modern Library), 21, 22.

10. Basil Davenport, *Inquiry into Science Fiction* (New York: Longmans, Green), 9, 10; Richard Lupoff, "What's left of the Science Fiction Market," *The Writer,* LXIX (May, 1956), 165.

11. Antonina Vallentin, *H. G. Wells, Prophet of our Day* (New York: John Day), 325.

Comparative Aspects

PROBLEMS OF SCIENCE TEACHING IN THE UNITED STATES*

James R. Killian, Jr.

Special Assistant to the President for Science and Technology

In speaking as an ardent advocate of better science teaching and greater emphasis on science in U. S. schools, I must start with some personal observations about education. Our overriding objective today must be to elevate standards of performance and enlarge the intellectual content of the secondary school program. There needs to be a greater interest in matters of the mind, a weeding-out of the trivial, peripheral, narrowly vocational subjects, and a more general acceptance by parents, teachers and students of the importance of intellectual qualities and high standards in all parts of the secondary school program. If we are to have better science education, we must have better over-all education and if we are to have better education, we must have a shift in values so that intellectual interests and performance are not played down and socially denigrated. We must cultivate in all of our education a distaste for the "take-it-easy" and anti-intellectual attitudes, and a positive taste for what is excellent in intellect and spirit.

These same observations can be made about college education. The emphasis on quality must run through the whole spectrum of education. We need to bring down into the undergraduate college program more of the spirit of independent and creative work that now marks our good graduate schools, and we need to bring down into the high school more of the depth and the sense of individual responsibility for intellectual effort that is to be found in college. In both secondary school and college, we must provide both opportunity and incentive for high talent; especially in the secondary school must there be an unremitting search and enlarged opportunity for talent and intellectual giftedness.

In the development of our public school system, we have concentrated in recent years on making it universally available and of the greatest help to the greatest number. The next phase—the next great mission of our educational system—should be to introduce more extensively into our

*Text of Remarks given on March 23, 1958, at the 50th Anniversary Celebration of St. Alban's School in Washington, D. C.

system of mass education the opportunities and means for differentiation in order to permit the fullest encouragement and development of our high talent.

In emphasizing the importance of intensifying the cultivation of talent and the raising of standards as objectives of top priority, I do not mean to suggest that our great secondary school system should cease to provide effectively for all of the varying needs of our young people. In discussing education, it is a frequent error to think of one's own son and, in the broader sense, of one's own kind of people when discussing school curricula. I am struck by the number of presumably intelligent college graduates who apparently think the sole purpose of the public schools is to prepare students for college. I am equally shocked by enthusiasts of another aspect of education who apparently forget that one extremely important function of schools is to prepare youngsters for college. Too many college professors think of the high school only in terms of its responsibility to prepare students to do well in the freshman subjects they teach.

Our schools must be designed to help all children, and the needs of children of different ability, different background and different aspirations must constantly be kept in mind. The diversity of educational needs in America is very great. The special needs of the community must, of course, be taken into consideration in planning the secondary school program. Regional differences, economic differences, differences in the desires of the people, all these things and many others make it difficult to generalize about the high school curriculum.

This necessity of thinking of the whole list of things schools should attempt to accomplish brings us hard up against the thorny question of priorities. It is easy to argue that almost any kind of instruction is at least potentially helpful to students—that is one reason why the curricula of schools and colleges, like a perfect gas, seem to have an infinite capacity for expansion. The list of school objectives tends to proliferate almost endlessly.

The question we must ask ourselves is not whether this course or that course of instruction has any good in it; what we must ask ourselves is what school objectives are the most important for a given community, a given time and for the nation. There are limits to what the schools can attempt, but most important, there are limits on the student's time. Schools can't attempt to offer students every useful kind of instruction in the world. Instead, school administrators must help each student to use whatever years he has for education most advantageously. Time is the most precious ingredient of all in education. The average student has only a few pennies of time to spend on education, and he can't buy everything in the store. It is up to his elders to help him spend his time as wisely as possible—to purchase not just a lot of little educational trinkets, but something that will sustain him all his life. Let us never forget the preciousness of the student's time. As long as we remain fully appreciative of that, I believe that we will find ourselves thinking in terms of educational priorities, rather than in terms of omnibus lists; of excellence rather than coverage.

With the expression of these general views on education, let me now turn more specifically to the problem of priority for science teaching and science education. During the last quarter century our schools have gone through a phase during which languages, mathematics, and science have been far too generally neglected or ostracized. I think it futile to try to assess blame for this; it has occurred in part because the attitudes and values of the American people resulted in a low value being placed on these subjects. But clearly the time has come for a redress of this imbalance. The needs of the nation today require that these subjects be restored to a priority at least as high as other principal subjects in the high school curriculum, and that they be taught, not superficially, but thoroughly and well—and imaginatively. It is not that we want to make scientists of all our young people—far from it. Rather, it is that science courses have come to be taught much more poorly in many schools than have the humanities and social sciences, and they need to be brought up to par. Up until now, we have done little—save in our best schools— where science is probably taught as well as anywhere in the world. We have been blocked by the baseless fear that if we strengthen our science education, we might run the risk of distorting the emphasis on other subjects. I hold that we have extraordinary opportunity and unique incentive now to strengthen science education, and that in doing so it can serve to strengthen other parts of the curriculum. Science can be the flagship in leading to a deepening and strengthening of the high school curriculum. It also may well do much good by serving as a sorting-out factor of excellence, since scientific courses are likely to give better mental discipline and a better test of student mettle than descriptive courses. In the longer view, it is well also to remember that it is usually easier to make a good businessman out of a scientist than it is to make a good scientist out of a businessman.

At the institution from which I am now on leave, we have been sponsoring an inter-institutional project for preparing a new approach to the teaching of secondary school physics. It early became clear to the teachers conducting this project, after examining extensively the present approach to the teaching of physics, that it has largely failed to keep pace with the progress of physics, and that far too little of the exciting and important new concepts had found their way down into the high school program. It also became clear that too much of the science instruction was either descriptive or technological and that it did not illuminate the most fundamental concepts and views of the universe, which makes physics so basic and powerful a subject—both culturally and in terms of science. I cite this as an example of the importance of re-thinking the content of our school subject matter at the secondary level, not only in the sciences but in other fields, and of achieving a deeper grasp of the great fundamentals which underlie our culture and our professions. In a subject which changes so rapidly as science, especially is this important. As the Physical Sciences Study Project continues with its tryouts of new materials it becomes increasingly clear that students of more than average ability readily respond to the excitement and invigoration and the penetration that comes from grappling with some of the profound and basic concepts of physics, even though they are intellectually quite demanding.

Another requirement for the improvement of science education is the correction of some of the strange notions about science.

There is a widespread view, for example, that science is "vocational," that it is "materialistic," and "anti-humanistic"—that it contributes only to the practical needs and the defense but not to the quality of our society.

It is my own deep conviction that the liberal arts cannot be liberal without including science, and that humanism is an indispensable ally of science in a sound scientific education. In the face of the practical responsibilities which rest in science and engineering for our security and our material welfare it is all too easy for people to conclude that science is inimical to the spiritual ends of life and for them to fail to understand that in reality it is one of man's most powerful and noble means for seeking truth and that its driving force is the thrust of man's curiosity to discover more of the beauty and order of the universe. Scientists have an obligation to make this true character of science better understood, not by an arrogant advocacy of science and technology as the only objective means to increase our understanding and well-being, but by the balanced and tolerant presentation of science as one of the powerful means by which man can increase his knowledge and understanding and still remain humble and ennobled before the wonder and the majesty of what he does not understand. As George R. Harrison has eloquently written: "There is no evil, no inhumanity, in the primary task of science, to forward man's love and desire for truth. An increased awareness of truth has often made men uncomfortable, but seldom has it made them less human. Science increases the areas of spiritual contact between man and nature, and between man and other men."

Let us not forget this complementary character of science, which deals with nature, and of the humanities which deal with man. To be most effective the scientist must study nature in a world of men and the humanist must study man in a physical environment dominated by science. Neither can achieve optimum effectiveness without working in harmony with the other and without the benefit of a harmonious and understanding reaction one with the other.

Let it also be noted that we have a far better chance of producing great scientists in the U. S. if we have an educated community that understands science and values it as well as the humanities and that views them both as essential parts of our common culture and the wholeness of learning.

Our progress in science will be greatly affected by our achievement of a high degree of scientific literacy among the rank and file of Americans. A man cannot be really educated in a relevant way for modern living unless he has an understanding of science. Our young people, whether they become scientists or not, need some of the intellectual wealth of science, some of its excitement and adventure, some of its special vision for interpreting nature—some of the understanding which our citizens should have if they are to deal intelligently with the great issues of our time arising out of science.

Let me give an amusing example of the pitfalls of scientific illiteracy. Many years ago a parson-naturalist became interested in deer flies which are justly noted for their speed. He made some observations, unhappily colored by innocent enthusiasm, and published his estimate that the deer fly cruises at 800 miles per hour. This estimate was accepted as gospel truth for years. But a few years ago the great chemist, Irving Langmuir, who was also a nature lover, became worried as to whether such a tiny insect could store enough energy or fuel in his body to fly at such a rate. He also perceived that at such speeds the deer fly, like the ballistic missile, would have nose-cone problems because of the heat generated by air friction. Troubled by these questions, he hunted up some deer flies and made careful measurement of their speed. It proved to be about 70 miles per hour.

Despite this correction I am told that a recently published table of flying speeds, circulated to an estimated fifty million readers, reported the male deerfly as cruising at 818 miles per hour, while the female struggled along at 800. Thus a titillating sex angle was introduced which will probably preserve the 800-mile estimate for years to come.

Finally, and in summary, one cannot fail to ask whether we Americans in our drive to make and acquire things have not been giving too little attention to developing men and ideas. If we are to maintain leadership in this century of science we must be sure that we devote an adequate amount of our energy and resources to the cultivation of talent and quality and intellectual accomplishment.

If we fulfill our potential for skill, talent, education, and quality; if we can give full recognition in our national life to the importance of emphasizing quality and of achieving intellectual pre-eminence, both for our internal benefit and our external position, there would appear to be no real impediment to our steady scientific and technological advance.

With our own American prized pattern of education, with the laboratories and factories and advancing skills and freedom of our industrial society, we may well show the way to a nobler level of living for all men and enhanced freedom and dignity for man the individual.

DR. JAMES R. KILLIAN, JR., Special Assistant to the U. S. President for Science and Technology, was born in Blackburg, South Carolina (1904), and studied at Trinity College (Duke University, 1921-23), and received his B.S. degree from Massachusetts Institute of Technology

(1926). He became Assistant Managing Editor of the *Technology Review* (M.I.T.) (1926-1927), Managing Editor (1927-30), and Editor (1930-1939); in 1939 he was appointed Executive Assistant to the President of the M.I.T., Executive Vice-President in 1943, Vice-President in 1945, 10th President of that Institution from 1949-1959, and Chairman of the Corporation in 1959. He has also served on numerous governmental and private advisory committees, and especially as Chairman of the President's Science Advisory Committee (1957-   ).

SCIENCE EDUCATION IN GREAT BRITAIN

KENNETH LAYBOURN
City and County of Bristol

The Stages of Education

Statutory education in England begins at the age of five although nursery schools and classes may be provided for still younger children. Primary schooling continues to the age of eleven, in Infant departments (ages 5-7) and in Junior departments (ages 7-11). Beyond this every boy and girl must stay at school at least to the age of fifteen in one or other of the various Secondary establishments.

The commonest organization of Secondary education is the tri-partite system, under which, following a Selection Test at the age of eleven, pupils pass to either a Grammar school, a Technical school or a Modern school—the latter being called in Scotland a Junior Secondary school and in Northern Ireland an Intermediate school. Technical High schools, accepting pupils between the ages of 11 and 18, are still comparatively few in number and differ from Grammar schools mainly in two respects: they offer a more vocationally-biased curriculum, and they attract pupils most of whom would have accepted a Grammar school place had one been offered. Roughly a quarter of all boys and girls pass either to Grammar or Technical schools, and these will be expected to remain at school at least to the age of 16, when they will be eligible to take papers for the General Certificate of Education (Ordinary Level). Up to 20% of pupils in Grammar schools (fewer in Technical schools) will continue for a further two or three years at school, and may take Advanced Level subjects in the General Certificate examination at about the age of 18. From here, they may proceed to Universities, to Technical Colleges or to Teacher Training Colleges in order to continue their formal education.

An alternative to the tri-partite system is found in the Comprehensive schools now being developed in London and under a number of other local education authorities. These schools are staffed and equipped to cater for children of all ranges of ability—from the most intellectually able, who elsewhere would go to Grammar schools, to the slowest. In these schools and in Modern schools there are signs that many more pupils are prepared to stay beyond the age of 15 if facilities for education are geared to their interest in practical methods, real-life situations and future careers.

In the Maintained schools, accounting for something like 93% of all boys and girls in Great Britain, education is entirely free. Especially in England, a number of fee-paying schools persist—from the so-called Public Schools, through the various Independent and Direct-Grant Grammar Schools, to lesser Private Schools.

The Development of Science in Education

It was as a result of pressure from a generation immensely interested in the spectacular achievements of industry, commerce and medicine that Science gradually penetrated the old-established Grammar Schools (then the *only* secondary schools) in the latter half of the nineteenth century. Although the subject had appeared earlier in a number of Public schools (including Eton and Rugby), a classical education was still generally considered to be the only one worth having, and where Science did get a foothold it was seldom on cultural grounds: the country needed chemists, engineers and geologists. Right up to the end of the century, and often far beyond, if Science *was* taught, it was Chemistry and Physics for boys and only Botany for girls.

The Education Act of 1902 empowered local education authorities, for the first time, to build Secondary schools and thereafter the proportion of 'free places' gradually improved. In these "maintained" schools, Science teaching established a pattern which has remained essentially the same in many Grammar schools down to the present day. Pupils commonly begin Science—either General Science or separate sciences—at the age of eleven. In a school week of $27\frac{1}{2}$ hours they may begin with $2\frac{1}{2}$-3 hours of Science, increasing the time somewhat (depending on the course they elect to pursue) up to the first public examination at 16, when 5 hours total Science time is not unrepresentative. In Girls' schools, Botany has often given way to Biology and a proportion of girls might add a second science.

Of course in the early years of the present century the great majority of boys and girls were in Elementary schools. Before 1861, teacher training colleges had included some instruction in Natural Science in their curricula, and ten years later there was a recommendation that some Science teaching should be given in all elementary schools: in London and elsewhere large boxes of apparatus were sent round the schools and peripatetic science demonstrators were employed. In the closing years of the century some of the larger School Boards developed Higher Grade departments (thirteen was then the statutory leaving age) and the upper standards of such schools could earn grants which enabled them to equip laboratories. So it came about that the Thomson Report of 1918 was able to say that, in such schools, "a master with exceptional character and qualifications" might give "an amount of science teaching that may be of great value"! With the increasing reorganisation of "all-age" elementary schools into Primary and Senior departments, after 1926, the Board of Education's Inspectors could report (in 1932) that "the new (Senior) schools as a whole have decided that a course of work in Science of some kind is desirable as part of the normal curriculum for all children." But the Senior schools had neither the staff, the accommodation nor the equipment to do much: if a boy wanted to take up any serious study of Science, he must attend a Grammar school.

The General Science Movement

In 1916 the Association of Public School Science Masters (the parent of the Science Masters' Association) launched a campaign to draw attention to the danger in which the country stood in the First Great War through the neglect of Science teaching. This led directly to the Prime Minister's Committee to enquire into the position of Natural Science in the educational system of Great Britain, with Sir J. J. Thomson as Chairman. Meanwhile, the A.P.S.S.M. was looking into the question of the *kind* of science that should be taught to boys, with particular reference to the non-specialist, and its report "Science for All" was the forerunner of all our modern schemes of General Science in this country. Briefly, the report suggested that the basis of science teaching should be broadened, so as to cover all the main fields; and that at the same time content and method should be brought much more into line with the pupils' own interests and experience.

The Thomson Report of 1918 gave official backing to the new outlook. It urged that more attention should be paid to those aspects of the sciences that bear directly on everyday life, that work for pupils under sixteen should include, besides Physics and Chemistry, some study of plant and animal life, and that all through the Science course stress should be laid on the accurate use of the English language. But the most striking thing about the Thomson Report is the way in which it anticipated the very problems that we face today:

". . . real progress in education depends on a revolution in the public attitude towards the salaries of teachers and the importance of their training";

". . . a larger number of students in training colleges should be encouraged to take Advanced Courses in Science";

". . . concerted efforts should be made by employers, teachers, local education authorities and the State to increase the flow of capable students to the Universities and higher technical institutions with a view to securing the large supply of trained scientific workers required";

". . . and those specialising in Science should continue some literary study, and those specialising in literary subjects should give some time to Science";

". . . increased attention should be given to the teaching of Science in girls' schools."

All this was forty years ago!

The General Science movement made but slow progress until the nineteen-thirties, when the Science Masters' Association published two reports strongly supporting the new outlook. Through the years since their publication, every conceivable argument for and against General Science (as distinct from the individual sciences) has been debated. Can General Science be taught effectively by the average teacher? Is it science or merely information? Can it, in any case, be dealt with adequately in the time usually available? The number of candidates taking the subject at the First School Certificate rose rapidly and today the desirability of a wide first view for average pupils and non-specialists is generally accepted, although many Science teachers in Grammar schools still prefer the branches to be taught by separate subject specialists.

Science in Secondary Schools

The aims of science teaching in grammar schools have been described as follows in a policy statement ("Science and Education," November, 1957) by the Committee of the Science Masters' Association:—

(a) To lead pupils to observe, and to solve problems by controlled experiments, to draw conclusions from observations, and to appreciate the systematic laws and principles of science.

(b) To give knowledge and understanding of the origins and development of science, of the achievements of scientific pioneers and of the implications, now and in the future, of modern scientific and technological developments.

(c) For Science specialists, to provide a suitable preparation for further scientific or technological education.

With these aims in mind the Committee recommends that *all* pupils in the Grammar school should follow the same course in Science up to the end of the fifth-form year (age 16; "Ordinary" Level G.C.E. standard), depth of study being determined by the abilities of the pupils and not at all by considerations related to later specialist subjects. There should be no division into arts and science 'sides' at this level. The Committee further recommends that Science should be studied by *all* pupils in the sixth-form, as a humanistic and cultural subject, and that existing "Advanced Level" syllabuses (for those who will become Science specialists) should be reduced in width of content, though not in depth.

During the first two years of the grammar school course, it is suggested that a broadly-based syllabus taught on a topic basis is appropriate. The main aim here is to arouse curiosity and develop observation, with training in simple laboratory techniques and in the use of good descriptive language. In the second phase, extending over the next three years, the work should be composed of physics, chemistry and biology and all pupils should devote the same total time to science, the more able reaching G.C.E. 'O' Level in all three branches. The keynote of the course should be experiment as a means of solving problems, but practical work should lead to the formulation of empirical laws and hypotheses, and, eventually, to simple ideas of some of the great generalisations of science.

In the Sixth Form, two courses are proposed—one intended for all students and the other for Science specialists only. The former course would embrace the history and philosophy of science, its present-day social and technological consequences and its future possibilities. It might include topics such as cosmology, evolution, heredity, man's command over sources of energy, and the effects of technology on society. There might be a place for some introduction to psychology and the social sciences.

How does the present position differ from that outlined by the Committee? Largely in the early beginnings of specialisation (by restriction of subjects); in the absence of Science from the curricula of most Sixth-Form 'Arts' pupils; and in the restrictions placed upon practical work by accommodation and staffing problems.

The major difficulties in the way of a general expansion of Science teaching in the Independent Schools have been removed during the last three years by the action of leading British industrialists in contributing $3\frac{1}{4}$ million pounds for "expanding, modernising and equipping Science buildings in Direct Grant and Independent schools." In the State schools the many competing demands for money have meant that science facilities are still far from ideal in many places, although improvements are constantly being made. The principal complaints are of laboratories inadequate in number and size, and of lack of adequate laboratory assistance. These handicaps are infinitely greater in the older Modern schools, where it is the exception to find more than a single Science room, but the enormous programme of Secondary school building now being undertaken should rapidly bring about improvement and Britain already has many hundreds of fine new schools with well-planned Science accommodation.

It is increasingly recognised that, if Britain is to get the technicians and craftsmen required by her expansion programme, Science must be made an effective part of education for boys and girls other than those in attendance at Grammar schools. But Science teachers in Modern schools are even more concerned that *every* pupil should be introduced to ideas and ways of thinking that are an essential part of modern life. They are concerned that tomorrow's citizens should feel at home in tomorrow's world.

Development of Science and Technology in the Universities

The late Professor Alexander of Manchester, a great philosopher and university teacher, defined the purpose of a university in the following terms:

". . . an association or corporation of scholars and teachers engaged in acquiring, communicating and advancing knowledge, pursuing *in a liberal spirit* the various Sciences which are a preparation for the professions or higher occupations of life . . . not that its purpose or object is the professions, but that it does in fact prepare for the professions by pursuing the Sciences on which those professions are founded."

Not everyone would accept such a definition as sufficient; yet it has a peculiar relevance to the developments now taking place in the English universities.

The "natural philosophy," which was the door through which Science first entered the universities a century ago, soon gave place to the system of separate subject schools which is the basis of the modern university structure. In Britain the undergraduate course in science is commonly of three years' duration, during the first of which the student usually studies a group of related sciences (often three). In the following two years he will concentrate upon the various aspects of his Honours subject (Chemistry, Physics, Engineering, etc.) and will specialise in some particular branch of that subject. There are variations upon this pattern in different universities and faculties. Post-graduate students may undertake research work in the university, and they will in due course present themselves for higher degrees.

There are, of course, many critics of the system. It is certain that departmentalization often goes too far—that subjects are taught too often without reference from one to another, and that each faculty appears to try to work independently of the others. There are complaints that the constant pressure to add more and more to the content of study (as scientific knowledge is extended) is seldom accompanied by an equal determination to root out material that is no longer relevant or essential. There is severe criticism of the system of entrance scholarships which effectively prevents the broadening of the curriculum of the Grammar school Sixth Forms and leads to a wild scramble for University places. But the talk of 'crisis' is untrue in fact and unfair to a great system under which academic freedom remains the bulwark of truth. Present stresses arise from the impact of two revolutions—the democratic and the technological. The rapid growth of student numbers may tend to destroy the common life and personal relationships which are the best element in university education. Every university knows the danger and great efforts are being made to combat it. On the technological side, the Government is very much alive to the vital need to greatly increase the output of scientists and technologists upon which Britain's future as an industrial nation depends.

Universities in Britain are now very largely supported directly from state funds, administered by the University Grants Committee. The grant has never had any strings attached to it; it has never been a subject of political dispute; and parliament deliberately entrusts financial control to men and women whose sympathies are almost entirely academic. Since the war many millions of pounds have been allocated for major building projects (60% of the total being for science in some form or another), and on top of this the Government has embarked upon a great programme of university technological expansion. The development of the separate sciences, pure and applied, is being specially encouraged at particular universities—chemistry at Birmingham, Newcastle, Sheffield and Leeds; chemical engineering at Birmingham, Cambridge and Manchester; fluid mechanics at Cambridge; and so on. Immense building projects are being carried forward in Manchester and Glasgow where Colleges of Science and Technology will form an integral part of the University. And in London a massive expansion of the Imperial College, costing fifteen million pounds, will provide for 1,500 undergraduate students and a similar number of post-graduate students, with perhaps one half of the latter engaged in research. All of these projects are well on the way and the universities are already formulating plans for further advance.

The number of graduates in science and engineering from British universities rose from 3,000 in 1939 to over 7,000 in 1957. To these must be added graduates (and those with equivalent qualifications) from technical colleges, making a total for 1957 of about 12,500. The Scientific Manpower Committee has defined the national objective as that of increasing this figure to 20,000 in the next ten years, and the task will probably be shared almost equally between the universities and the colleges of technology.

Technical Colleges

In 1956 the Government announced its intention of spending about £70,000,000 within five years on the expansion of technical education throughout Great Britain; this is only a first instalment. It is recognised that our future prosperity and perhaps even our National survival depend upon a vast increase in the number of our trained scientists and technologists.

The technical colleges and institutes of Great Britain are nearly all maintained by local education authorities. They cover work at every level and they vary from small local colleges to large regional institutions. There is a tremendous volume of evening and part-time day work, but the Government has now designated eight Advanced Colleges of Technology (and intends to increase the number) in which the bulk of full-time technological education will be concentrated. Such Advanced Colleges will offer courses to the highest levels: the newly-established Diploma in Technology is broadly equivalent to a University Honours degree.

A boy who leaves school and enters industry (or commerce) at age 15 may attend a part-time one-year course which will improve his general education and help him to start at 16 on an industrial apprenticeship. At that stage, joined by others who have left school at 16 (probably with 'O' level G.C.E.), he will start on a senior part-time course lasting two or three years. Many of these senior courses are geared to the needs of operatives and craftsmen, but parallel with them or following them are courses aimed at intermediate technician level, offering qualifications such as Ordinary National Certificate.

Advanced part-time courses begin at age 18 or 19, may last from two to four years, and often lead to Higher National Certificate or to a London University (external) degree. Professional qualifications demand industrial experience and further study, so that it is not until he is 25 or more years old that the qualified technologist will emerge via this part-time route.

Boys who have stayed on at school to age 18 and who hold Advanced Level subjects in G.C.E. may either enter industry direct, as Student Apprentices who will pursue their academic education part-time or in sandwich courses, or may start on full-time courses in one or other of the major colleges. They will generally proceed either to a Higher National Diploma, to an external Degree, or to the Diploma in Technology; and beyond this post-graduate work is available in the Advanced colleges. It should be noted that courses for the Higher National Diploma are much more broadly based than are those for Higher National Certificate; their scientific content is deeper, their outlook much more liberal, and they usually cover the full academic requirements for exemption from examinations of the professional bodies.

The need to ensure that technological education generally shall be broadly and liberally conceived is reflected in the attention which is being given to the need for residential accommodation, library and recreational provision, and general education as part of a technical course.

The 'sandwich' course is probably the key to the massive expansion of advanced technical education in Great Britain. Under this scheme a student spends alternate periods (usually of six months) in a college and in training in industry, over three to five years. Such courses are planned jointly by colleges and firms, and co-operation between management, trade unions and college staff is necessary if the works periods are to be used to best advantage. Many industrialists are finding that in addition to university-trained scientists and engineers, they need men and women who have 'grown up' in industry—who have that immediate grasp of shop-floor problems that can only be acquired through first-hand acquaintance with production processes and the people engaged on them. Part-time courses of the old kind (evening study) are no longer a sufficient answer. The Government's present plan is to increase the output of students with *advanced* qualifications from the 1955 figure of 9,500 to about 15,000, the bulk of the increase being in respect of full-time or sandwich courses.

In the technical colleges the physical sciences are of course very well established; except in the largest colleges the biological sciences are not yet developed to anything like the same extent and advanced work in the biological field is usually undertaken via the Grammar School VIth Form/University route. The social sciences have in the past contributed "a mere trickle to the flood of technical education, almost wholly through part-time courses—disguised as management and administration." But Education for Management now forms an important element in many sandwich courses and there is a growing number of advanced short courses in the subject.

At the less advanced level, the aim is that of greatly increasing the number of day-release students. The traditional method of training craftsmen and technicians in Great Britain is apprenticeship, of which the characteristic feature is practical training on-the-job over a period of five years. But the need for understanding the principles behind the machine, and for flexibility and versatility in a swiftly-changing scene, make it imperative that the education of apprentices and other young people in industry should not stop short at 16 or rely solely upon (voluntary) attendance at evening classes. For this reason many employers release their apprentices for one day a week for attendance at technical classes and there has been a great increase in facilities for day-release courses since the war. But as expansion proceeds many more industries will be brought into the net and more attention will be focused upon the suitability of courses to meet the demands of particular workers. This kind of development should go far towards reducing the serious wastage that still takes place at about the age of 16, wastage which is accounted for by "reliance upon evening classes, the need to work overtime, being away on a job, shift work, travelling difficulties, ill-health, a change to another job, home conditions which make study impossible, or lack of encouragement from the employer." The 1954/5 figure of day-release students was 355,000: the aim now is to double this.

In May, 1957, the Willis Jackson Report on the supply and training of technical college teachers was published. It showed that an average net annual increase of 1,400 full-time teachers would be required over the next five years and it made suggestions about recruitment and training. Industry must be willing to encourage the transfer to full-time College work of experienced staff (whom it could ill-afford to lose on the short-term view), as the only way of ensuring a sufficient supply of young teachers of high quality; it must support the release of staff for part-time teaching; and selected senior staff must be brought into close and responsible association with the academic activities of colleges through part-time appointments carrying special status. On the training side, it was recommended that a residential staff college should be established to study the teaching, administrative and technical problems of further education, and that a permanent advisory committee on the supply and training of teachers should be set up. This committee was in fact formally constituted by the end of 1957.

Technical education in Great Britain is co-ordinated through the activities of Regional Advisory Councils (nine in England and five in Scotland). Associated with them are Regional Academic Boards for ensuring close co-operation between the universities and technical colleges in the provision of advanced courses. At the centre, a National Advisory Council on Education for Industry and Commerce advises the Minister of Education on national policy.

The Supply and Training of Science Teachers

The overall problem of teacher supply is one that has been faced realistically since the war and thousands of extra teachers have entered the schools to ensure that the effects of the post-war birth-rate 'bulge' did not get out of hand. The National Advisory Council on the Training and Supply of Teachers has just (July, 1958) advised the Minister of Education that 16,000 additional Training College places should be provided forthwith in order to counter the reduction in output that will follow the expansion of teacher-training from two to three years in the early 1960s. Even this will leave many desirable reforms untouched—the raising of the leaving age, reduction in the size of classes, and the extension of further education for young men and women who have entered industry or commerce.

One of the most pressing problems, however, is the shortage of teachers for Science and Mathematics. In 1938 all Science posts in Maintained Grammar schools were filled, almost entirely by graduates (60% of whom held either First or Second Class Honours degrees). But in 1957, 1,100 Science posts out of a total of 7,000 in these schools were either unfilled or unsatisfactorily filled—in the sense that they were not filled by men or women with good graduate qualifications. The reasons include the vastly increased numbers in Science Sixth-Forms, the doubling of the size of University Science staffs in the last decade, and the competition of industry and the Government for Science graduates. Steps have been taken to increase the supply: today a Senior Science Master gets £1,400 a year as against £800 in 1948, and he can work until he is 70; but it is certain that he is still seriously underpaid in relation to men with similar qualifications in industrial or university posts. It seems clear too that a much more serious effort must be made in the schools and universities to publicise teaching as a career and to organise training along attractive and effective lines.

In the Training Colleges, which supply the great bulk of teachers—usually non-graduates—for Primary and Secondary Modern Schools, the position is quite as difficult. One of the most encouraging developments in Science education in England is the interest now being taken in it in many Primary schools. Here it is felt that every child, from a very early age, should be given a great range of practical experience which will enable him to start his Secondary school Science on a basis of real understanding. Whilst, therefore, we shall not perhaps need scientists in our Infant and Junior schools, we *shall* need teachers who are interested in Science and sympathetic towards it. When the general three-year training course starts in 1960, there seems to be an overwhelming argument for including a basic course in Science as part of the personal education of all students. In the Modern schools we need men and women who are both trained to teach Science and trained to understand the pupils in these schools. Again the three-year course will do much to enable the colleges to develop the kind of Science teaching that they pioneered after the war—teaching which took into account the need to relate Science to the child's everyday environment, to his interest in a career, and to his ability to learn through using his hands. The total number of students taking Science as a main subject in training colleges has substantially increased since 1953, but there is still a great lack of men and women—and especially women—in the schools. Part-time and supplementary full-time courses are helping, but nothing short of an all-out effort by everyone concerned from the Government downwards, will be sufficient to solve this problem.

Selected Bibliography

B. V. Bowden, *Proposals for the Development of the Manchester College of Science and Technology* (Manchester, 1956). An account of the aims and organization of technological education in one of Britain's largest colleges.

Science Masters' Association, "Provision and Maintenance of Laboratories in Grammar Schools," *School Science Review* (London: John Murray), CXXXIX (June, 1958), 438-446. Report of a sub-committee on the existing provision of school science laboratories and laboratory assistants in Maintained Grammar Schools.

British Association for the Advancement of Science, *Science in Schools* (London: Butterworth, 1958). Report of a conference on Science education in Britain, including the problem of supply and training of teachers.

P. F. R. Venables, *Technical Education* (London: G. Bell & Sons, 1955). A detailed account of the organization of technical education in Britain.

J. D. Bernal, *The Social Function of Science* (London: Routledge & Sons). A critical examination of the function of Science in society.

Sir Edward Appleton, "Science and the Nation," *Reith Lectures, 1956* (Edinburgh University Press, 1957). A discussion of the kind of Science the nation needs and how it is to be provided.

Science Masters' Association, *Secondary Modern Science Teaching* (London, John Murray, Parts I and II, 1957). Report on the teaching of Science in Secondary Modern schools.

Laybourn and Bailey, *Teaching Science to the Ordinary Pupil* (London: University of London Press, 1957). A source book for teachers on classroom and laboratory methods, teaching techniques and practical work, illustrating the modern trends in England.

Board of Education, *Report of the Consultative Committee on Secondary Education* (London: Her Majesty's Stationery Office, 1947 reprint). A report on organisation, curriculum, and examinations in Grammar and Technical High schools in England and Wales.

D. S. L. Cardwell, *The Organisation of Science in England* (London: Heinemann, 1957). An analysis of the history and development of English Science and Technical education.

KENNETH LAYBOURN, Ph.D., M.Sc., Chief Inspector of Schools of the City and County of Bristol (England), graduated at Durham in 1930, with First Class Honours in Chemistry, and then carried on research in physical chemistry for three years and subsequently taught in grammar school for fifteen years. During this latter period he strongly supported the development of the General Science movement in England. He was appointed Head of the Science Department in a London school and then Deputy Headmaster of a Yorkshire Grammar School, before joining the Manchester Education Authority in 1948 as District Inspector of Schools. *Teaching Science to the Ordinary Pupil* (University of London Press), of which Dr. Laybourn is co-author, was the first book in England to demonstrate that every part of the Science syllabus for boys and girls of average ability can be treated practically by the pupils themselves using familiar everyday materials. In 1955 he was appointed Chief Inspector of Schools for the City and County of Bristol. At present he is also engaged in a campaign for the widening of the curriculum of primary schools, from the nursery stage upwards, so as to ensure that later Science studies are based upon first-hand experiences in every branch of the subject. He is Chairman of the Secondary Schools Examination Committee of the Union of Educational Institutions.

SCIENCE EDUCATION IN THE U.S.S.R.

L. A. D. DELLIN
The University of Vermont

Introduction

At the Twentieth Congress of the Communist Party of the Soviet Union, Stalin's successor Nikita Khrushchev boasted that "no capitalist country has so many schools, *tekhnikums,* higher educational institutions, scientific-research institutes, experimental stations and laboratories, theaters, clubs, libraries, and other cultural and educational institutions as the Soviet Union."[1] This statement was made within the framework of another often repeated boast that illiteracy had been completely wiped out among the odd sixty nations of the 200 million Soviet people.

These impressive achievements, although marred by the usual Soviet manipulations of statistical criteria, did not generate any noticeable reaction among the American public, except among a few for whom the Soviet strides were not, however, a novelty. It was thought by the public that either the Soviet Union had remained where it was forty years ago or that mass education was only a matter of quantity, largely offset by primitive quality as well as by the connotations of a Communist system of government and education. *Sputnik* and the perfectioning of the intercontinental ballistic missile came therefore as a violent shock. It became painfully evident that mass education was coupled with quality, that the system of government and education did not foreclose creativeness, and that the Soviet Union had moved ahead of the United States in fields of science and technology which were synonymous with American ingenuity, know-how, and material power. This time we moved, however, to the other extreme. We plunged into almost fatalistic generalizations and became inclined to overstate today what we understated yesterday. It took us quite a while to sober and embark upon the only sensible way: to find and analyse dispassionately the hard facts, to recognize the challenge in its true dimensions, and to act quickly but rationally, in order to meet that challenge.

There can hardly be any argument any longer about the fact that Soviet science and science education present a tangible challenge, more— a real threat to the United States and the free world as a whole, not because of the academic strides in themselves but because of their actual use as an instrument of aggressive Soviet military and economic power and their potential use for Soviet communism's ultimate goal—world domination. The U.S.S.R. is admittedly ahead in many aspects of modern rocketry and ballistics as well as in its output of scientists, engineers and technicians, and the theory and practice of Soviet communism need not be recalled. Therefore, the matter is bluntly one of survival.

Realizing this, one should not, however, assume that Soviet education, in general, and science education, in particular, are either faultless or not vulnerable. In fact, they have decidedly both assets and liabilities as it will be attempted to demonstrate. Even the very concept of education may be challenged on the ground that its Soviet equivalent is not really education but rather training and indoctrination, so that the Soviet youth "have been denied the privilege of education"[2] altogether.

Still, no matter what Soviet education in reality is, it is undoubtedly a deadly-serious affair in form as well as in contents. In this connection it seems appropriate to do away with an erroneous notion, i.e. that the Soviet approach to formal education with its rigid and uniform curriculum and the Soviet attitude towards learning and the learned is a new and typical Soviet phenomenon. It is not. It is the well-established continental approach which was also dominant in Russia of Tsarist memory. Of course, substantial modifications were made by the Soviet regime in terms of an all-permeating political indoctrination, of "science for politics' or applied science's sake," of "democratization," "secularization," and "polytechnization" of education, and of several other innovations, inherent in the Soviet drive for creating a "new man" and an advanced economy. Yet the continental setting is there with the heavy load for the student and the respect, even veneration, for the learned. This is in fact one of the many contradictions between Communist theory and practice: the classless society of the workers, on the one hand, and the privileged class of the Party elite *and* intellectual elite, on the other hand.

This article is limited not only in space but also in contents. It purports to survey the general trends of science education in the U.S.S.R. It cannot avoid, however, the broad framework of general education or of the primary-secondary school for the simple reason that science education is part and parcel of the general school and that its real dimension will appear best by measuring its relative share within general education. Furthermore, the fragmentation of higher education will not allow much more than a summary of trends and features with an emphasis on the university (as against the institute) to which theoretical science education is almost exclusively confined. Advanced study because of its non-instructive character is only briefly discussed while science work outside the educational framework such as in the Academy of Sciences is omitted altogether. No attempt has been made to compare Soviet science education with any other system due not only to the immensity of a similar undertaking but also to the valid objection raised by almost all specialists that the values involved are hardly comparable. Such references to American education are only illustrative, any conclusive comparison being left to the reader.[3]

This article is based mainly on Soviet sources, in the processing of which the author has years of experience. It was fortunate that for the first time since 1939 the Soviet authorities published only recently voluminous statistical abstracts on Soviet economic and cultural developments, listed in the bibliography, which despite the known manipulation filled in a considerable gap. Moreover, in 1957, appeared another Soviet standard book on education, summarizing the most recent trends from the Soviet point of view.[4] All these publications, alongside with other better-known Soviet sources, have been used for updating statistical reference and Soviet interpretations.

Englsh language sources, especially in this country, have grown in geometric progression over the past years, so that only the most recent and most pertinent have ben listed in the bibliography. Among these Alexander G. Korol, *Soviet Education for Science and Technology* and Nicholas DeWitt, *Soviet Professional Manpower* have been most useful for this limited topic.

The Educational Framework

Literacy. Since 1952, the Soviets have repeatedly claimed of having eradicated illiteracy completely. Although this claim is validly disputed,[5] the Soviet achievements in this respect are impressive, especially so far as the non-Russian nationalities are concerned.[6] However, there are reasons and explanations of a particular character which place these achievements in proper perspective. Lenin himself clearly suggested the main reason: "An illiterate person stands outside [the reach] of politics; we must first teach him the alphabet. Without this there can be no politics."[7] Thus, the three R's—and not necessarily much more—were considered a mandatory medium for political indoctrination—not necessarily for learning as such. And the semi-literates whose number is still overwhelming in the U.S.S.R., exposed at that to a concentrated unilateral propaganda, are an easy prey for securing loyalty and even enthusiasm in favor of the regime.[8]

There is the additional reason suggested by the expansion of contemporary military service and by the ambitious goals of modernizing Russia's economy which require at least a minimum of technological preparedness and consequently a minimum of literacy. This reason accounts, of course, with major force for some type of formal education, as will be evident later on. An explanation for the rapid success in the struggle against illiteracy which holds true for formal education as well is the very nature of the Soviet state which can and did muster all available resources toward the realization of this aim.[9] Therefore, the over-all achievements must be evaluated with a view of the ultimate motivation

and the force and effort applied, all of which has no counterpart anywhere else and still illiteracy has had a similar fate in many a democratic
country.

The Setting of Formal Education. Soviet education is compulsory for
the first seven grades of the ten-year general-education course which embraces over 30 million students. An additional two million study in the
secondary "specialized" schools and the same number attends the higher
educational institutions. A total of about 50 million people are said to
attend some kind of school which amounts to about 2,500 students in all
types of schools and 93 students in higher educational institutions per
10,000 inhabitants.[10] 6,253,000 "specialists," of whom 2,631,000 with
higher and 3,622,000 with secondary education, i.e. professionals and
semi-professionals, compose the Soviet "socialist intelligentsia" and 239,-
000 of them are "scientific workers," or the cream of that intelligentsia.[11]
53 per cent of the "specialists" with higher education are women and
87,000 women make part of the cream.[12]

The officially announced aim of Soviet education is "to develop ver-
satile, active, and conscientious builders of the Communist society" and
it includes "intellectual, polytechnical, moral, esthetic, and physical education."[13] The emphasis throughout is on "productive practice" as
against "pure theory," although within this framework several shifts
have occurred since the first Program on Education was adopted by the
Eight Party Congress of March 1919. At first it was all "labor education"
for the ignorant Party enthusiast who could enter a higher institution
without having gone through much more than the three R's. Then,
during the middle 1930's, after it was realized that there is no substitute
for knowledge, formal education regained its place. Still, the vocational
orientation of Soviet education prevails, which is evident not only from
the large number of vocational schools but within general education
itself.

What matters most is complete control. In fact, the Government, run
by the Party, owns and operates all schools. It determines their number
as well as the number of students and teachers on the basis of the state
economic plans, embodying the estimated needs of the country. It writes
the rules and quotas for admission and graduation, the curricula, and
the textbooks. It even prescribes the "daily regime" of the student (and
of the teacher) not only in school but also for all the remaining hours of
the day (even how long to sleep). Such an all-embracing and minute
control has not been known anywhere anytime.

There are, however, compensations of importance. There are no
tuitions and students in higher institutions receive scholarships.[14] Only
textbooks and notebooks must be purchased as a rule. Schools are co-
educational. And the teachers, especially of the higher educational institutions, are paid well by Soviet standards, much more than the industrial
worker or technician.[15] The expenditures on education comprise about

12.8 per cent of all budget expenditures.[16] All this immensely facilitates mass education and provides strong incentives for the gifted and laborious to aspire to the intellectual elite, especially if his social origin and political standing are not objectionable. For the individualist the field least permeated with politics becomes the best alternative, and this was and is science.

The general-education school is the basic unit within the Soviet educational framework. Since 1932-33, its full course is ten years,[17] comparable to our twelve-year elementary-high school course, but Soviet general-education schools can be either elementary (first four grades), incomplete, or junior, secondary (first seven grades), i.e. the now compulsory period of Soviet education, or complete secondary, providing the full ten-year course.[18] Formal education starts with the age of seven and the full general-education course can be completed at age 17. Curricula and syllabi are identical for each corresponding grade of these three types of school, so that they are in fact superimposed and interchangeable. Most of the general-education schools are regular residence schools which offer, however, evening and correspondence courses, mainly for adults. There are also incomplete and complete secondary schools for young workers as well as incomplete secondary schools for young peasants which offer condensed part-time courses outside of the working time. To the general-education schools belong also military schools, schools for musically gifted and convalescent children, and special schools for physically or mentaly handicapped children.

The "specialized," i.e., vocational schools are open to persons with completed compulsory education and are either complete secondary special schools, with a four-year course, or incomplete secondary schools, with a six-months to two-year course. The complete secondary special schools, called mostly tekhnikums, train "junior specialists" or semiprofessionals who may enroll in higher educational institutions after a three-year employment in their profession. Outstanding students are released from this obligation. The incomplete vocational schools, called labor reserve or factory-plant schools, turn out skilled workers for the national economy who may complete their education in the secondary schools for young workers and peasants without leaving employment.

Graduates of the secondary general-education schools may enroll without employment requirements[19] in either the regular higher educational institutions with four to six-year courses or in the tekhnikums or, since 1954, in technical schools with a six-month to two-year course which train highly skilled workers and "junior technical personnel" for occupations requiring completed secondary education. Secondary school graduates may obtain special skills in various vocational-technical schools by taking short courses. Evening and correspondence courses are also open at the secondary and higher level for those gainfully employed.

Administratively, the individual republic ministries of public education together with the local Soviets manage the general-education schools within their borders. The federal republic Ministry of Higher Education is responsible for the higher educational and the secondary special institutions, although some of them are under the direct jurisdiction of the respective governmental agency. The Chief Labor Reserve Administration at the Council of Ministers manages the labor reserve, factory-plant, and technical schools. Of course, "the policy in the field of education and the leadership of all organs of public education is exercized by the Communist Party of the Soviet Union,"[20] which has its representatives at every level and in each school.

The school authorities are assisted by the Communist youth organizations, the trade unions, enterprises, kolkhozes, sovkhozes, machine-tractor stations, and offices—the latter five business units "patronizing" an educational institution, by parents committees, and other public bodies, but the actual responsibility lies with the head of the school (in Communist parlance, "one-man responsibility"). This device for strengthening the authority of the competent is a long way from the initially-established authority of the ignorant yet faithful Party member and has had positive results for students and teachers alike.

Soviet pedagogics is, of course, based on dialectic and historic materialism and the methods of such Soviet pedagogists as A. S. Makarenko, but increased reference is made to the well-known educators of the Russian, non-Soviet, past like M. V. Lomonosov, V. G. Belinsky, N. G. Chernishevsky, A. I. Khertsen, and even Lev Tolstoi. The initial theory of "withering away of the school," like the "withering away of the state," has long been abandoned and the present top Soviet pedagogic institution, the Academy of Pedagogical Science of the Russian Republic, is entrusted with standardizing the scientific-methodological aspects of education along less orthodox lines.

General Education

Schools, Attendance, Character, Trends. The general-education school is the backbone of Soviet elementary-secondary education. It not only embraces the overwhelming majority of Soviet students but also provides the higher educational institutions with their largest contingent, and is the only breeding ground for future scientists. There are now over 30 million students in the more than 200,000 general-education schools of all types. But regular residence students number 28.2 million, attending 196,600 schools.[21]

Elementary, seven-year, and complete secondary schools comprise 55.6, 30.1, and 13.3 per cent of the total number of schools respectively. Attendance follows the opposite direction, with students enrolled in complete secondary schools five times as many as in elementary schools and about 1.6 times as many as in seven-year schools. With regard to grade distribution, most of the students, about 13 million, are in grades 1 to 4, about 10 million in grades 5 to 7, and about 5.2 million in the upper grades 8 to 10.[22]

The trend reveals a decrease of elementary schools (peak year 1931-32), a relative standstill in seven-year schools, and a considerable increase in the ten-year school. The total school population was largest immediately preceding the Soviet involvement in World War II (1940-41) and war casualties as well as a declining birth rate have seriously affected the elementary school population, have begun to indent grades 5 to 7 enrollment, while grades 8 to 10 attendance is at its peak, due to the expansion of secondary education.[23] Although upper grade enrollment could be expected to level off as the deficit moves upward, this is in fact offset by the goal of making the entire ten-year course compulsory. The whole future of the Soviet educational system depends on the solution of this increasingly pressing problem as will be discussed in more detail at several instances.

Now attendance in rural areas accounts on the average for about half of total attendance and the decline in lower grade enrollment has occurred at the expense of the peasants which, in spite of unquestionable strides, indicates the still existing cleft between town and village in the U.S.S.R., where the rural population outnumbers the urban one by 26.2 million.[24]

In 1956, over 1,300,000 students obtained secondary education and about seventy per cent of them had to be employed in farms and factories or otherwise barred from immediate ascendency to higher education. Some of them were even sent to till the Siberian virgin lands and the steppes of Kazakhstan or to work in mines and collieries.[25] The sixth five-year plan, now stripped and in process of being replaced by a seven-year plan, not only set as a 1960 goal to accomplish universal secondary education but also to produce 6.3 million secondary-school graduates between 1956 and 1960, i.e. to roughly maintain the 1956 output.[26]

Thus, historically Soviet general education points in quantitative terms to a continuous, although not always smooth, trend upwards, with a saturation point to be reached in the immediate future.[27]

For the time being the greatest shortcoming in this respect seems to be the shortage of school buildings. Soviet sources admit that about half of the students now attend schools built in "Soviet time" and that the shortage makes two and more shifts not unusual.[28] The Soviet press complains continuously about overcrowding and unhygienic conditions. Insufficient seems to be also the supply of textbooks and other school facilities, although complete standardization is applied.

General education in Communist understanding is conceived as a combination of mental knowledge, working habits, technical training as well as moral, esthetic, and physical education. It passed from a predominantly labor-technical stage through a more conventional formal education with college-preparatory function and is now being molded into a combination of academic-subject schooling and labor-technical training. Since Stalin enunciated the new character of general education at the Nineteenth Congress of the Soviet Communist Party, a vocational reorientation is being sought by again stressing "poly-technization," i.e. acquainting the students with the main branches of production, this time for the obvious reason that the outflow of secondary school graduates is far too large to be chanelled into higher institutions and must instead be increasingly employed in production, for which industrial training is more necessary.[29] The sight of a high-school graduate as an agricultural or industrial worker seems a waste of means and effort, aside from the natural reaction and disillusionment of the graduate himself, but the Soviet system is not known as a model for economizing or for its regard toward human feeling and ambition.

In 1955 was elaborated and began to be applied the new plan for the general-education school which introduced the following new subjects: manual work (grades 1 to 4), practical work in school experimental plots and shops (grades 5 to 7), and practical study *("praktikum")* in agriculture, machine elements, and electrotechnique (grades 8 to 10), at the expense of reduced hours in the humanities "which do not have an educational importance."[30] The implementation of this plan is in full swing and it remains to be seen how, how much, and how soon the "greatest shortcoming" of present-day Soviet general education, namely "a certain alienation of education from life, the insufficient preparation of the graduates for practical work"[31] will be eliminated. Officially enunciated is, however, a "considerable change in the organizational form and contents of the general-education school, the structure of the school period and the internal school setup, the contents and volume of the lessons and the relationship between lessons and laboratory work, the practical and independent work of the students, and in many other respects."[32] Debates and experimentations to adjust to the hard reality are still inconclusive but the trend is toward more practical industrial training and specialized science.[33]

No matter what precise shape Soviet general education will eventually take, it is now and will basically remain a serious and systematic impartation of knowledge with a strong emphasis on science. This is not only a legacy of the past and the European tradition but is also complemented by the Communist materialist philosophy which glorifies science (not always for its own sake) as the key to the understanding of the laws of nature and society and a weapon against religious belief and other "scientifically unfounded" superstitions. The emphasis on science is also dictated by practical considerations arising from the commitment to a short-cut modernization and industrialization of economy and society. Last but not least, Soviet Russia's ultimate goal—world domination—requires aggressive strength, best exemplified by modern destructive weapons which only an advanced science and technology can deliver.[34]

Curriculum. Following is the most recent available curriculum of the typical Soviet general-education school:

Ten-year school curriculum (1956-57)

| | Weekly hours per grade | | | | | | | | | | Total hours | | |
Subject	1	2	3	4	5	6	7	8	9	10	Weekly	Yearly	Per cent
Russian language and literature	13	13	13	9	9	8	6	5.5	4	4	84.5	2,788	28.8
Mathematics	6	6	6	6	6	6	6	6	6	6	60	1,980	20.5
History	....	....	....	2	2	2	2	4	4	4	20	660	6.8
Constitution of the U.S.S.R.	....	....	....	....	....	....	....	....	....	1	1	33	0.3
Geography	....	....	....	2	3	2	2	2.5	3	....	14.5	479	5.0
Biology	....	....	....	2	2	2	3	2	1	....	12	396	4.1
Physics	....	....	....	....	....	2	3	3	3	4.5	16.5	544	5.6
Astronomy	....	....	....	....	....	....	....	....	....	1	1	33	0.3
Chemistry	....	....	....	....	....	....	2	2	3	3.5	10.5	347	3.6
Psychology	....	....	....	....	....	....	....	....	....	1	1	33	0.3
Foreign language....	....	....	....	....	4	4	3	3	3	3	20	660	6.8
Physical culture......	2	2	2	2	2	2	2	2	2	2	20	660	6.8
Drawing	1	1	1	1	1	1	....	....	....	....	6	198	2.1
Drafting	....	....	....	....	....	....	1	1	1	1	4	132	1.4
Singing	1	1	1	1	1	1	....	....	....	....	6	198	2.1
Manual and practical work	1	1	1	1	2	2	2	....	....	....	10	330	3.4
Practical study in agriculture, machine elements and electrotechnique	....	....	....	....	....	....	....	2	2	2	6	198	2.1
Excursions...............												198	n.c.
Total	24	24	24	26	32	32	32	33	33	33	293	9,857	100.0

Sources: M. M. Deyneko, *op. it.*, 134; per cent distribution which does not consider the hours for excursions, in *Narodnoe Obrazovanie*, No. 7 (July, 1955), 4. The curricula of the non-Russian republic schools contain minor modifications, the most important being the study of the respective native language and literature alongside Russian. This accounts for an additional hour in the schedule for grades 3 and 4, one and one-half hours in grade 8, and two hours in grades 5, 6, 7, and 10. The total weekly hours are 306.5.

It is evident that the entire curriculum embraces 18 subjects (counting excursions but not counting as separate subjects the subdivisions of mathematics, biology, history, and geography, which would make the total 28) taught in 9,857 or 9,669 hours, if excursions were excluded. It is furthermore evident that after Russian (and the native language and literature) follows mathematics with three times as many hours devoted to it as to the next-closest subjects and that the natural-mathematical sciences, together with the practical work of all types comprises 44.5 per cent of the curriculum, or up to 55 per cent in grades nine and ten. Science and mathematics alone occupy 34.2 per cent and the trend is still further away from the humanities. The new emphasis is on the practical and technological application of science, as evidenced from the introduction of manual and industrial training, now occupying 5.5 per cent of the total course.[35]

This curriculum, even as presently diluted, is overwhelming, especially in terms of the American school.[36] It should not be forgotten that this is a universal curriculum, that there are no electives, and that failing, or better—lack of excellency, is almost equivalent to economic and social suicide as it bars the way for joining the élite except perhaps if political merits of the students, and more so, of the parents are exceptional. Six weekly hours of mathematics each year for ten years, starting with arithmetics (grades 1 to 4), continuing with algebra and geometry (grades 6 to 10) and concluding with trigonometry (grades 9 and 10) is undoubtedly impressive. Not less impressive is the time devoted to the natural sciences: 544 hours of physics for four years and including mechanics, acoustics, heat, electricity, optics, and molecular physics with some time devoted to the structure of the atom; 347 hours of chemistry for four years and including inorganic as well as organic chemistry with the basic theories. Then 396 hours of biology for five years and including botany (grades 5 and 6), zoology (grade 7), anatomy and human physiology (grade 8), and principles of general biology (grade 9). Astronomy is being taught 33 hours in grade ten. And all science subjects are supplemented by application in class as well as out of class in various "science circles."[37] Still it is the rest of the curriculum that comprises over one half of the total schedule with the humanities and social sciences accounting for nearly 50 per cent of the entire curriculum.[38] This is a fact that those who try to press an all-out drive in this country for science education alone should not overlook.

The school year itself lasts 34 weeks for grades 1, 2, 3, and 10, and 35 weeks for the remaining grades. [It starts on September 1 and ends in May, for grades 1 to 4, or June, for the remainder.] It embraces 213 school days (for the lower grades) and 230 for the upper ones, [has four quarter terms and two recesses: 12 days in January and 10 days in March.] The students go to school six days a week for four to six hours daily. [Each class lasts 45 minutes.]

Besides the long and concentrated school day with the additional required out-of-class and out-of-school activity, the student must devote considerable time to daily homework assignments, the duration of which is fixed by the authorities but which in reality require much more time and make relaxation and play almost nonexistent. Soviet sources continuously complain that homework cuts even the hours for sleep and imperils the students' health in the long run.[39] Great part of the summer vacation is also taken up with "socially-useful labor," mainly in agriculture.

Checking the knowledge of the student via recitations and written homework occurs almost daily but formal final examinations for passing the grade are held only twice: at the end of the seventh year and at the end of the ten-year course, the latter entitling the student to a *matura* diploma.[40] Students passing the *matura* examinations with excellent grades and excellent behaviour are awarded gold or silver medals which entitle them to preferential treatment for admittance in higher educational institutions.

Appraisal. This is undoubtedly a heavy schedule, too heavy for every child and teenager between the age of seven and seventeen, because of the mental but also because of the physical work and the additional burden of extra-curricular activities.[41] Not that the Soviets do not realize this themselves. But they are still acting under the obsession of "lagging behind" which coupled with the must of industrialization and the ultimate goal of world supremacy, drives them relentlessly to try the short-cut, hoping that on balance the advantages will outweigh the disadvantages. And they may be able to afford this overload because of the vast reservoir of human stock which in spite of the known Communist disregard toward man, may still allow an ample supply of survivers for the scientific and intellectual elite. The positive results of this selectivity system can hardly be questioned now, although in the long run the stresses and strains will have to manifest themselves.[42]

How good is Soviet education, in general, and Soviet science education, in particular, at the general-education level? There is no question about the quantitative saturation of the curriculum and the syllabi and their advanced contents. There also is no question about the mutilation of most of those disciplines which allow politically and doctrinairily-motivated bias, the social sciences in the first place and the humanities, to a lesser extent. The re-writing of history textbooks is a perpetual phenomenon. Even the natural sciences, and biology in the first place, are not unaffected. But no regime can modify even if it wished to, the basic principles of physics, chemistry or mathematics.[43] And it is mathematics and science that occupy such a prominent place in Soviet general education. Therefore, with all the qualifications that follow, one may conclude that the average Soviet secondary school graduate is well-versed in the science subjects and without counterpart among his average American colleague graduating high school. Such a realization takes, of course, into account many difficulties and discrepancies of various nature. We lack many pertinent data for a complete evaluation of the actual width and depth of science knowledge of the Soviet graduate. We know that overload, rigidity, and memorization do not further quality of achievement. We are aware of the discrepancy between one school and another, between one science subject and another, between urban and rural education, and between regular, evening, and correspondence students. Thus, there is a wide variation of quality. But on the over-all the quality of science teaching has undergone substantial improvements, and other related criteria, such as student-teacher ratio, textbooks and other facilities, syllabi, examination standards, etc., are without too serious flaws. It should not be forgotten that all Soviet students in the general-education school are required to study the same amount of mathematics and science, which is overwhelming, and that the relatively low success rate of the three upper grade students and of the tenth grade graduates, estimated at averaging between 30 and 50 per cent and about twenty per cent, respectively is an indication for achievement. In short, the quality of Soviet science education, in spite of several environmental and some intrinsic shortcomings, is to be regarded at present as more than satisfactory in academic terms and coupled with the broad numerical basis presents perhaps the most dangerous challenge to our high schools and—why not—to our survival.

Before discussing the top of the pyramid—higher education and advanced degree study, a gap must be filled by sketching the role and training of the teacher and the state of the professional-vocational school as they relate to science education.

Teachers

One of the most serious shortages of Soviet education in the past was the great shortage and poor training of teachers. Due to the initial disdain toward learning and the learned, reinforced by the Communist hatred for the intellectuals—"the bourgeois remnants"—the overwhelming number of qualified teachers were either liquidated or put in the dog-house and no new contingents were existent, nor even desired. When formal instruction began to be rehabilitated, there was a frightening vacuum of teaching personnel and all that could be done in the short run was a rushed output of obviously deficient educators. Yet gradually were not only some of the old, mostly science teachers, rehired but the familiar planned drive for training new teachers was initiated and pursued with great vigor and determination.[44]

Shortcomings still beset the quality of teaching not only because the deficiencies in the schooling of the recent past cannot yet be fully overcome, but also because of the inherent bias of Communist pedagogy.[45] Still, the supply seems now adequate and is expected to improve further, due to the stationary and even decreasing number of students except in the upper grades which will present a problem not to be minimized.[46] With regard to quality, the arguments brought forward in connection with general and science education hold with equal validity when applied to the conveyer, not only to the recipient. One point must be made, however, namely that the teacher is charged with tremendous responsibility for molding the outlook of the "new Soviet man" and is carefully trained for this task, checked upon, and held accountable if unsuccessful. This "mission" makes even the science teacher not entirely immune, although his is the most scholarly field.

Elementary school teachers (for grades 1 to 4) are trained in pedagogical schools of four-year duration, for seventh grade graduates and, since 1954, of two-year duration for secondary school graduates, the latter now predominating. Secondary school teachers (for grades 5 to 10) are trained mostly in higher pedagogical institutes whose course of study has been extended since 1956-57 to five years and which offer now three "unified" groups of specialties: history-philosophy, physics, mathematics, and natural science-geography.[47] This measure aims at creating specialists in a related group of subjects. The "polytechnization" of general education has affected the training of the teachers in the two latter groups of specialization by adding the new approach and subject to the anyway heavy curricula which include also teaching experience and research.[48]

In 1956-57, there were 282 pedagogical schools, about 30 secondary special schools for teachers in drawing, singing and music, and physical culture, 204 higher pedagogical institutes, and 14 pedagogical institutes for foreign language teachers. A total of 620 higher and secondary teachers' establishments is reported, the difference being made up of schools for training teachers for pre-school children and the remaining non-general education institutions. Also the 35 state universities train up to eighty per cent of their graduates to become teachers in the upper grades, in accordance with a government decree.[49] They are, in fact, the best teachers but represent only about one-fifth of the total number of teachers coming from other establishments.[50]

The total number of graduates from all teacher-training establishments was reported as around 100,000 in 1956, the state universities excluded. In 1957, there were 1.8 million teachers in all general-education schools, which gives a student-teacher ratio of about 17 to 1.[51] The number of science teachers was 370,000 and the grand total of teachers in all educational institutions was slightly over 2 million.[52] About 70 per cent of the teachers in grades 5 to 7 and 96.2 per cent of the teachers in grades 8 to 10 were reported as having completed higher pedagogical education, obtained in many instances by correspondence.[53] Although these figures and their real meaning must be taken with caution, there is no doubt about the strides made in the training of sufficient and adequate teaching personnel, due last but not least to the attraction of the profession in terms of prestige and material rewards.

Indeed, the Soviet teacher has become a "central figure" in Soviet society. His position has been enhanced to the point of having become a member of the privileged class. His good work is rewarded with titles and decorations, his salary is comparatively high, his overtime and many other assignments are paid additionally, his vacation and pension guaranteed, and his advancement facilitated.[54] Although their daily instruction hours are four, in grades 1 to 4, and three, in grades 5 to 10, the Soviet teachers are busy all the day and very much overworked, due also to tremendous paper work which the authorities request, in order to keep a thorough check on their teaching. Inspectors check on their efficiency and the Party on their political reliability.

In conclusion, there is little doubt about the quality of the Soviet science teacher with all the qualifications that were made in relation to the appraisal of science education in the general-education schools.

Professional Vocational Schools

The variety of professional-vocational schools below the higher educational level deserve only little attention within our terms of reference, as their curricula are saturated with vocàtional training subjects and their graduates have little opportunity for further formal education. In fact, these schools aim primarily at the training of skilled workers and "specialists" for the various branches of production who as a rule get employed immediately and leave the educational stream. Only about five per cent of the top *tekhnikum* graduates are permitted to enter higher educational institutions of their major immediately after graduation and another negligible percentage ever have the chance to do so afterwards.[55]

This should not minimize their importance for the Soviet economy, for they provide an answer to the pressing need and quest for qualified manpower at the semi-professional level.[56] Nor should it minimize the fact that with the mass and compulsory seven-year education preceding the entry in the four-year professional school or even more so with the completed secondary general education, the students in and graduates from those specialized schools have now a much larger and more solid general education and science background than not long ago. Still, the *tekhnikum* curriculum centers on an extremely narrow specialization and although the authorities claim that the graduates are not less prepared in science and mathematics than the secondary general-education school graduates, this can be validly disputed.[57]

The recent and continuous increase in the number of secondary general-school graduates who will have to be barred from higher education, coupled with the decline in the influx of new workers,[58] presages a renewed emphasis on vocational training in the professional schools, in the general-education schools, and on the job, in order to meet the demands of the state, as visualized by the Soviet rulers.

For our purpose, it will be even more evident that the vast reservoir for the higher educational institutions, in general, and for higher science education, in particular, is increasingly limited to the graduates from the general-education, and not the professional-vocational schools.

Higher Education

Entering the door of a higher educational institution is almost equivalent to being admitted to the ante-chamber of the Soviet paradise and completing it means to be admitted to the paradise itself. Thus rigorous is the screening and thus rewarding the achievement that higher education has become a supreme goal for most of the Soviet youth. One should, of course, never forget the Soviet setting, and among other things, the fact that a college graduate is assigned to work where he may not like it at all, but still he will receive his relatively high income, be favorably discriminated from the rest, hope for advancement, and, most of all, he is convinced of having chosen one of the best alternatives that Soviet society offers to a selected few. The inexistence of private business precludes any success outside of education and, of course, of Party career.

It should be stressed that the belief and claim that the Soviet system permits a mass higher education is wrong. It allows larger segments of the population to aspire to higher education than was the case in Tsarist Russia and has, of course, expanded the higher institutions and their enrollment. But the road blocks to higher education are many and by no means exclusively limited to achievement. First of all it is again the state plan which establishes the over-all and specific admission quotas. Political and social discrimination still plays a role, although much less than in the past. Now the children of the élite are favored. Then, unlike in the secondary institutions, wanting financial status is a handicap as state scholarships are not generally available nor sufficient.[59] But what is more important is the fact that the Government simply bars the entry to higher education by shifting the mass of secondary school graduates to immediate employment in industry and agriculture or to professional-vocational-education, in accordance with the needs of the moment. And the present and future needs for such semi-professional manpower are increasing. This discriminatory power of the Party Government, coupled with the benefits expected from higher education, make the latter such an exclusive and precious good.[60]

Although achievement remains the most important criterion, the limited admission quotas and the hard and competitive admission examinations require outstanding performance and more than a grain of luck in order that an applicant who has passed the examinations does not still fail to qualify and thus jeopardise his entire career.[61]

The stresses and strains for entering a higher institution and also for staying in and graduating from it, must then be evident. No wonder that the average Soviet college student must give all he has to maintain his standing and that the average college graduate must be thoroughly prepared in his field.

Character of Education, Schools, Enrollment

Soviet higher education is specialized and considerably interwoven with practice. All students are prepared for a specific professional career and general-education subjects are given only to the future scientists and engineers. There is nothing comparable to our liberal arts college which makes Soviet higher education appear even narrower and more utilitarian than it is. Here again the European influence is felt. Thus, it should not be forgotten, that broad and intensive general education has already been administered in the secondary school so that the university curriculum is meant to concentrate on a specialized field. In this connection a deep dividing line should be drawn between universities and other special higher educational institutions. Only the former offer science and mathematics majors and their course is thoroughly impermeated with theory as well as application, while the latter are built around a specific vocation or industry and are pronouncedly restrictive and practical, with the possible exception of the enginering institutes where general subjects and science theory are part of the curriculum.[62]

Lately, the authorities are trying to overbridge the gap between "too much theory" in the universities and "too much practice" in the special higher institutions by introducing "polytechnization" and more vocational training in the former and "scientific background" in the latter.[63] Recognizing the narrowly differentiated and vocationally-oriented character of Soviet higher education as compared with our higher institutions, still even Soviet engineers carry a heavier load not only in their main field but also in the sciences, i.e., in their "service courses" than students at roughly corresponding American schools.[64] This "magic" is possible by nothing more than tremendous work load, almost beyond the possibilities of a youngster.

According to official data, 1,225,000 regular students were graduated in 1956 from secondary schools and 458,000 freshmen were admitted to the higher educational institutions, of whom about one-half as regular and the other half as extension students.[65] This shows that less than 38 per cent of the graduates are permitted to continue in higher institutions and that less than 20 per cent do so as resident students who are in fact those mainly eligible for post-graduate study. If one considers the fact that an additional 650,000 students were graduated as extension students from secondary school, the above percentage figures shrink considerably more. Furthermore, the 1956 freshmen were with 3,400 less than those admitted a year earlier.[66]

In 1956-57, there were 767 higher educational institutions of all types, i.e. universities and special higher institutions, the latter being called institutes, higher schools, schools, academies, and conservatories, with a total enrollment of 2 million students of whom 1.3 million were residence students.[67] This enrollment was the highest in Soviet history but the number of estabishments has been reduced, due mainly to the elimination of most of the two-year teachers institutes and their consolidation into higher pedagogical institutes, but also to the increasing emphasis on extension students. In fact about one-third of all students were extension students who are regularly employed, and their number is planned to increase by 1960 to about one million.[68] In view of the already discussed trends, this increase in extension students will have to be made at the expense of residence enrollment and it is known that a non-resident student's training and quality are generally deficient.

Of all higher educational establishments, technical and pedagogical schools are by far the most numerous, i.e. 230 and 225, respectively, followed by agricultural schools, 108, which clearly reveals the production-oriented character of higher education, alongside with the stress on training teaching cadres. There are also 35 universities, 30 economics and law schools, 77 medical schools, 47 arts academies and conservatories, and 15 schools for physical culture. By far the highest enrollment was in the first-mentioned category, namely 550,000 students in industrial and civil engineering schools and nearly 100,000 in transportation and communication schools. The agricultural schools add another 200,000 students engaged in technical education. This concentration on technical education has led to the doubling of the number of students in industrial, civil engineering, and agricultural schools during the fifth five-year plan.[69]

Of the 35 universities which among others are also the top science educational institutions, 12 were established in Tsarist times. Their enrollment was reported as 166,200, of whom three-fourths are resident students. Moscow University is the largest with 12 departments and about 22,000 students, of whom about 18,000 are residents, with 210 chairs and about 2,500 faculty, among whom are 89 members and candidate members of the Soviet Academy of Sciences. It is followed by the Leningrad and the Kiev universities. The Rusian and Ukrainian Soviet Republics alone account for most of the student body and of the higher educational institutions which reveals the familiar feature of concentration around the larger cities of European Russia at the expense of the vast remaining territory and not negligible population of the country.[70]

A more radical shift has occurred in the male versus female ratio of students. Thus, female students are in majority in the medical schools— 69 per cent—and in the universities and pedagogical institutes—72 per cent which is also reflected in the majority of women teachers and physicians throughout the country. In the various technical schools their share is about 37 per cent on the average.[71]

The 1956 graduating class of all higher educational establishments numbered 260,000, of whom about 70,000 were extension-correspondence students. Only about 5 per cent graduate in the humanities. Graduates in engineering and science account for 55-60 per cent of the total. The major share falls on the non-university establishments, the universities graduating only about 10 per cent of the total.[72]

Higher Science Education

The special appeal of science is best evidenced by the fact that about 65 per cent of the applicants to higher institutions wish to be science majors.[73] The main reasons for this otherwise unusual occurrence must be sought in the prestige and rewards as well as in the relative political immunity, connected with this field of knowledge, rather than in natural inclination and urge. This should not suggest, however, that, once accepted, the science student is allowed to move ahead as an average student. To the contrary: he must put as much additional effort as he may lack in talent, although as a rule he would have been thoroughly screened for both before having been admitted.

As was mentioned before, majoring in science is possible only at a university. In fact, about three-fourths of the university graduates major in science and mathematics, the rest majoring in humanities. There are no engineering, medicine, law or economics majors at the universities.[74]

Space would not allow the reproduction of science curricula or syllabi because each science comprises a separate and distinguished major, not to mention the fact that science-subject curricula vary considerably for the many types of engineering schools and majors. Moreover available curricula refer only to specific institutions, which would require further detail and elaboration. Last but not least DeWitt's and particularly Korol's volumes on Soviet science and professional education contain the latest available curricula, to which little could be added.[75]

The length of the university study is five years, and the yearly course hours range from close to 1,000 to 1,300. About 90 per cent of the time is devoted to the major field, although for the first two years the course is somewhat diluted by general-science subjects, and the political and social science subjects comprise about 6 per cent of the total curriculum. The often cited physics course at the Kharkov University reveals that the total course hours are 4,290, of which 3,356 are spent on physics, mathematics and chemistry. This is nearly twice as much as the number of hours for the basic subjects of the corresponding physics course at the Massachusetts Institute of Technology. Bearing in mind that Soviet science courses are standardized for all higher institutions, while M.I.T. is our top institution in the field, the disparity between Soviet and American instruction as far as the amount of exposure of the average science students to basic subjects is concerned becomes even more obvious.[76]

Before passing a judgment upon the quality of higher science education, one must add that class attendance is compulsory for six hours, six days a week, ten months a year, for four and one half years in class attendance, that applied work extends between 12 and 30 weeks, that preparing the home work requires as much time as the instruction hours, that about 50 examinations and other reports and checks are given during the course of study, and that extra-curricular activities such as participation in scientific circles and societies and scientific contests, not to mention political obligations, are a part of the heavy schedule of the science student. Considerable stress is laid on the student's independent work and the last semester is entirely free from formal instruction, being devoted to the preparation of the diploma project—a small but original research piece, to be defended at the graduation examination, which comprises also the mandatory major subjects. No degree is conferred at graduation.[77]

This thorough and extensive schedule leaves little doubt that the

Soviet science graduate is generally a well-trained specialist. It is obvious that the deficiencies in the humanities and social sciences and also in the politically-manageable science fields like biology, encountered in secondary education are magnified at the higher educational level. But so are the virtues of the sciences. It is also true that indoctrination subjects such as Marxism-Leninism, political economy, and philosophy are given in every university department and that the Party, Komsomol, and trade unions do the same outside of the university but in the science field this is taken seldom seriously by the students and they as well as the authorities are mostly concerned with academic achievement. Thus, making the necessary allowances for non-scientific influences as well as for differences between resident and non-resident students, between one field of science and another or, to a lesser degree, between one institution and another, competent analysts are generally agreed that Soviet higher science education is by no means inferior to ours and in many respects more extensive and thorough than at the corresponding level of our higher institutions.[78] This realization, coupled with the estimated yearly output of about 150,000 scientists and engineers of all types, of whom 70,000 engineers and 12,000 university-trained science majors represents a serious challenge to our colleges and universities and to our future as well.[79]

Advanced Degree Training

As the Soviet advanced-degree training can hardly be considered an educational pursuit as understood in this country, only a brief sketch should suffice. As was mentioned before, the bulk of the higher-school graduates must enter immediate employment in their respective fields and up to 90 per cent of the graduates of universities and pedagogical institutes become teachers, This leaves only a few thousands who could aspire yearly to an advanced degree. In fact, in 1956, there were 29,400 advanced-degree students of whom 16,800 were trained for "scientific and teaching activity" at the higher educational institutions.[80] Besides in the universities and other institutes, students are trained also in the institutes of the Academy of Sciences and of the various Ministries. There are two academic degrees: "candidate of sciences" and "doctor of sciences." The outstanding characteristic of Soviet advanced-degree training which differentiates it clearly from our practice is that there is no course study connected with it. According to an August 1956 special ordinance by the Party and Government, aiming at the improvement of the quality of the candidates, only such persons should be admitted to advanced-degree study who have at least a two-year experience in their given field and show abilities for scholarly work. Direct admission of university graduates is permitted only as an exception in the theoretical disciplines, such as mathematics, theoretical physics and a few other fields. All applicants must pass an admission examination and present a paper or published work in the chosen field of specialization in order to qualify.[81]

This reveals the particular nature of Soviet advanced-degree training which centers exclusively on independent contributions without instruction and presumes an already proven scholarly productivity by the applicant. The length of time required for advanced study varies from three years upward. A dissertation or equivalent piece of original research may be defended only after having been published.[82]

The annual award of advanced degrees mainly "candidates" is reported as 8,500, of which 7,000 or 82 per cent are in sciences and engineering, with the latter by far most numerous.[83] But, as far as defended dissertations is concerned, 75 per cent of them were in the field of science.[84]

The concentration on science and the requirement of outstanding achievement rather than formal study, points to the conclusion that the advanced degree holder in the USSR and especially the "doctor of sciences," if not honoris causa, must be a highly qualified scientist.

Then, he reaps exclusive rewards. Professorship and advanced research with all the personal satisfactions deriving from mental and material compensation and recognition by society are open to him. The scientist is now a full-fledged member of the Soviet paradise.[85]

The Party and Government back science with huge expenditures, which make this recognition possible. And still there is a handicap. The authorities establish priority of research and in many fields there is the emphasis on practical and most immediate application. And one should never forget the political atmosphere which, although rarefied at this high level, is still present and potentially menacing.

Summary and Conclusions

We have examined the pyramid of Soviet education, trying to follow the seam of science education first within the compact structure of general education and then in the exclusive science majoring institutions of higher education. Reference to science teaching in the secondary and higher technical schools had to be kept to a minimum.

The balance sheet reveals a combination of European thoroughness and rigidity and a numerically broader basis closer to the American pattern. What is typically Soviet is the total and centralized control, the concentration of resources in priority fields, and the whole Communist philosophy which subordinates the individual to the state and which places science on a pedestal.

The recognized good quality of science education coupled with the 2 to 3 times numerically-superior output of Soviet scientists and engineers as compared with our own would not be alarming were it not for the aggressive aims of the Soviet rulers who train their specialists as if they were weapons fighting a war with the United States. This realization leaves or should leave us little choice but to adjust ourselves accordingly. This does not mean that we either should or could ape their pattern. Our system and our values are radically different and we do not wish them to be otherwise. But it is imperative that we too devote a larger portion of our resources to meet the challenge as in time of war. We are able to expand and deepen science education by providing at least the same incentives and rewards to students—teachers and scientists alike. Our resources and our values are so much superior than the Soviet's that, once determined, we could not fail to obtain our goal.

We should not forget that Soviet education has also its many weaknesses and vulnerabilities. Right now the Soviet school is at a crossroad. Khrushchev himself revealed in his speech to the Thirteenth Congress of the Komsomol, in April, 1958, that in 1957 alone at least 700,000 secondary school graduates could not enter the higher educational institutions, not even the *tekhnikums,* and that between 1953 and 1956 their number amounted to 2.2 million. It is apparent that not only is higher education to become even more discriminatory but that general education is to be limited to a seven or eight grade course which will undo the glamorously-announced goal of a ten-year compulsory education. In fact, while concluding this article, Khrushchev made known a set of proposals of the Party which imply that the overwhelming majority of the young people must quit school at the age of 15 or 16 and be immediately employed as farm or factory workers and that a handful of the most gifted will be allowed to continue their education and aspire to the elite. Such a predicament unmasks the allegedly democratic character of Soviet education and confirms the class stratification of Soviet society with all the favoritisms for the children of the privileged class. Although science education *per se* would not be directly affected, this restrictive reorganization of the Soviet educational system with all the stresses and strains accompanying it and with the many other shortcomings mentioned in the text should guard us from considering that system as approaching perfection.

A balanced and impartial view is necessary to make us take the most appropriate measures. The recent National Defense Education Act is a beginning. Much more is expected, however, if we are to meet our responsibilities.

DR. L. A. D. DELLIN was born in Bulgaria and educated in East and West European and American Universities; he holds certificates from the University of Sofia (Bulgaria), Vienna and the Polytechnical School of Milan; he also received a M.A. degree from New York University and a J.S.D. from the University of Genoa. While in Europe, he was foreign correspondent for European newspapers and periodicals. In the United States, he was associated for seven years with the Mid-European Studies Center in New York as Research Superviser. He was also Lecturer in the Program on East and Central-European Affairs at Columbia University and served as guest lecturer in the graduate courses of Temple University, New York University, and Brooklyn College. He has worked on scholarly projects of The Bookings Institution, Georgetown University, and The United States Congress, and appeared as an expert before the United Nations Ad Hoc Committee on Forced Labor. Among his latest publications is *Bulgaria* (New York: F. A. Praeger, 1957). He is today Assistant Professor of Economics, College of Technology, The University of Vermont, where, alongside his regular courses, he conducts a seminar on Eastern Europe.

NOTES FOR SCIENCE EDUCATION IN THE U.S.S.R.

1. N. S. Khrushchev, *Otchetniy Doklad Tsentralnogo Komiteta Kommunistcheskoy Partii Sovetskogo Soyuza XX Suezdu Partii* (Moscow: Gospolitizdat, 1956), 93.

2. Alexander G. Korol, *Soviet Education for Science and Technology* (New York and London: The Technology Press of Massachusetts Institute of Technology and John Wiley & Sons, 1957), 411.

3. For most recent statistical comparisons, see Nicholas DeWitt, "Basic Comparative Data on Soviet and American Education," *Comparative Education Review,* Vol. 2, No. 1 (June, 1958), 9-11.

4. M. M. Deyneko, *40 Let Narodnogo Obrazovaniya v SSSR* (Moscow: Gosuchped, 1957). An abbreviated English translation, M. Deineko, *Forty Years of Public Education in the U.S.S.R.* (Moscow: Foreign Languages Publishing House, 1957), is also available and is revealing for its omissions and additions, compared with the original. We have kept our transliteration, in order also to easier distinguish the cited source.

5. UNESCO, *World Iilliteracy at Mid-Century* (Paris, 1957), estimates a five to ten per cent illiteracy among the adult population.

6. See A. M. Ivanova, *Chto Sdelala Sovetskaya Vlast po Likvidatsii Negramotnosti sredi Vzrolykh* (Moscow: Uchpedgiz, 1949). It should be noted, however, that with the introduction of the Cyrillic alphabet among the non-Russian nationalities, Russification became a concomitant aim. (See George L. Kline, "Education Toward Literacy," *Current History,* Vol. 35, No. 203, July, 1958, 19).

7. V. I. Lenin, *Sochineniya,* Vol. 33 (Moscow, 1951), 55.

8. George S. Counts, *The Challenge of Soviet Education* (New York: McGraw-Hill Book Company, 1957), 185, remarks that the campaign for liquidating illiteracy is waged "not to liberate the mind of the individual, but to hold it captive."

9. The "liquidation" of illiteracy was officially decreed in 1919, ordered again in 1925, and again in 1929, before the ball started rolling in the middle 1930's, and a score of mass and special organizations, backed by Party and Government, were employed in this drive, such as special schools, the trade unions, the Komsomol, the armed forces, the "Down

with Illiteracy" society and publishing house, commissions of the local Soviets and kolkhozes, as well as literate individuals comprising the so-called "culture army." An inside account of this drive and its difficulties is given by Nina Nar, "The Campaign against Illiteracy and Semiliteracy in the Ukraine, Transcaucasus, and Northern Caucasus, 1922-1941," in G. L. Kline, ed., *Soviet Education* (London: Routledge & Kegan Paul Ltd., 1957), 139-159.

10. *Narodnoe Khozaystvo v SSSR v 1956 Godu* (Moscow: Gosstatizdat, 1957), 224, 250, and 17. The mean number of years of schooling completed per capita is estimated at about 4, and 1 out of 10 persons have completed secondary education (DeWitt, *op. cit.*, 9).

11. *Narodnoe Khozaystro v SSSR v 1956 Godu,* 208 and 257.

12. *Ibid.*, 211 and 261. See also V. Bilshay, *Reshenie Zhenskogo Voprosa v SSSR,* (Moscow: Gospolitizdat, 1956), 119.

13. M. M. Deyneko, *op. cit.*, 23.

14. The tuition for the three upper grades of the secondary school and the higher institutions was definitely abolished only in 1956. That same year were established the first boarding schools with expenses to be charged according to the parents' income and up to a limit.

15. Estimates vary, yet the multiplier ranges from about three to five times the pay of a worker; in case of university professors, their average earnings are estimated at 16 times the pay of an average unskilled worker (George L. Kline, "Education," *New Leader,* June 16, 1958, 7).

16. S. P. Partigul, *Statistika Materialnogo i Kulturnogo Urovnya Naroda* (Moscow: Gosstatizdat, 1956), 107. But the budget figure usually includes expenditures on education and culture, so that the nearly 79 billion rubles or 400 rubles per capita for 1957 are misleading (*O Gosudarstvennom Byudzheta SSSR na 1957 God,* Moscow: Gospolitizdat, 1957, 24).

17. It was once nine and even seven years. Eleven-year schools function in the Baltic states and Georgia. Now various new projects are being considered for reasons to be discussed later on.

18. Most ten-year schools are in the larger cities. Plans for compulsory ten-year education have been announced for 1960.

19. Only recently preference is given to applicants with a previous 2-3 year employment.

20. Deyneko, *op. cit.*, 42.

21. Deyneko, *op. cit.*, 93.

22. Figures rounded up from *Narodnoe Khozaystvo SSSR* (Moscow: Gosstatizdat, 1956), 223 and *Kulturnoe Stroitelstvo SSSR* (Moscow: Gosstatizdat, 1956), 80-81, 84-85, and 122-123.

23. According to Hans Rogger, "The Russian Ten-Year School," *Current History,* XXV (July, 1958), 13, upper grade attendance is still about one-third of all the 14 to 17 year-olds.

24. Even seven-year attendance in the villages does not seem to be

completely enforced and the majority of Soviet youth discontinues academic training after only seven years in school. (H. Rogger, *op. cit.,* 14.)

25. M. Deineko, *op. cit.,* 32.

26. M. Postolovsky, *U.S.S.R. in 1960* (Moscow: Foreign Languages Publishing House, 1957) , 80.

27. Eric Ashby, *Scientist in Russia* (Harmondsworth Middlesex England: Penguin Books, 1947) , 63, remarked that nineteen out of twenty children left school before completing the ten-year course due to family needs or the needs of the state for labor reserve trainees.

28. M. Deineko, *op. cit.,* 130.

29. To quote Khrushchev himself: "The program of the secondary school must be reshaped with an emphasis on greater production specialization so that the ten-year graduates have a good general education, opening the road to higher education and at the same time be prepared for practical work because the greater part of the graduates must be immediately included in the work of the various branches of the national economy." (N. S. Khrushchev, *op. cit.,* 94.)

30. Deyneko, *op. cit.,* 130.

31. *Rezolyutsii XX Suezda Kommunisticheskoy Partii Sovetskogo Soyuza* (Moscow: Gospolitizdat, 1956) , 18.

32. Deyneko, *op. cit.,* 131.

33. The experimentation with the boarding school, to reach 1 million enrollment in 1960, as well as with a twelve-year course, is discussed in its latest forms by Hans Rogger, *op. cit.* However, of even greater importance seem to be most recent developments. Thus, in a Moscow Radio broadcast, reported in *The New York Times* (May 19, 1958) , the leading Soviet pedagogist, Nikolai Goncharov, outlined a draft plan for eight years of formal schooling, to be followed by three or four years of further study but combined with work in industry and agriculture, while only the gifted should be made to continue strictly academic subjects in the sciences, mathematics, and humanities. While concluding this article the eight-year school is reported to have received greater attention, especially by the Soviet manpower tycoon, G. I. Zelenko, head of the Chief Labor Reserve Administration. Thus, according to *The New York Times* (September 11, 1958) , the majority of eight-year school graduates would be sent to work, although they could continue their study in the evening or by correspondence. Other students would be chanelled to the various types of specialized schools and only the most promising ones would be allowed to continue formal residence schooling.

34. Counts, *op. cit.,* 174, remarks that "To the Bolsheviks science is power, and they are the foremost students of power in the contemporary world." Or, as Ashby, *op. cit.,* 202, puts it: "Russia has endowed science with the authority of religion."

35. Aiming at a more thorough "polytechnization" of the general-education school, the authorities intend to put greater stress on manual work, practical work in experimental fields and shops, practical work and

studies in enterprises, kolkhozes, sovkhozes, and machine-tractor stations, as well as physical culture, with a view of almost doubling the hours now devoted to them at the expense of literature, history, geography, the constitution of the U.S.S.R., and even of physics and chemistry, whose hours would be reduced. Psychology will be eliminated altogether. (Deyneko, *op. cit.*, 137.)

36. Partial confrontations and comparisons with the American school are given by Korol, *op. cit.*, 61-3 and the U. S. Department of Health, Education, and Welfare, Office of Education, in *Education in the USSR* (Washington, 1958), 67-70, and "Education in the USSR," *School Life* (December, 1957), 6-7. The most striking difference is that less than one third to one fourth of the American high school graduates have taken a year of chemistry or physics, and less than one seventh have had advanced mathematics.

37. For individual syllabi the reader is referred to: Ashby, *op. cit.*, Appendix 4; DeWitt, *op. cit.*, Appendix to Chapter II; Korol, *op. cit.*, Appendix B, and Office of Education, *Education in the USSR*, 74-83.

38. Variations are due to the presence or absence of native language and literature, in addition to Russian. Note also the emphasis on foreign language study, accounting for 6.8 per cent of the total course.

39. Some details are contained in Office of Education, *op. cit.*, 89-90.

40. Detailed procedure and contents of *matura* examinations in some subjects are given by Korol, *op. cit.*, 82-98 and Appendix D and E. It should be recalled that during the initial stage of Soviet rule there were no examinations at all, that later on there were final examinations after each grade from grade 4 on, and that a final examination at the end of the fourth grade existed until September 1956, since when also the two remaining final examinations have been drastically reduced in the number of required subjects. Now the *matura* examination includes: Russian language and literature, geometry, algebra, trigonometry, physics, and chemistry (See H. Rogger, *op. cit.*, 14).

41. *The New York Times Magazine* (June 22, 1958) correctly remarked that the school is not a playground and that a Soviet child entering it says good-by to his childhood.

42. See Harrison E. Salisbury, "Tensions Mount for Khrushchev, Too," *The New York Times Magazine* (August 24, 1958).

43. In those subjects textbooks date as far back as the 1880's, although brought up-to-date. See Korol, *op. cit.*, 74-5 and Appendix C.

44. A more detailed account based on personal experience is Vladimir D. Samarin, "The Soviet School, 1936-1942," in G. L. Kline, ed., *op. cit.*, 25-52.

45. Brush-up courses in the Moscow Central and other institutes for improving the qualification of teachers, as well as in pedagogical study rooms are still very much alive. Soviet teachers are trained mostly in subject knowledge, little in methodology.

46. Many elementary school teachers are simply moved together with

the students to the upper grades without having the necessary full qualification.

47. Deyneko, *op. cit.*, 105-116, *passim.*

48. Curricula are reproduced and discussed in Korol, *op. cit.*, 47 and 276 ff and Office of Education, *op. cit.*, 205-207. See also the personal experience of Ivan Rossianin "Teachers Colleges in the Soviet Union" in G. L. Kline, ed., *op. cit.*, 79-93.

49. Deyneko, *op. cit.*, 39, 110, and 111. About 20 per cent of the university graduates in the sciences go into teaching in the secondary schools (Office of Education, *op. cit.*, 204).

50. Korol, *op. cit.*, 51 and 271. The same source gives a break down of the teachers according to type of graduating institution, as follows: about 60,000 come from higher pedagogical institutes, about 30,000 from two-year pedagogical schools, and about 12,000 from universities (272).

51. Deyneko, *op. cit.*, 113 and 114. If only the regular teachers in the academic subjects were considered, the ratio would be 22:1 (Korol, *op. cit.*, 43).

52. DeWitt, *op. cit.*, in *Comparative Education Review*, 11.

53. Deyneko, *op. cit.*, 115.

54. It is the college and university teacher, however, who is paid much higher, about five times more than the general-school teacher (Korol, *op. cit.*, 52). There are also wide differentiations within the progressive wage scale, based on location and type of school, grade taught, education and experience. The monthly salaries range from 441 to 759 rubles for the first four grades teachers, from 544 to 874 rubles for teachers in grades 5 to 7, and from 635 to 935 rubles for the three upper grades' teachers. The teachers are entitled to three raises over a period of 25 years, conditioned upon passing specified university courses. The final raise increases the basic salary by 25-35 per cent. Teachers with over 25 year service receive a pension amounting to 40 per cent of their basic salary. Rural teachers are given free quarters, heat, and electricity, as well as other benefits. (See Deyneko, *op. cit.*, 116.)

55. Deineko, *op. cit.*, 103.

56. According to Soviet sources, all labor reserve schools graduated between 1940 and 1955 over 8 million people, and in 1956 alone an additional 665,000 (*Narodnoe Khozaystvo SSSR*, Moscow: Gosstatizdat, 1956, 216). The *tekhnikums* and other secondary professional schools graduated for the same period over 3 million "junior specialists," and in 1955 alone 387,000 (Deyneko, *op. cit.*, 248). Still, the Soviets are not content with the present ratio of professional to semi-professional manpower, which is estimated at 1 to 1.8, and aim at ratios of 1 to 2, 1 to 3, and even 1 to 4, in some fields. See Office of Education, *op. cit.*, 168. Pertinent and comparative data are contained also in Korol, *op. cit.*, 109-112, and DeWitt, *op. cit.*, 84-5, and Chapter Six, and cited article in *Comparative Education Review*, 10.

57. Of course, this excludes the graduates from the ten-year general-

education schools, who then attend professional institutions. Curricula are contained in and discussed by Korol, *op. cit.*, 123 ff., and DeWitt, *op. cit.*, 81 ff.

58. See Solomon M. Schwartz, "Education for Russian Industry," *Current History*, XXXV, 203 (July, 1958), 30 and 38.

59. Tuitions were abolished in 1956. The scholarships affect about 75 per cent of the students, are highly discriminatory and may be withdrawn in case of unsatisfactory performance or violation of discipline. (See Korol, *op. cit.*, 177-8, and DeWitt, *op. cit.*, 143.) Ashby, *op. cit.*, 74, remarks that "From talking to students I had the impression that they can just manage to live on their state grants, but it needs self-denial and good luck to do so."

60. An additional road block is the recent preference for such applicants who have labor experience. Furthermore, preferred also are those with highest qualifications for a given major and those with military service behind them.

61. See Korol, *op. cit.*, 171-2. All, except the medal holders, must pass the admission examinations, consisting of five subjects: Russian language and literature, written; mathematics, oral and written; physics; chemistry; and foreign language, all three oral (*ibid.*, 178). A person may apply to only one institution at a time, so that if rejected he must wait for one more year, if he can afford it. Examples of examination questions are contained *ibid.*, 180-182.

62. See Korol, *op. cit.*, 135, 190, and 232, DeWitt, *op. cit.*, 19, and Counts, *op. cit.*, 35. Many of the departments and courses at American colleges and universities would be considered in the U.S.S.R. as below the level of higher education.

63. See Korol, *op. cit.*, 256 and 271.

64. See Korol, *op. cit.*, 253. In a 1954 Ordinance of the Party and Government the "overdifferentiation of specialties" was assailed and an "integration of departments and chairs" invoked. As a result some subjects were eliminated and some disciplines made elective. Also the time devoted to production practice was increased and the examinations decreased.

65. Deineko, *op. cit.*, 32 and 108.

66. See *Kulturnoe Stroitelstvo*, 203 and Postolovsky, *op. cit.*, 80.

67. *Kulturnoe Stroitelstvo*, 201-2.

68. Deyneko, *op. cit.*, 264.

69. *Ibid.* As we do not intend to pursue the non-university technical establishments, it can only be specified that they are of various types and with numerous majors and that only a relatively small portion of the technical schools are really engineering schools, comparable to our technological and engineering institutions.

70. *Ibid.*, 265-7. The three largest universities accommodate about one-third of the total student population.

71. *Kulturnoe Stroitelstvo*, 205.

72. *Deyneko, op. cit.,* 269. Also DeWitt, *op. cit.,* 88-9.

73. See John O'M. Bockris, "Higher Education in the Soviet Union," *Current History,* XXXV, 203 (July, 1958), 24.

74. Korol, *op. cit.,* 204-5.

75. The reader is referred to Korol, *op. cit.,* in particular Chapter 8, and DeWitt, *op. cit.,* 110 ff., and 276 ff. Also Ashly, *op. cit.,* in his pioneering study of Soviet science education reproduces curricula and comments on them, in particular 51-59, 78-86 and 91-3.

76. Data are from Korol and DeWitt, *op. cit., passim.* Korol presents and analyses the Kharkov University curriculum and compares it with that at the M.I.T. (261 ff.)

77. Examination figure from Korol, *op. cit.,* 221. Remaining information from Deyneko, *op. cit.,* 271.

78. This evaluation refers to the science graduates, the engineering graduates coming from too many institutes with too many specializations, to allow a generalized judgment. Teachers, facilities and textbooks for the science students are also considered generally adequate (see Korol, *op. cit.,* 284 ff., and DeWitt, *op. cit.,* 147 ff).

79. Figures from DeWitt, cited article in *Comparative Education Review,* 11, and Korol, *op. cit.,* 204-5. The latter calls Soviet science education "comparable to the very best." (*Ibid.,* 356).

80. Deyneko, *op. cit.,* 272.

81. Ordinance discussed *ibid., loc. cit.* The examination subjects are the major field, foreign language, and Marxism-Leninism.

82. *Ibid., loc. cit.*

83. DeWitt, *op. cit.,* in *Comparative Education Review,* 11.

84. DeWitt, *Soviet Professional Manpower,* 210.

85. The senior faculty members receive up to 15 times the salary of a beginning instructor. If administrative position, multiple jobs, overtime, royalties from books, and advanced degree are considered the salary increases by a multiple. A full professor with a doctor's degree receives 5 to 6.4 thousand rubles monthly and a full member of the Academy of Sciences from 10 thousand rubles upward. These estimates made by Korol, *op. cit.,* 303-309, *passim,* speak for themselves.

CLASSICAL EDUCATION, SCIENCE AND THE WEST

WERNER HEISENBERG
Nobel Prize-Winner for Physics (1932)
Max Planck Institute for Physics, Munich

The Traditional Reasons for the Defense of a Classical Education

Many people have asked whether a classical education is not too theoretical or unwordly, and whether in our age of technology and science a more practical education would not be much more suited to equip us for life. This bears directly on the frequently discussed question of the relationship between the humanities and contemporary science. I cannot deal with this question fundamentally, for I am not a pedagogue nor have I been overmuch concerned with educational problems. I can, however, try to recall my own experiences, since I myself had a classical education and later on devoted most of my work to science.

What are the arguments that defenders of the humanities have produced, time and again, in favour of concentrating on ancient languages and ancient history? In the first place, they rightly point to the fact that our whole cultural life, our actions, our thoughts and our feelings, are steeped in the spiritual roots of the West, *i.e.*, in that attitude of mind which in ancient times was initiated by Greek art, Greek poetry and Greek philosophy. With the rise of Christianity and the formation of the Church great changes took place, and finally, at the end of the Middle Ages, there occurred the tremendous fusion of Christian piety with the Greek spirit of enquiry, and the world, as God's world, was radically altered by voyages of discovery, by science and by technology. In every sphere of modern life examination of the root of things, whether methodological, historical, or philosophical, brings us up against the concepts of antiquity and Christianity. Thus we may say in favour of a classical education that it is always a good thing to know these roots, even if they may not always be of practical use.

Second, we must stress the fact that the whole strength of our Western culture is derived, and always has been derived, from the close relationship between the way in which we pose our questions and the way in which we act. In the sphere of practical action other people and other cultural groups were just as experienced as the Greeks, but what always distinguished Greek thought from that of all other peoples was its ability to change the questions it asked into questions of principle and thus to arrive at new points of view, bringing order into the colourful kaleidoscope of experience and making it accessible to human thought. It is this link between the posing of questions of principle and practical action which has distinguished Greek thought from all others, and which during the rise of the West at the time of the Renaissance, the turning-point in our history, was responsible for the rise of modern science and technology. Whoever delves into the philosophy of the Greeks will encounter at every step this ability to pose questions of principle, and thus by reading the Greeks he can become practiced in the use of the strongest mental tool produced by Western thought. Hence, in this respect, we can fairly say that a classical education teaches us something very important.

Finally, it is justly said that concern with antiquity gives us a judgment in which spiritual values are prized higher than material ones. It is precisely in Greek thought, and in all the traces of it that we have inherited, that the pre-eminence of the spirit clearly emerges. True, people of today might take exception to just this fact, for they might say that our age has demonstrated that only material power, raw materials and industry are important, and that physical power is stronger than spiritual might. It would follow that it is not in the spirit of the times to teach our children to attach greater importance to spiritual than to material values.

In this connection, I am reminded of a conversation which I had some thirty years ago in the forecourt of our University. At that time Munich was in the throes of a revolution. The inner town was still occupied by the Communists, and I, then seventeen years old, had been assigned with some school comrades as auxiliaries to a military unit which had its headquarters opposite the University, in the Theological Seminary. Why all this happened is no longer quite clear to me, but it is probable that we found these weeks of playing at soldiers to be a very pleasant interruption of our lessons at the Maximilian Gymnasium. In the Ludwig Strasse there was occasional, if not very heavy, shooting. At noon we fetched our meals from a field-kitchen in the University courtyard. On one such occasion we had a discussion with a theology student on the question whether these struggles in Munich had any meaning, and one of us younger ones said emphatically that questions of power could never be settled by spiritual means, by speeches or by writing; only force could lead to a real settlement of our conflicts with others.

The theology student replied that in the final analysis even the question of what was meant by 'we' and 'the others,' and what distinguished the two, would obviously lead to a purely spiritual decision, and that in all probability we should have gained a great deal if we could settle this question more reasonably than by the usual method. We could hardly object to this. Once the arrow has left the bow, it flies on its path, and only a stronger force can divert it; but its original direction was determined by him who aimed, and without the presence of a spiritual being with an aim it would never have been able even to start on its flight. In this regard we could do far worse than teach our youth not to rate spiritual values too low.

The Mathematical Description of Nature

However, I have strayed too far from my proper theme, and I must revert to my first real encounter with science at the Maximilian Gymnasium in Munich, since, after all, I am speaking of the relation between science and a classical education. Most schoolboys are introduced to technology and science when they begin to play with apparatus. Emulating the example of a fellow pupil, or perhaps because of a present received at Christmas—or even through school lessons—they begin to have a desire to handle small engines, and perhaps even to build one. This is precisely what I did with great enthusiasm during the first five years of my life at high-school. This activity would probably have remained a mere game and would not have led me to real science, if another event had not occurred.

At the time, we were being taught the basic axioms of geometry. At first, I felt this to be very dry stuff; triangles and rectangles do not kindle one's imagination as much as do flowers and poems. But then our outstanding mathematics teacher, Wolff by name, introduced us to the idea that one could formulate generally valid propositions from these figures, and that some results, quite apart from their demonstrable geometric properties, could also be proved mathematically. The thought that mathematics somehow corresponded to the structures of our experience struck me as remarkably strange and exciting.

What had happened to me was what happens only too rarely with the intellectual gifts we are handed at school, for school lessons generally allow the different landscapes of the world of the mind to pass by our eyes, without quite letting us become at home in them. According to the teacher's ability these landscapes are illuminated more or less brightly, and we remember the pictures for a shorter or a longer time. However, very occasionally, an object that has thus come into our field of view will suddenly begin to shine in its own light, first dimly and vaguely, then ever more brightly, until finally it will glow through our entire mind, spill over to other subjects and eventually become an important part of our own life. This happened in my case with the realization that mathematics fitted the things of our experience, a realization which, as I learnt at school, had already been gained by the Greeks, by Pythagoras and by Euclid.

At first, stimulated by Herr Wolff's lessons, I tried out this application of mathematics for myself, and I found that this game between mathematics and immediate perception was at least as amusing as most other games. Later on, I discovered that geometry alone was no longer adequate for this mathematical game which had given me so much pleasure.

From some books I gleaned that the behaviour of quite a few of my home-made instruments could also be described mathematically and I now began to read voraciously in somewhat primitive mathematical text-books, in order to acquire the mathematics neded for the description of physical laws, *i.e.*, the differential and integral calculus. In all this I saw the achievements of modern times, of Newton and his successors, as the immediate consequence of the efforts of the Greek mathematicians or philosophers, and never once did it occur to me to consider the science and technology of our times as belonging to a world basically different from that of the philosophy of Pythagoras or Euclid.

Although, in my youthful ignorance, I was not fully aware of it, this enjoyment of the mathematical description of nature had introduced me to the basic trait of all Western thought, namely, to the inter-relationship between the way in which we pose questions and the way in which we act. Mathematics is, so to speak, the language in which the questions are posed and answered, but the questions themselves are concerned with processes in the practical material world; thus, geometry, for instance, was designed for measuring agricultural land. Because of all this, I remained far more interested in mathematics than in science or apparatus during most of my life at school, and it was only in the two upper classes that I acquired a special liking for physics—oddly enough because of a fortuitous encounter with a fragment of the modern physics.

Atoms and Classical Education

At that time we were using a rather good textbook of physics in which, quite understandably, modern physics was treated in a somewhat off-hand manner. However, the last few pages of the book dealt briefly with atoms and I distinctly remember an illustration depicting a large number of them. The picture was obviously meant to represent the state of a gas on a large scale. Some of the atoms were clustered in groups and were connected by means of hooks and eyes supposedly representing their chemical bonds. On the other hand, the text itself stated that according to the concepts of the Greek philosophers atoms were the smallest indi-visible building-stones of matter. I was greatly put off by this illustration, and I was enraged by the fact that such idiotic things should be pre-sented in a textbook of physics, for I thought that if atoms were indeed such crude structures as this book made out, if their structure was com-plicated enough for them to have hooks and eyes, then they could not possibly be the smallest indivisible building-stones of matter.

In my criticisms I was supported by a friend from my youth club with whom I had gone on many hiking expeditions, and who was much more interested in philosophy than I was. This friend, who had read some essays on atomic theory in ancient philosophy, had also unexpectedly come across a textbook of modern atomic physics (I believe it was Som-merfeld's *Atomic Structure and Spectral Lines*) where he had seen visual models of atoms. This had led him to the firm conviction that the whole of modern physics was false, and he tried to convince me that he was right. At that time our judgments were obviously very much rasher and more dogmatic than they are today. I had to agree with him that these visual models of atoms were indeed false, but I reserved the right to look for the mistakes in the illustrations rather than in the theory.

In any case, I had gained the wish to become better acquainted with the case for atomic physics, and here another accident came to my aid. At the time we had just started reading one of Plato's *Dialogues,* but school lessons were irregular. I have already told how I, as a young boy, had been a member of a military unit during the Munich revolution and that we had been stationed in the Theological Seminary opposite the University. We had no rigid plan of work at the time—far from it. The danger of lounging about was very much greater than that of over-exertion. In addition, we had to be prepared to be called even at night, and thus we were without any control by parents or teachers.

It was then July, 1919 (a warm summer), and there were hardly any military duties, particularly in the early mornings. Thus it came about that frequently, shortly after sunrise, I would withdraw on to the roof of the Theological Seminary and lie down there to warm myself in the sun, any old book in my hand; or I would sit on the edge of the roof and watch the day beginning in the Ludwig Strasse.

On one such occasion, it occurred to me to take a volume of Plato on to the roof, for I wanted to read something different from the books we were supposed to study in school. With my somewhat modest Greek knowledge, I came upon the dialogue called *Timaeus,* where for the first time and from the original source I read something about Greek atomic philosophy. This lecture made the basic thoughts of atomic theory much clearer to me than they had been; or at least, I believed that now I had an inkling of the reasons that had in the first place caused Greek philosophy to conceive of these smallest indivisible building-stones of matter. True, I did not feel that Plato's thesis in *Timaeus—i.e.,* that atoms are uniform bodies—was fully convincing, but at least I was happy to learn that they did not have hooks and eyes. In any case, at that time I was gaining the growing conviction that one could hardly make progress in modern atomic physics without a knowledge of Greek natural philosophy and I thought that our illustrator of the atomic model would have done well to make a careful study of Plato before producing his particular illustration.

Thus, without properly knowing how, I had become acquainted with that great thought of Greek natural philosophy which links antiquity with modern times and which only came to full fruition at the time of the Renaissance. This trend in Greek philosophy, typified by the atomic theory of Leucippus and Democritus, used to be described as 'materialism'. Historically this is a correct description, but today it is easily misunderstood, since the word 'materialism' was given a very one-sided bias in the nineteenth century by no means in accordance with its meaning in Greek natural philosophy. We can avoid this false interpretation of ancient atomic theory if we remember that the first modern investigator to return to the atomic theory in the seventeenth century was the theologian and philosopher Gassendi, who, we may be sure, did not use the theory in order to combat the dogma of the Christian religion; indeed, even for Democritus atoms were merely the letters by which we could record the events of the world, but not their content. In contradistinction, nineteenth-century materialism was developed from thoughts of quite a different kind, thoughts which are characteristic of the modern age and are rooted in the division of the world into separate material and spiritual realities, as proposed by Descartes.

Science and Classical Education

We have seen that the great stream of science and technology of modern times springs from two sources in the fields of ancient philosophy. Although many other tributaries have flowed into this stream, and have helped to swell its current, the origins have always continued to make themselves felt. Because of all this the sciences cannot but benefit from classical studies. People who are concerned with the more practical schooling of youth for their struggles in later life will continue to assert that the knowledge of this spiritual foundation has little relevance for practical activities, and that they should rather acquire the necessities of modern life: modern languages, technical methods, accounting and commercial practice. These (they say) will set the youngsters on their feet, but a classical education, being, so to speak, merely of decorative value, is a luxury which only those few can afford for whom fate has made the struggle for life less exacting. Perhaps this is true for the many people who will do nothing in their later lives but carry on a purely practical business, and who themselves will have no wish to influence the spiritual climate of their age. Those, however, who find this inadequate, and wish to get to the root of things in their chosen vocation, whether it be in technology or medicine, are bound sooner or later to encounter the sources of antiquity, and their own work can only benefit if they have learnt from the Greeks how to discipline their thoughts and how to pose questions of principle. I believe that in the work of Max Planck, for instance, we can clearly see that his thought was influenced and made fruitful by his classical schooling.

Perhaps I may here cite yet another personal experience which occurred three years after I left school. While a student at Göttingen, I discussed with a fellow student the problem of the model of the atom that I had found so disturbing while still at school. This question was obviously the basis of the puzzling phenomena of spectroscopy which were still unsolved at that time. This friend defended perceptual models, and he believed that all that was needed was to enroll the help of modern technology in the construction of a miscroscope with a very great resolving power—for example, one employing gamma-rays instead of ordinary light. We should then be able to see the structure of the atom, and so my objections to perceptual models would finally be dispelled.

This argument disquieted me deeply. I was afraid that this imaginary miscroscope might well reveal the hooks and eyes of my physics textbook, and once again I had to resolve the apparent contradiction between this envisaged experiment and the basic conceptions of Greek philosophy. Here the education in disciplined thought that we had received at school was to help me a great deal, and make me wary of accepting unproved solutions. In this I was greatly helped also by what little acquaintance with Greek natural philosophy I had made at that time.

In contemporary discussions about the value of a classical education one can no longer maintain that the relationship between natural philosophy and modern atomic physics is a unique case. For even if we rarely meet such questions of principle in technology, science or medicine, these disciplines are basically connected with atomic physics and thus, in the final analysis, lead to similar questions of principle. Chemical structure is explained on the basis of atomic physics. Modern astronomy is connected with it most closely, and can hardly make any progress without it. Even in biology, many bridges are being built towards atomic physics. The connections between the different branches of science have become much more obvious in the last decades than at any previous time. There are many signs of their common origin, which, in the final analysis, must be sought somewhere in the thought of antiquity.

Faith In Our Task

With this conclusion I have almost returned to the point from which I started. At the root of all Western culture there is this close connection between our way of posing questions of principle and our actions; this we owe to the Greeks. Even today the whole force of our culture rests on this connection. From it springs all our progress, and in this sense a declaration of faith in a classical education is an avowal for the West and for its culture.

However, do we still have a right to this faith when the West has lost so terribly in power and prestige in the last decades? Our answer is that all this does not involve questions of right, but questions of *will*. For the activity of the West does not stem from the theoretical insights—our ancestors did not base their actions on theories—but from quite a different origin. What is, and always has been, our mainspring, is faith. By faith I do not mean only the Christian faith in a God-given and meaningful framework of the world, but simply faith in our task in this world. Here, faith obviously does not mean that we hold this or that to be true. If I have faith, it means that I have decided to do something and am willing to stake my life on it. When Columbus started on his first voyage into the

West, he believed that the earth was round and small enough to be circumnavigated. He did not think that this was right in theory alone, but he staked his whole existence on it.

In a recent discussion of this aspect of European history, Freyer has rightly referred to the old saying: *'Credo ut intellegam'*—'I believe so that I may understand'. In extending the application of this idea to the voyages of discovery, Freyer introduced an intermediate term: *'Credo, ut agam; ut intellegam'*—'I believe so that I may act; I act so that I may understand'. This saying is relevant not only to the first voyages round the world, it is relevant to the whole of Western science, and also to the whole mission of the West. It includes classical education as well as science. And there is no need to be over-modest: one half of the modern world—the West—has gained immeasurable power by applying the Western idea of controlling and exploiting natural resources by means of science. The other half of the world, the East, is held together by its faith in the scientific theories of a European philosopher and political economist. Nobody knows what the future will hold and what spiritual forces will govern the world, but our first step is always an act of faith in something and a wish for something.

We wish that spiritual life may once again blossom here, that here in Europe thoughts may continue to grow and shape the face of the world. We stake our existence on this, and in so far as we remember our origins, and recover the harmonious interplay of Western influences, we shall make the external conditions of life in the West happier than they have been for fifty years. We wish that, despite all outer confusion, our youth will grow up in the spiritual climate of the West, and so draw on those sources of vitality which have sustained our continent for more than two thousand years. Let us not worry about the detailed ways in which this might be brought about. It does not matter whether we prefer a classical or a scientific education. What alone matters is our unshakable faith in the West.[1]

(For Bibliography see the chapters in this Section and Section II)

DR. WERNER HEISENBERG was born in Würzburg (1901), son of Dr. August Heisenberg, then teacher at the Gymnasium. He studied physics at Munich under Sommerfeld, Wien, Pringsheim and Rosenthal and under Born, Frank and Hilbert in Göttingen; in 1923, working under Sommerfeld in Munich, he obtained his Doctorate in Philosophy, and was appointed an assistant of Born in Göttingen; in the summer of 1924 he was given the *venia legendi* there. In the winter of 1924-25 he worked as a Rockefeller Scholar under Niels Bohr in Copenhagen. In 1926 he was appointed Lecturer in theoretical physics at the University of Copenhagen, and in 1927 became Professor-in-Ordinary of theoretical physics at the University of Leipzig. In 1929 he went on a long lecture tour of the United States, Japan and India; in 1932 and 1939 he again lectured in the United States. In 1933 he was awarded the 1932 Nobel Prize for Physics. In 1941 he was called to the University of Berlin and also became Director of the Kaiser Wilhelm Institute for Physics in Berlin-Dahlem. At the end of the war, he was captured by American troops and taken to England. After his return to Germany in the spring of 1946 he reorganized, with some of his former colleagues, the Kaiser Wilhelm Institute for Physics in Göttingen; this institute, which since 1946 has carried on its work within the framework of the Max Planck Society (founded as successors to the Kaiser Wilhelm Society), and which since 1948 has been known as the Max Planck Institute for Physics, devotes its scientific work chiefly to the investigation of cosmic radiation and thus to the atomic physics of the most energetic elementary particles. During the spring of 1948, Heisenberg lectured at Cambridge, and in the autumns of 1950 and 1954 in the various American institutions; he also gave the Gifford Lectures at the University of St. Andrews in 1955-56, and is now Director of the Max Planck Institute for Physics, which has recently moved from Göttingen to Munich. In 1958, Heisenberg announced that he had developed a new, all-embracing theory of the elementary particles, which he hopes will offer a thorough mathematical analysis leading to a derivation of the properties of these particles, and thereby to a unified field theory (see: Earl Ubell, "West Germans Press Study of the Atom," New York *Herald Tribune,* August 17, 1958, Section 2).

NOTES ON CLASSICAL EDUCATION, SCIENCE AND THE WEST

1. From *The Physicist's Conception of Nature* by Werner Heisenberg (c 1955 by Rowohlt Taschenbuch Verlag, GmbH, c 1948 by Hutchinson & Co. (Publishers) Lits). Reprinted by permission of Harcourt, Brace and Company, Inc.

INDEX

DATE DUE